MODERNISM AND THE CULTURE OF EFFICIENCY:
IDEOLOGY AND FICTION

EVELYN COBLEY

Modernism and the Culture of Efficiency

Ideology and Fiction

UNIVERSITY OF TORONTO PRESS
Toronto Buffalo London

Toronto Buffalo London
www.utppublishing.com
Printed in Canada

ISBN 978-0-8020-9957-0 (cloth)

Printed on acid-free, 100% post-consumer recycled paper with vegetable-based inks.

Library and Archives Canada Cataloguing in Publication

Cobley, Evelyn Margot, 1947–
Modernism and the culture of efficiency : ideology and fiction / Evelyn Cobley.

Includes bibliographical references and index.
ISBN 978-0-8020-9957-0

1. Industrial efficiency – Social aspects. 2. Technological innovations – Social aspects. 3. English fiction – 20th century – History and criticism. 4. Modernism (Literature). 5. Technology in literature. I. Title.

PN56.M54C63 2009 306.4′6 C2009-902132-3

This book has been published with the help of a grant from the Canadian Federation for the Humanities and Social Sciences, through the Aid to Scholarly Publications Programme, using funds provided by the Social Sciences and Humanities Research Council of Canada.

University of Toronto Press acknowledges the financial assistance to its publishing program of the Canada Council for the Arts and the Ontario Arts Council.

University of Toronto Press acknowledges the financial support for its publishing activities of the Government of Canada through the Book Publishing Industry Development Program (BPIDP).

To David

Contents

Acknowledgments

This project was made possible by the support and encouragement of friends, colleagues, and students who kindly provided comments and insights at all stages of its development. The greatest debt of gratitude, however, goes to David Thatcher, my patient partner, most assiduous volunteer research assistant, and most critical reader of the manuscript. I am also most grateful to my friends and colleagues Pam McCallum, Linda Hutcheon, Len Findlay, Smaro Kamboureli, R.B.J. (Rob) Walker, Stephen Ross, J. Hillis Miller, Ed Pechter, Ann Pearson, and Jim Ducker for their encouragement and assistance. Special thanks go to my research assistants Kate Wellburn, Tara Thomson, and Paisley Mann.

I have greatly benefited from the valuable suggestions offered by the three anonymous readers for the Press who were expert and attentive in their consideration of the manuscript.

I would also like to acknowledge the efficient way in which the University of Toronto Press dealt with my manuscript. Jill McConkey, Daniel Quinlan, and Richard Ratzlaff were highly supportive academic editors, while John St James was exemplary as a meticulous copy-editor.

Preparation of this book was greatly facilitated by the generous support of a Standard Research Grant from the Social Sciences and Humanities Council of Canada and by the study leave granted by the University of Victoria.

MODERNISM AND THE CULTURE OF EFFICIENCY: IDEOLOGY AND FICTION

Introduction

My interest in efficiency began many years ago during a visit to the concentration camp at Dachau, near Munich. At age twenty, I was moved, not unexpectedly, by the sight of the partially reconstructed camp; however, what I had not anticipated, and what was most shocking to me, was the documentation in the small museum showing the rational efficiency with which the Nazis had gone about the irrational goal of eliminating the Jewish population. On a visit to Auschwitz-Birkenau in the summer of 2007, I had my earlier sense of shocked incredulity once again confirmed and intensified. *Modernism and the Culture of Efficiency* thus grows out of an awareness that the everyday investment in efficiency as a rational means for achieving aims we deem desirable also carries with it risks that are too often disavowed. Although efficiency need not lead us straight to the concentration camp, we may want to consider its dark underside, especially in our own times when it increasingly threatens to become an end in itself.

Efficiency in itself is neither a positive nor a negative quality; the value we attribute to it depends on how we use and assess it in our lives and society. Without a certain commitment to efficiency, we would endure material hardships and social chaos. There can be little doubt that most of us like efficient organizations, offices, governments, workers, machines, and gadgets. How could we not? We appreciate trains that run on time and cars that carry us to our destinations on well-organized and signposted roads. Efficiency is undoubtedly closely allied to what we appreciate as progress in the material and social conditions of our existence. Indeed, efficiency strikes us as the very cornerstone of modern civilization. It is our commitment to efficiency that has driven inventions like the wheel, the steam engine, the car, the airplane, and the computer. Our daily comforts are made possible by myriad large and small advances in the efficiency of the production and distribution of the goods and services we enjoy. Most of us would be loath to give up our dishwashers, central heating, increasingly sophisticated kitchen ranges and fridges, televisions, cars, and digital cameras.

What frustrates us above all is inefficiency: we are frustrated by machines and appliances that break down, and we disdain lazy and slovenly people. Our sense of human dignity is to no small degree dependent on our self-image as an efficient homemaker, office worker, carpenter, or medical expert. It is often not just what we do to earn a living but how well we think we are performing our assigned tasks that matters. In short, most of us take great pride in being efficient in most areas of our lives. Even our leisure time is frequently dominated by an efficient use of time or talents. We want to be the best recreational tennis player we can be. Or we want to excel as amateur musicians, gardeners, or solvers of crossword puzzles. Being idle makes us feel guilty. For our vacations, we feel obliged to improve our cultural knowledge through travel to foreign places or, at the very least, spending quality time with our families. The lure of doing things as efficiently as possible is even perceptible in the socialization of our children; their days are so filled with organized activities that they constantly rush from pillar to post. On her birthday in August, the eleven-year-old daughter of a friend informed me that she and her friends could not fit a birthday party into their busy schedules until December. Efficiency has taken hold of us and has penetrated into our innermost thoughts. Those who refuse to be efficient are dismissed as either romantic dreamers or social outcasts. Whenever we catch ourselves being lazy or inefficient, we may put on a defiant pose, which simply reinforces the importance we attach to efficiency.

Yet we also watch with considerable dismay how efficiency is taking over our lives. While enjoying all the material advantages of an efficiently organized existence, we worry that we may have sacrificed some of the values that make us human. We are discomfited when human workers are replaced by machines, when employees are rendered redundant by more efficiently organized corporations, when family and friends are sacrificed to the efficient time management demanded by our career ambitions. We often feel overwhelmed by the rapidity with which our conditions of existence are transformed and made increasingly complex. Time has become a commodity that has to be managed with optimal efficiency. We train our children early in the efficient use of time; rarely left to while away an afternoon, they spend their days in activities organized by schools and after-school programs, and in parentally arranged and supervised pursuits. There are undoubtedly moments in all our lives when we ask ourselves if the price we pay for our material comforts may not be too high. Newspapers are full of articles dealing with the stresses of juggling the demands of work, family, and play. Some people opt out of the rat race to live simple lives no longer heavily dependent on consumer goods. Not surprisingly, perhaps, we find that books documenting or advising resistance to the efficient life are finding an eager public. Tom Hodgkinson's *How to Be Idle,* Al Gini's *The Importance of Being Lazy,* Corinne Maier's *Bonjour paresse* (*Hello Laziness*), and Carl Honoré's *In Praise of Slow* are only the most obvious examples.

Instead of contributing to the growing literature on the 'slow movement,' this study concentrates on the ideological implications of our cultural commitment to efficiency by focusing on its most salient features in both socio-economic and literary registers. Although efficiency became an issue in the nineteenth century, it was during the first three decades of the twentieth that it generated a host of cultural anxieties most dramatically captured in modernist novels such as Aldous Huxley's *Brave New World* (1932). While Joseph Conrad's *Heart of Darkness* (1902), D.H. Lawrence's *Women in Love* (1920), Ford Madox Ford's *The Good Soldier* (1915), and E.M. Forster's *Howards End* (1910) are not primarily concerned with the conflicted logic of efficiency, this logic nonetheless informs the personal and cultural anxieties manifested by characters whose attitudes and behaviours invite critical scrutiny. Conceived as an ideology, efficiency can be seen to be 'working by itself'; it is an 'obviousness' that deserves to be analysed and deconstructed. Without drawing attention to itself, it has infiltrated the consciousness of individuals and suffused the social fabric. As we will see, the broader social implications of our socio-economic investment in technical efficiency speak to preoccupations with cultural shifts from capitalism as a 'free' to a regulated market, from the 'free' to the socially constrained individual, from contingent to socially engineered communities, and from liberal-democratic to totalitarian political organization. In short, these novels reflect, in various registers, an almost imperceptible cultural slide from the desire for the perfectibility of machines to the perfectibility of society.

The culture of efficiency informing the modernist literary canon has its source in socio-economic transformations that can be traced to the invention of the steam engine, first manufactured by Watt and Boulton in 1794. This external combustion engine prepared the way for the internal combustion engine powering first locomotives and then cars. The proliferation of machines made possible by the combustion process was most memorably celebrated at the Great Exhibition of 1851 in London. Throughout the nineteenth century, improvements in the construction of locomotives spawned a growing network of railroads that required the kind of massive financial resources that were beyond the capabilities of predominantly family-owned companies. When the well-documented rush to buy railway stocks resulted in a bubble that eventually burst, investors lost not only the invested capital but all their assets. Having to reassure a now risk-averse public, the state introduced 'limited liability' legislation to encourage people to keep investing in the market. With this legislation, the modern form of the corporation came into being. The nineteenth century thus saw the more or less simultaneous emergence of the concept of efficiency in its modern sense, of the modern age of efficient machines, and of the efficient modern corporation. At the beginning of the twentieth century, these three aspects of efficiency combined to establish the Ford Motor Company not only as the quintessential American corporation but also as the graphic symbol of rapid sociocultural transformations in the Western world.

In *The Mantra of Efficiency: From Waterwheel to Social Control* (2008), Jennifer Karns Alexander captures the nuances of efficiency as 'a slippery concept' (Alexander 4) and indicates its complex history: 'Efficiency had been a philosophical concept describing agents and causes of change, yet it dropped out of sight in the eighteenth century, only to resurface in the nineteenth in a different form, as a technical measurement of the performance of machines. It moved into economics and then, early in the twentieth century, into more common use, as an efficiency craze swept through Europe and the United States' (3). With the invention of the steam engine, the word 'efficiency' took on its modern sense of increasing the output of a machine or organization through the elimination of waste. Closely associated with nineteenth-century industrial developments, the concept took its meaning from thermodynamic laws invoked to measure the performance of machines as these converted useless into useful energy. An 'interest in output-input relationships' (Alexander 9) does, of course, precede the emergence of the concept itself; it can be traced as far back as antiquity. The term 'efficiency' also has a long history; it was first used in the philosophical sense of *efficient cause,* that is, as 'an agent that brings something into being.' When something was deemed to be 'efficient,' it had simply caused '(a thing) to be what it is' (*Oxford Dictionary of Word Histories*). As an early philosophical concept, it was 'based upon the ancient Aristotelian system of four causes, in which the efficient one was the active and immediate principle that produced change' (ibid. 9). In the medieval tradition, it was God who was not only the 'efficient cause' of change but also the guarantor of the proper functioning of the cosmos he had created and was now governing.

As a historian of science, Alexander focuses on case studies ranging from nineteenth-century technical debates on how motion should be measured in the performance of waterwheels to broader social controversies surrounding the 'efficient' management of people – from Darwin's principle of 'natural selection' to the 'efficient' seating of workers at the Siemens factory in Weimar Germany and to the 'efficiency' of slave labour in the United States. Focusing on case studies, Alexander is able not only to cut a long historical swath but also to suggest a multiplicity of meanings and uses of efficiency that point beyond 'one local historical context to the larger phenomena of industrial modernity itself' (Alexander 14). As her subtitle announces, she seeks to shed light on how a concept denoting technical measurement and quantification came to signify techniques of social surveillance, control, and exploitation. In its emphasis on social control, *The Mantra of Efficiency* covers territory also central to *Modernism and the Culture of Efficiency.* Yet the two texts approach efficiency from significantly different perspectives. I should perhaps mention that I had already completed my manuscript when I first saw an advertisement for Alexander's just published book. From an interdisciplinary perspective, it was interesting to discover how, independently of each other, the historian of science and the cultural critic focus on different materials and pose different questions. Alexander's case studies are snapshots of various aspects of efficiency at specific moments in history from the eighteenth century to the twenty-first. For her analysis of efficiency's contradictory meanings and its competing social roles and uses, she relies on material that the general reader might consider obscure or unfamiliar.[1] In contrast, I concentrate on the first three decades of the twentieth century, focusing on well-known emblematic figures and events so as to document the impact that an increasing technical, social, and personal investment in efficiency had on human consciousness at a moment of cultural crisis.

Instead of introducing readers to rediscovered illustrative material, I seek to uncover the logic of efficiency, as well as its underlying ideological assumptions, by drawing on the extensive scholarly literature that has already been devoted to well-known figures such as Frederick Winslow Taylor, Henry Ford, or Max Weber as well as to phenomena such as the Great Exhibition of 1851 or the Nazi concentration camp at Auschwitz. The case studies in *Modernism and the Culture of Efficiency* are not intended to acquaint readers with the unfamiliar but to invite a critical re-examination of the supposedly 'obvious' through the 'defamiliarization' of the familiar.

Although *The Mantra of Efficiency* and *Modernism and the Culture of Efficiency* agree that efficiency has both static and dynamic aspects, they differ in the evaluation of this opposition. Devoting several chapters to how the performance of machines was measured, Alexander extrapolates from the 'interconvertibility of energy, heat, and motion' (Alexander 7) specific to the thermodynamic laws to argue that efficiency in this technical sense was a *static* concept. The laws of thermodynamics are for her 'bounded by natural, fixed limits'; since quantities of 'motion, heat, and energy' can only be 'transformed' but not 'created,' she concludes that 'conservation laws provided a theoretical upper limit to the effectiveness of a machine' (13). In other words, for her, a machine always exists in a state of balance; she approaches and measures it as an end product or a static totality. I would argue, however, that this totality is always an illusion. In the process of transforming useless into useful energy, some energy always escapes or remains useless. It is wasted. What counts for us is how successfully an engineer is able to minimize waste and to increase the ratio of useful energy to be squeezed out of a mechanism. It is not the static balance between output and input that fires the cultural imagination but the dynamic desire to increase the total output of machines at all cost. Efficiency is thus marked by the lure of a perfectibility remaining always out of reach.

It is not that Alexander denies that friction creates resistances to be overcome. Indeed, as she repeatedly stresses, the two 'distinct senses of the word' she invokes are also 'woven together in a paradoxical rhetoric': 'One was an efficiency of balance, a static efficiency, the highest measure of which accounted for the conservation of measured elements. The other was creative and dynamic efficiency, which allowed growth through careful management and brought as its reward not merely conservation but growth' (Alexander 2–3). Throughout her case studies, she associates the *technical* sense of efficiency (thermodynamic laws) with the 'conservation of measured resources' (13); in other words, efficiency is here identified with the maintenance of a homeostatic condition. Mechanisms of change arise for her only with 'techniques of agency, especially planning and discipline' (11); in other words, her dynamic conception of efficiency involves social agents attempting 'to direct the course of change' through 'the surveying and apportioning of resources to achieve a desired end' (12). Her recourse to paradox to explain the relationship between static and dynamic conceptions of efficiency suggests that she cannot maintain a distinction on which her study nevertheless relies. Initially tempted to adopt a similar static/dynamic opposition, I soon discovered that I could not sustain it. It is only by isolating a machine from its social context that it can be deemed a static totality.

Yet even in its technically most restricted sense, efficiency is always a comparative measure. As Alexander herself acknowledges, 'Efficiency allowed comparisons between machines or systems of widely different design and function and had come to describe the general goal of machine design: to approach *as nearly as possible* to a perfect correspondence between output and input' (80; my emphasis). In both its technical and social manifestations, efficiency is, for me, marked by a dynamic *totalizing desire* intent on achieving an always receding static or perfected totality. If we keep in mind that the thermodynamic process can never entirely overcome waste or resistance, then the desired totality always remains open to disruption and transformation. Where Alexander posits a static balance or equilibrium, I see a utopian model mobilizing ceaseless efforts to contain and close off a process that has neither beginning nor end. From my ideological perspective, the crucial problem with efficiency is that it is no longer just a means to an end but has become an end in itself.[2] In the social arena, the relentless pursuit of efficiency gains animates social-engineering utopias, which tend to deteriorate into nightmare scenarios of social control. At stake in *Modernism and the Culture of Efficiency* is the impact that our increasing commitment to efficiency has on our individual consciousnesses and on our cultural self-understanding.

The most graphic illustration of the total-output model of efficiency is undoubtedly the assembly line introduced at the Ford Motor Company in 1913. In the process of pioneering a car for the masses, Henry Ford inaugurated a revolutionary mode of production and a new consumer society. The logic of the *technical* efficiency Ford sought to obtain seeped into the *social* organization of the factory floor. Although Ford vehemently denied having been influenced by Frederick Winslow Taylor, his actual practices agreed with Taylor's contention that human beings had to be 'managed' if waste was to be minimized or eliminated. In the popular imagination, workers were reduced to 'cogs in the machine' both by Ford's assembly line that chained them to machines and by Taylor's 'stopwatch' that exposed them to the surveillance of the 'efficiency expert.' But Ford's ability to assimilate Taylor's ideas into his total-output model of efficiency in effect obscures the utopian dimension of Taylor's *The Principles of Scientific Management* (1911).

Central to Taylor's ambition to increase productivity was his attempt to alleviate the 'wasteful' class conflict pitting management against labour. His system was predicated on the assumption that both workers and bosses would be better off if they could be persuaded to cooperate with each other rather than harm each other. His principles depart from the total-output model of *technical* efficiency to instal a measure of *social* efficiency indebted to a formula developed by the Italian mathematician and social theorist Vilfredo Pareto, who is credited with 'applying the concept of efficiency to social institutions' (Heath 22). Pareto efficiency or 'optimality' stipulates that 'maximum efficiency has been achieved when it is impossible to improve anyone's condition without harming someone else' (23). It is not the *total output* but the *optimal distribution* of resources that is at stake in this utopian equilibrium model of efficiency. But in Taylor's version of Pareto efficiency neither labour nor management can be trusted to cooperate spontaneously; it is only if they are presented with a rational plan, drawn up by an 'efficiency expert' who is an 'objective' outsider, that they will be prepared to act in their own and the company's collective interests and thereby increase the system's overall efficiency. It was the risk of spontaneous cooperation succumbing to the logic of the Prisoner's Dilemma central to game theory that compelled Taylor to introduce the figure of the efficiency expert. Unfortunately, once the optimal distribution of input and output is entrusted to an outside expert rather than negotiated by participants, Taylor's utopian vision tends to be haunted by the spectre of totalitarianism. It is this aspect of the Taylorization of the Ford Motor Company that inspired Huxley to associate the utopian world of 'Our Ford' in *Brave New World* with Mustapha Mond's totalitarian political system.

Through Ford's pragmatic assimilation of Taylor's utopian ideal, his company provides us with an illustration of the often-conflicted logic of efficiency in its technical and social dimensions. But Ford workers were not only oppressed by machines and manipulated by efficiency experts; they were also seduced by the prospect of the good life in the consumer society. What separates Ford most decidedly from Taylor was 'his explicit recognition that mass production meant mass consumption' (Harvey 125–6). Ford was intent on constructing not only a new type of worker suited to the new industrial process but also a new type of man or woman prepared to spend rather than save money. For the modern corporation to increase its productivity, it had to market its goods by creating a desire for affluence. At the Ford Motor Company, efficiency was thus contradictorily predicated on both the elimination of waste in the production process and the encouragement of waste at the consumer end. In addition to the two incompatible models for technical and social efficiency, Henry Ford introduced a third conception that complicates the emphasis on thrift in the first two by stressing a profligate expenditure of 'energy.' In short, the Ford Motor Company operates at the overlapping intersection of the drive for increasing the technical efficiency of machines, the social productivity of workers, and the desires of consumers; the internalization of these conflicting notions of efficiency contributed to the reified consciousness of the modern individual and to the new rationalized self-understanding of modern society. My ideological investigation of the ways in which efficiency was increasingly being internalized by individuals and infiltrating the culture represented in the modernist literary canon thus touches on economic and social theories, on Continental philosophy and critical theory, and on aesthetics and politics.

It is, of course, in *Brave New World* that the Ford Motor Company is shown to invite precisely the kind of speculation Huxley engaged in when he associates the assembly line with the utopian dream of the perfect society that devolves into the dystopian nightmare of the totalitarian state. Although this novel inevitably drew my attention to the culturally symptomatic importance of Ford, Huxley's direct attack on the corporation's homogenizing effect on society is so broadly satirical that it no longer attracts the critical scrutiny it deserves. While the provocative title reverberates in the popular press ('Scientists predict brave new world of brain pills,' 'The Brave New World of Scalar Electromagnetics,' 'Berlin's Brave New World of Public Transport,' 'Brave New World for Public Media,' 'Bush's Brave New World of Torture,' 'Brave New World: Virtual Law and Order'), the academic community tends to shrug the novel off as a more or less prescient anticipation of technological innovations we now mostly take for granted. We seem more intrigued by modernists who preferred more subtle modes of representation to capture a social crisis Huxley mocked so openly. Standing alone in his attempt to subject the ideology he associates with Fordism to intellectual analysis, Huxley offers the most encompassing and provocative dramatization of its troubling features but also the most overstated social and political commentary. Although *Brave New World* quite obviously *insists* on being included in a study of efficiency, here it will be considered *last* as the culmination of cultural anxieties expressed in less concentrated fashion in other modernist texts.

Eschewing Huxley's utopian fantasy mode, other writers of his day resorted to what were often considered to be experimental narrative techniques that, nevertheless, kept characters and events firmly anchored in the everyday realities of the times. No matter how disjointed the narrative plot and how unreliable or limited in perspective the narrator may be, it is always possible to reconstruct the events of the story and the sociohistorical contexts of the characters in Conrad's *Heart of Darkness,* Lawrence's *Women in Love,* Ford's *The Good Soldier,* and Forster's *Howards End.* One could, of course, analyse the formal experiments typical of modernist poets and novelists for further signs of a cultural preoccupation with efficiency. As Suzanne Raitt shows convincingly, the methods of modernist art were indeed 'aimed at adjusting the economy of the art-work to the economy of the world' (Raitt 89). Focusing on imagism and stream of consciousness, she argues that 'art had to be modern and to be modern meant to be accurate, stream-lined, and efficient' (90). Concerned with forms of literary writing that 'did not waste words,' poets such as Ezra Pound offer the 'image' as a privileged site for the 'modernists' quest for precision and compression' (89). 'Short forms such as the Japanese *haiku*' illustrate one economical approach to literary expression while stream of consciousness, with its apparent proliferation of words through the 'inclusion of the smallest details in novels' (89), signals not 'waste' but 'an unfamiliar manifestation of accuracy and completeness' (101).[3] In 'Picturing Efficiency: Precisionism, Scientific Management, and the Effacement of Labor,' Sharon Corwin similarly focuses on the 'relationship between scientific management and paradigms of visual representation,' exploring how 'the aesthetics of efficiency, standardization, simplification, and economy' entered into 'dialogue with other contemporaneous forms of visual representation, in particular, the formal project of Precisionist art' (140). In contrast, *Modernism and the Culture of Efficiency* is self-consciously *thematic* in its analysis of the terms in which characters in the novels under discussion both object to modern civilization and reinforce its investment in efficiency. Overt preoccupations with sexual relations, family structures, identity crises, social aspirations, and moral dilemmas in these novels conceal from the reader how much space is often devoted to the efficiency calculus surfacing in depictions of industrialization, urbanization, speed, utilitarianism, materialism, commodification, instrumental reason, and capitalism. Critics have, indeed, commented on the efficient Wilcoxes in *Howards End,* the efficient Leonora in *The Good Soldier,* the efficient Gerald Crich in *Women in Love,* and the efficient Kurtz in *Heart of Darkness;* however, there has been no sustained critique of the ambiguous, slippery, and often contradictory ideological work performed by efficiency. The deliberately exaggerated features of an emergent investment in efficiency in *Brave New World* are thus perceptible in more muted tones in other novels of the period. Once we become alert to the broader ideological ramifications of such features, our interpretations of these novels will have to be modified or reconfigured.

The reader may of course ask why these particular novels have been singled out for inclusion in this study. The selection is neither carelessly arbitrary nor entirely motivated. Reactions by colleagues to the project at conferences and in conversations repeatedly alerted me to other cultural emblems and novels that could have strengthened my case or added to the thematic territory to be mapped out in this study. The question of efficiency generated animated discussions, leaving me with the hope that *Modernism and the Culture of Efficiency* will encourage others to pursue avenues I had to leave unexplored. In the interest of the narrative I am trying to construct around the emergence of efficiency as an ideology during the first three decades of the last century, I had to make some pragmatic decisions. Instead of analysing Ford's overt treatment of the efficiency movement in *Parade's End,* for instance, I consider *The Good Soldier* precisely because here the ideology of efficiency has seeped unannounced into the novel's whole social fabric. In an effort to avoid unnecessary repetition, each chapter devoted to the novels is meant to illustrate different but interrelated aspects of efficiency in overlapping cultural contexts. Most broadly speaking, the literary engagement with efficiency during the first three decades of the twentieth century ranged from public debates on eugenics and the social evils of rationalization to depictions of the most personal internal struggles and the most private interpersonal relations. Where the eugenics debate overtly emphasizes a link between efficiency and social control that can be traced back to Goethe's *Faust* and forward to Auschwitz by way of the Ford Motor

Company, the novels of Conrad, Lawrence, Ford Madox Ford, and Forster exemplify in more oblique terms how the drive for efficiency can exact unanticipated social and personal costs.

What Goethe's *Faust* inadvertently illustrates is a sinister possibility or unintended consequence lurking in the promise of improving the world through effort, application, and planning. We may recall that Faust is presented as the quintessential 'modern' who enters a pact with the devil in order to free humanity from its premodern shackles. Unlike Marlowe's Faust, Goethe's protagonist is ultimately saved because, as the prototypical enlightened humanist who works to reclaim land from the sea for the betterment of humanity, he replaces capricious self-indulgence with responsibility to the human community. However, as Marshall Berman has pointed out, this triumphant narrative of enlightened progress contains a generally overlooked warning. In his zeal to reclaim land, Faust sacrifices an old couple who refuse to sell their land to make room for Faust's project. These two old people 'are the first embodiments in literature of a category of people that is going to be very large in modern history: people who are in the way – in the way of history, of progress, of development; people who are classified, and disposed of, as obsolete' (Berman 67). The social benefits of Faust's efficient reclamation of land from water are thus implicated in a social-engineering project that eliminates those standing in the way of progress. In his attempt to create as much agricultural land as possible, Faust treats both the sea and the old couple as 'waste' to be eliminated.

In Berman's reading of Goethe's play, the Enlightenment narrative contains the seeds of the death camps in Nazi Germany. The trajectory from good, 'enlightened' intentions to abhorrent, 'barbaric' consequences is not so much the story of a moral failure as it is an illustration of troubling implications inherent in, among other things, the pursuit of efficiency for its own sake. In Goethe's story, Faust is clearly not making a moral judgment; he is neither an evil nor an irresponsible figure. But he is driven to act against his own moral sensibility by an investment in efficiency. Once he embarks on his engineering project, he becomes obsessed with the blind pursuit of converting always more water into land. The removal of the old couple is for him simply an efficient solution to a technical problem. By instituting the category of dispensable people, Faust raises the question posed by Carl Schmitt and Georgio Agamben: 'Who decides on the exception?' On what grounds does Faust presume to judge who lives and who dies? Where Goethe 'sees the modernization of the material world as a sublime spiritual achievement,' Berman contends that Faust, this 'archetypal modern hero,' anticipates the barbaric fate of displaced people like the Jews who perished at Auschwitz (66). While Auschwitz stands as a worst-case scenario long left behind us, the troubling link between a dedication to efficiency and a desire for control over both nature and human beings manifests itself in the modernist novels under discussion in often surprising ways.

On the surface, at least, our modernist novels are primarily interested in dramatizing the detrimental effects of the internalization of the efficiency calculus on the consciousness of individuals and on traditional cultural values. What Conrad, Lawrence, Ford, and Forster deplore above all is the loss of a nourishing inner life, the failure of meaningful personal relationships, the separation from nature, and the decline in the appreciation of cultural traditions. Modern life seemed to most of the authors and their characters to be superficial, utilitarian, materialistic, and rootless. Forster's Margaret Schlegel encapsulates the conflicted attitude of intellectuals towards their own bourgeois comforts when she points out that, without efficient men like Henry Wilcox, there would be 'no trains, no ships to carry us literary people about in, no fields even. Just savagery. No – perhaps not even that. Without their spirit life might never have moved out of protoplasm. More and more do I refuse to draw my income and sneer at those who guarantee it' (Forster 164). Margaret here voices Forster's own ambivalence about the benefits and costs of efficiency, an ambivalence shared by the other novelists analysed in this study. But even if they mostly 'sneer' at efficiency and sometimes unequivocally express their hostility to this 'evil of modernity,' they find it difficult to locate any viable sites of resistance to its infiltration of the social fabric. In their search for alternatives, they most often retreat into nostalgic longings for a past they know to be lost. In the early twentieth century, then, efficiency was not yet taken for granted but manifested itself as a contested source of anxiety. From a distinctly Western perspective, efficiency raises the question of what constitutes not the 'bare life' (Agamben) but the meaningful life according to the cultural elite of the bourgeoisie.

It should perhaps be stressed that, methodologically, Fordism and Taylorism are of sociocultural interest in their own right rather than serving simply as a paradigm to be applied to the interpretation of novels. Instead of exact correspondences between the socio-economic and literary registers, this analysis of the Ford Motor Company and predominantly British novels is intended as a first step in the cognitive mapping of an ideology that has received scant attention from literary critics in spite of its broad sociocultural implications. Although Fordism and Taylorism were specifically American manifestations, their impact was perhaps felt to be more traumatic in a society deeply rooted in specifically English traditions. The socio-economic transformations wrought by an investment in efficiency were experienced as more disruptive of the relatively stable British class system than of the more flexible conditions prevalent in American society. What is remarkable is the degree to which these British literary texts are, on close inspection, preoccupied with efficiency as a disorienting incursion into a previously comfortable cultural self-understanding.

Readers may well ask themselves why *Modernism and the Culture of Efficiency* concentrates to a large extent on American theories and social practices in part 1 and then proceeds to examine literary works by British authors in part 2. The two-part structure constitutes an ultimately unavoidable methodological compromise. Since the theory provides the framework for the readings of novels, it had to take temporal precedence. But this precedence may, unfortunately, imply an unintended causal relationship, suggesting that the novels were merely passive reflections of a socio-economic crisis in which they played no active part. In reality, the debates on efficiency in the various disciplines invoked in this study took place simultaneously. The socio-economic registers influenced literary representations just as the novels shaped the public reception of the efficiency calculus. In short, fiction did not simply comment after the fact on the ideological impact of the emergent socio-economic investment in efficiency; on the contrary, it was an active participant in an ongoing debate about the social, cultural, and personal *meaning* of this often ambiguous, contradictory, and shifting concept. There are, of course, *American* novels preoccupied with the culture of efficiency so emblematically embodied in Henry Ford and Frederick Winslow Taylor. At least two American novels do in fact receive some attention here; however, they serve primarily to illustrate sociological arguments. John Dos Passos's *The Big Money* is mentioned for comments on Ford and Taylor, while Sinclair Lewis's *Babbitt* exemplifies the internalization of efficiency more prominently theorized by sociologists. Why then this privileging of specifically *British* literature in my critical readings of novels as novels? *Modernism and the Culture of Efficiency* has its genesis in my interest in British modernism, a genesis that, for some readers, might be obscured by a two-part structure that seems to suggest that the theoretical engagement with Fordism and Taylorism is the primary concern while the novels play the secondary role of illustrating sociological findings. It is precisely because British culture seemed traumatized by Fordism and Taylorism that the American contribution to an understanding of the efficiency calculus in British novels takes on such prominence.

Not surprisingly, perhaps, the scholarly literature on efficiency can be divided into a large number of studies devoted to improving the efficiency of machines, corporate and government bureaucracies, public and private organizations; some of these take the form of how-to manuals. With over 350 books listed under 'efficiency' in my university library catalogue, this literature focuses primarily on economic and industrial efficiency, efficient educational outcomes, and improvements in energy- and eco-efficiency. These areas of concern are repeated in the popular press, with articles encouraging us to meet the threat of global warming by cutting back on our use of energy, by resorting to alternative sources of renewable energy, and by replacing current technologies with more efficient ones. The primarily empirical approach in both the scholarly and popular literature unquestioningly accepts increases in the total output of efficiency as a desirable goal. It was only when we started to ask ourselves what would constitute an efficient society that the demand for preference satisfaction became an issue. Although GDP (Gross Domestic Product) continues to be used as the standard for measuring the size of a country's economy, other criteria are starting to be used to establish what makes a society livable or desirable. A website like 'Cities Ranked & Rated' lists cities based on demographics, economy and jobs, cost of living, climate, education, health and healthcare, crime, transportation, leisure, and arts and culture. In the social sphere, it is not only a question of how efficiently a country's market produces goods and services but also how efficiently the public good is being delivered. The market and the state now compete for the privilege of being accepted as the preferred provider of the common good.[4] The debate on how much state intervention either creates or hinders optimal conditions for ensuring a society's quality of life is particularly pointed in Canada, a country reputedly dominated by 'welfare-state capitalism.' To my knowledge, the only academic treatments specifically targeting efficiency as a social problem are by two Canadians. In 2001 Stephen Heath published *The Efficient Society: Why Canada Is as Close to Utopia as It Gets,* a cautious endorsement of Pareto efficiency as the path to a socially more responsive and ecologically more sustainable approach to the problems of our times. In the same year, Janice

Gross Stein published *The Cult of Efficiency* as a deliberate response to Heath's heavy reliance on market forces to achieve the public good. In her view, it is the state that is to be preferred because, unlike market forces, democratic institutions can be held accountable to citizens. *Modernism and the Culture of Efficiency* is not intended as a contribution to this Canadian debate; it sidesteps questions of public policy in favour of a more theoretically inflected concern with economic and political forces. As I have already indicated, my comments on the sociocultural impact of efficiency could perhaps most usefully be considered alongside Alexander's more technical approach in *The Mantra of Efficiency.*

Although this study opens with chapters devoted to the Great Exhibition of 1851 and the Ford Motor Company, this emphasis on the 'machine' accentuates the metaphorical significance of the engine as a *totalizing* concept. What follows is neither a history of technology nor an attack on its destructive potential. Any analysis of efficiency necessarily subsumes the exponential increases in the output capacities of both material and social practices by means of technical improvements. However, this study is only tangentially a contribution to the scholarship dealing with technology as a dangerous enemy. Instead of going over the ground already thoroughly covered by such illustrious authors as Lewis Mumford, Jacques Ellul, Herbert Read, Arnold Gehlen, Ivan Illich, and Neil Postman, I approach the reification of human consciousness from the perspective of a commitment to efficiency that touches on technology without, for all that, singling out specific innovations like the mechanical clock, the printing press, the railway, the radio, the camera, or the computer. The logic of unintended consequences in Postman's *Technopoly* (1992) in many ways complements aspects of the ideological implications of our commitment to efficiency at issue in my analysis of efficiency. The 'case studies' in his emphasis on technology could be used to reinforce my arguments, just as my 'case studies' might find a comfortable home in *Technopoly.* The example of the mechanical clock Postman borrows from Mumford could be used to demonstrate the disruptive potential of residues of 'waste' in the process of transforming useless into useful energy. The 'mechanical clock' that the Benedictine monasteries introduced to 'provide precision to the rituals of devotion' inadvertently became an instrument destructive of the contemplating life in that 'the clock is a means not merely of keeping track of the hours but also of synchronizing and controlling the actions of men' (Postman 14), which became central to Henry Ford's production process. If this study of efficiency does not accentuate technology, it is partly because this topic has already received extensive analysis. But it is also because my literary 'case studies' influenced the overlapping areas of interest taking centre stage in this study. With the possible exception of Huxley's *Brave New World*, the literary texts under discussion were only marginally concerned with the technological enemy as such.

Modernism and the Culture of Efficiency is, then, divided into two parts, the first consisting of four chapters and the second of six. Part 1, 'The Culture of Efficiency in Society,' focuses on some efficiency-driven objects, figures, institutions, and events that have become embedded in the imagination of Western culture. The Great Exhibition of 1851 symbolizes the arrival of efficient machines, the assembly line at Ford Motor denotes efficiency gains in the production process, and the stopwatch of F.W. Taylor's 'efficiency expert' signifies the efficient management of human beings. These cultural images suggest a connection between a drive for efficiency and the disciplining of 'bodies' both through violence and through ideology. This first part concludes with an analysis of the impact of efficiency on Western society by considering the extreme-limit case of Auschwitz as an efficient killing machine and, at the other end of the spectrum, the disciplinary cost of the pleasures afforded by the capitalist consumer society epitomized in the American suburb. These two instances are meant to stress my agreement with Alexander that efficiency 'cannot be an intrinsic value,' but ought to be viewed as 'an instrument for implementing the other values that define the context of its use' (Alexander 5). It is, then, not a question of choosing between desirable and undesirable forms of efficiency. In the first instance, the two forms are so interdependent that they tend to reinforce each other. Although the efficient management of people in factories and suburbs cannot be equated with the horrors of mass murder, there lurks in even the most benign applications of efficiency the dark underside of coercion. But even in the case of Auschwitz, efficiency retained many of its more positive attributes. At the same time, celebrations of the efficiently organized factory, office, or suburb tend to disavow the signs of coercion clinging to processes of normalization. It is only when efficiency turns murderous that we unambiguously recoil from it; in most other instances, we tend to remain ambivalent about its benefits and its costs. Far from seeking to determine some 'proper' use of efficiency, this study means to historicize the ideological tensions and contradictions of this complex concept. What is of interest is the transmogrification of efficiency from an *instrumental* function (or means) to an *ideological* investment (or end in itself) manifesting itself in often unacknowledged ways within certain historical contexts. This tracing of a pattern risks giving the false impression that efficiency is being elevated to a monocausal explanation for historical events that have many different causes. The significance of the Great Exhibition of 1851, for instance, far exceeds the narrow confines of the efficiency calculus; far from anticipating only the modernist anxieties central to this study, it was emblematic of a broad array of specifically Victorian preoccupations.[5] Along similar lines, the importance here attributed to the role efficiency played in the case of Auschwitz clearly brackets many other contributory factors, including the long history of anti-Semitism. And the American suburb is a far more complex phenomenon than can be accounted for by the framework of Taylorization alone. There is, of course, always more history to explain the history underpinning the emergence of efficiency as an emergent ideology at the beginning of the twentieth century. It can only be hoped that the delineation of the logic of efficiency in this study justifies whatever dangers there may be of historical oversimplification. In short, the analyses in part 1 draw on historical reference points that are of interest in themselves but also serve as contexts for the exploration of fictional representations of the efficiency calculus in the second part.

Part 2, 'The Culture of Efficiency in Fiction,' explores the logic of efficiency as it manifests itself primarily in a selection of novels written by culturally representative and influential 'canonical' modernists and published in Britain during the first three decades of the twentieth century. The interpretations of these texts seek to uncover a preoccupation with the emergence of efficiency as a dominant ideology that neither authors nor critics have generally noticed or acknowledged. It is only through a thorough analysis of the ideological complexities raised by Fordism and Taylorism that we can appreciate the full extent of the contributions the novels under consideration have made to our understanding of social change during the modernist period. The chapters on Conrad's *Heart of Darkness,* Lawrence's *Women in Love,* Ford's *The Good Soldier,* Forster's *Howards End,* and Huxley's *Brave New World* present close readings of individual novels that are informed by the historical contexts and theoretical analyses provided in part 1. One chapter departs from this focus on single novels as 'good examples' of a broader cultural trend; the discussion of eugenics deals with responses to efficiency in various essays, letters, novels, and (in the case of Shaw) plays of four figures (Shaw, Wells, Orwell, and Forster). The intent of part 2 is to suggest the extent to which efficiency infiltrated the social fabric treated in the novels, the various ways in which it surfaced in different novels, and the conflicting perspectives it generated. It is my hope that this interdisciplinary focus on the theme of efficiency will result in a reconsideration of both canonical literary figures and the socio-economic debates in which they participated more productively than has previously been acknowledged.

PART ONE

The Culture of Efficiency in Society

1 Efficiency and the Great Exhibition of 1851: Elation and Doubt

At the Great Exhibition of 1851 at London's Crystal Palace, the Machine Age openly celebrated the drive to increase the efficiency of everything from locomotives to household gadgets as evidence of human progress. Visitors to the Crystal Palace were not only amazed by spectacles of human ingenuity, but assumed that the Machine Age would usher in a future of greater leisure and prosperity for all. There were, of course, those who warned that a social cost would have to be paid; most famously, the Luddites (1811–16) burned the looms that threatened the livelihood of weavers while Marx and Engels documented the appalling working conditions endured by those labouring in factories. But the protests of workers were more or less drowned out by the jubilant endorsement of the production of wealth destined to empower the emergent bourgeoisie. Not surprisingly, this new commercialism was an affront to the refined sensibilities of the fading aristocracy; for the landed gentry, industrialization threatened the culture of polite society and the idyllic vistas of unspoiled countryside they enjoyed from their mansions. In the simplest terms, then, the machine symbolized the historical shift from an agrarian to an industrial mode of production. That the machine as a symbol of society's investment in efficiency proved to be ideologically more conflicted than meets the eye emerges from responses to the machine in general and the Great Exhibition in particular. In 1944, E.M. Forster commented on the changes wrought by the machine in a way that typifies an ambivalence that had been widespread among British intellectuals and artists:

> There is a huge economic movement which has been taking the whole

> world, Great Britain included, from agriculture toward industrialism. That began about a hundred and fifty years ago, but since 1918 it has accelerated to an enormous speed, bringing all sorts of changes into national and personal life. It has meant organization and plans and boosting of the community. It has meant the destruction of feudalism and relationship based on the land, it has meant the transference of power from the aristocrat to the bureaucrat and the manager and the technician. Perhaps it will mean democracy, but it has not meant it yet, and personally I hate it. So I imagine do most writers, however loyally they try to sing its praises and to hymn the machine. (Forster, 'English' 278)

The significance of this exhibition itself can be measured by the extensive scholarly literature it has spawned; no other has been discussed as often or at such depth. Where the earlier critical focus tended to be on the machine as a primary symbol of economic *production,* more recent analyses discuss the display of commodities as a space of *consumption.*[1] But, for my purposes, the most insightful reaction to this 'functioning microcosm of mid-Victorian capitalism' (Richards 38) and incipient symbol of twentieth-century modernity was produced by an outsider, the Russian novelist Fyodor Dostoevsky, whose apprehensions about the ideological implications of the Crystal Palace have been the subject of a most astute analysis by Marshall Berman in his seminal study *All That Is Solid Melts Into Air: The Experience of Modernity* (1982).

The Great Exposition of 1851 signalled that industrialization 'had come of age' (Arnstein 66), such that, in Macaulay's words, '1851 would long be remembered as a singularly happy year of peace, plenty, good feeling, innocent pleasure and national glory' (66). In addition to improvements in agricultural output, Britain, for instance, saw annual increases in coal and steel production that translated into an expansion of the railway network and a boom in the shipbuilding industry, all of which contributed to 'massive statistical evidence of progress and prosperity' (66). This age of prosperity seemed poised to fulfil the high expectations of Enlightenment philosophers like Descartes, d'Alembert, Condorcet, and Diderot. In the mid-nineteenth century, it must have looked as if the material achievements celebrated in the Great Exhibition of 1851 corresponded to the philosophical aspirations of Enlightenment thinkers, which Robert Pippin summarizes as follows:

> A view of nature as to be mastered, not contemplated; a 'mathematizable' and materialistic view of nature; a rejection of final causes in explanation;

> compared with antiquity, a much more 'realistic' view of the ends to be achieved by knowledge, ends such as health, pleasure, freedom from pain, and not, say, 'wisdom'; an expectation of great social benefits from the free and unimpeded pursuit of scientific knowledge, and a corresponding assumption that the fundamental cause of human injustice was scarcity, that this problem could be corrected; and a general belief in the progressive and politically ever more enlightened course of human history. (Pippin 20)

In the self-understanding of the nineteenth century, advances in the material conditions of existence were thought to be the prerequisite for advances in social arrangements and cultural accomplishments. Science and art would combine to ensure the material and spiritual emancipation of all humankind.

Initially planned as 'a Grand Annual Exhibition of Manufactures' (Hobhouse 3), the Great Exhibition only caught the public imagination when Henry Cole exerted his influence to bring 'the Art Manufactures' (Hobhouse 5) into the picture. At this time, the machine still required the cultural legitimacy that only art could provide. As Prince Albert put it, aesthetic sensibilities would both inform and ennoble the technological innovations of the age. The exhibition was to be housed in the Crystal Palace, a building that was itself considered a marvel not only of modern engineering but also of modern taste. Aside from celebrating art in a hall of sculpture, the exhibits were generally arranged so as to teach visitors an appreciation for aesthetic order and harmony.

The machines celebrated at the Great Exhibition were above all considered to be further evidence of the kind of human ingenuity previously reserved for artistic creation. This aestheticization of the machine reinforced ideological discourses touting the unification of classes and nations under the sound guidance of Britain's practical judgment and democratic sense of justice. Such lofty ambitions and claims were from the start subjected to criticisms pointing to the commercial ends that such aesthetic and utopian rationalizations tended to conceal. From our own retrospective position, the emphasis on economic production denoted by the prominence of machines distracts from the injunction to consume, an injunction symbolized by the display of 'manufactured objects' or commodities that appeared as 'autonomous icons ordered into taxonomies, set on pedestals, and flooded with light' (Richards 4). This first 'world's fair' thus initiated 'a stable system of representation for commodities' and constituted 'a pivotal moment in the history of advertising' (Richard 3, 53); it also ushered in the 'era of the spectacle'

(3) later theorized by the likes of Guy Debord, Jean Baudrillard, and Fredric Jameson.

In the first instance, the Exhibition of 1851 is remarkable in that the planners understood themselves to be organizing a display of international goodwill; the advances in the standard of living in Britain, made possible by the increased efficiency of machines, were presented as the prelude to similar advances throughout the world. Offering half of the exhibition space to foreign nations, Prince Albert prided himself on the promotion of international peace through commerce. In a speech that Yvonne Ffrench (*The Great Exhibition: 1851* [1950]) quotes in full, the prince points out that 'we are living at a period of most wonderful transition which tends rapidly to accomplish that great end, to which, indeed, all history points – *the realisation of the unity of mankind*' (51). No longer the domain of a small elite, knowledge is public, so that 'no sooner is a discovery or invention made than it is already improved upon and surpassed by competing efforts' (52). Endorsing free-trade principles, he continues: 'The products of all quarters of the globe are placed at our disposal, and we have only to choose which is the best and the cheapest for our purposes, and the powers of production are entrusted to the stimulus of *competition and capital*' (52). In Albert's view, God has given us reason to tame nature and turn it to our advantage: 'Science discovers these laws of power, motion, and transformation; industry applies them to the raw matter which the earth yields us in abundance, but which becomes valuable only by knowledge. Art teaches us the immutable laws of beauty and symmetry, and gives to our productions forms in accordance to them' (52).

In line with the prince's conception, the exhibits were first of all arranged by nation and then divided into Raw Materials, Machinery, Manufactures, and Sculpture. As Hermione Hobhouse glosses a report written by the organizing committee, these divisions 'made up a logically consistent progression from the "raw materials which nature supplies to the industry of man," to "the machinery by which man works upon those materials," then the "manufactured articles which he produces," and finally "the art which he employs to impress them with the stamp of beauty"' (Hobhouse 40). This organization corresponds to 'a four-fold Aristotelian arrangement of material, efficient, final, and formal causes: raw materials, machinery, manufactures, and fine arts' (Miller 52). In these divisions we detect the self-confidence of the nineteenth century as the fulfilment of the Enlightenment legacy, which held that the taming of nature through the application of reason would result in the perfectibility of humankind. For Richards, such utopian aspirations reinforced capitalism and empowered the middle classes by promoting a commodity culture thriving on 'the myth of the achieved abundant society,' a vision that 'was not so much of a classless society as of a society in which everyone was equal in the sight of things' (66, 61). Implicit in this utopia are the reduction of the human being or citizen to the consumer and the perversion of the Enlightenment notion of reason into its utilitarian dimension. The emphasis on the invention of efficient machinery to produce a plethora of consumer goods exemplifies the instrumentalization of reason that sociocultural critics and novelists target as evidence of the impoverishment of the human imagination and creative urge.

The official title of the Exhibition, 'The Great Exhibition of the Works of Industry of all Nations, 1851,' leaves no doubt that the focus was unabashedly industrial. However, debates within the committee and reactions in the press illustrate that aesthetic taste remained a major preoccupation. Designed by the architect Joseph Paxton, the Crystal Palace housing the Exhibition was itself a source of constant amazement. Entering through the spacious transept with its Fountain of Glass, the 'exhilarated millions who strolled the miles of galleries among greenery, flags, machinery, art works, amid visitors from all over the world, might well believe that they were witnessing the birth of an age of inconceivable progress and tranquility' (Mandell 9). Anticipating the emergent ideology of the commodity spectacle destined to culminate in the department store and the shopping mall, the Crystal Palace also instantiates an attachment to the residual ideology of aesthetic high culture and patriarchal stewardship.

The residual attachment to cultural notions of aesthetic taste and social responsibility was clearly overshadowed by the endorsement of the machine as the high road to 'Prosperity, Propriety, and Progress.'[2] Although the focus in both Richards and Miller on 'the things produced (commodities)' rather than on 'the means for producing the world (machines)' (Richards 57) illuminates the 'changing view of commercial activity' (39) in mid-Victorian England, it unnecessarily downplays the excitement generated by inventions that visitors to the Crystal Palace greeted with an amazement we can no longer muster in our own jaded time. Seen as a 'vivid symbol both of Victorian material progress and of the sense of self-satisfaction to which it gave rise' (Arnstein 64), the Exhibition was noted, as even Richards concedes, not for 'articles common to everyday life' but for the display of 'mechanical wizardry' (40).

With the technological 'wonders of the new industrial world' (Arnstein 64) on display, the main attractions were a 'sewing machine,' a 'medal-making machine that could produce fifty million medals in a single week,' an 'electric telegraph office'(Arnstein 64), steam engines, the highly popular Shepherd's electric clock, and ingenious contraptions such as 'an alarm bedstead' and 'the British equivalent of an automatic baseball pitching machine, a "cricket catapult"' (Arnstein 64). Based on contemporary reports, Ffrench finally surmises that 'enormous locomotives, rotating machines, and the subjugation of mysterious natural forces, were awe-inspiring sights. To stand in the machine-room and be deafened by the pandaemonic sound of hissing steam and clanging metal was, we read, a deeply moving experience' (Ffrench 215). The textile section seems to have been of particular interest, displaying various looms that were 'driven at 220 picks a minute' (225), easily outperforming the old machines, which could only manage a mere 60 picks a minute. The emphasis in the celebration of all this machinery was either explicitly or implicitly on the efficiencies achieved through the potential mass production of consumer goods. Whether we concentrate primarily on production or consumption, it is undoubtedly the case that the Exhibition owed its popularity to its ability to successfully integrate the 'paraphernalia of production into the immediate phenomenal space of consumption' (Richards 30).

In spite of some early opposition to the Exhibition, it turned out to be a great success, reputedly attracting over six million visitors. However, while it was clearly a testament to British industrial might and to the emergent commodity culture, it did not live up to some of its other promises. The brotherhood among nations rings today especially hollow in that we have become more and more critical of the exploitative practices in the colonies underpinning British industrial supremacy. From our postcolonial perspective, we are likely to accuse Britain of bad faith as it self-righteously preached a liberal order meant to empower the very people being kept subservient to British colonial interests. In addition, Prince Albert's self-congratulatory connection between technological advances and the flourishing of the arts seems to have been undermined by the artistic exhibits at the Crystal Palace itself. The art exhibits are most charitably described as overly ornate and less charitably as 'kitsch.' Even the generally over-enthusiastic Yvonne Ffrench castigates 'the fine arts of 1851' as 'examples of the bastardisation of taste without parallel in the whole recorded history of aesthetics' (230).[3] Ffrench even throws some cold water on the optimism of the age. Quoting from an article in a contemporary magazine that outlines the savings introduced by modern looms, she draws attention to the potentially damaging effect of mass production on working people: 'Wonderful mechanical result! What are the moral results?' (226). Indeed, as both Richards and Miller stress, the role of labour in the production of the commodities on display has been elided: with 'goods from all points in the globe' juxtaposed, the commodities 'appeared to have come out of nowhere' (Richards 20), so that 'their original contexts' were apparently eliminated (Miller 51). Moreover, fear that the presence of workers and foreigners visiting the Crystal Palace might endanger the middle-class spectators for whom the Exhibition was primarily mounted, testifes to the failure of the 'world's fair' to overcome class and national divisions. The prince's lofty social goals of universal cooperation and aesthetic celebration were undoubtedly compromised from the start by his promotion of 'competition and capital' pitting everybody against everybody else.

The emphasis on competition was reinforced by the awarding of medals and prizes at the Exhibition measuring the success and failure of individuals, companies, and nations in terms of efficient outcomes. As John Kemper's recent analysis[4] suggests, the Great Exhibition of 1851 is marked by close and contradictory links between free trade (competition), colonialism (exploitation), and liberal values (goodwill). It was through private business initiatives that Britain was in effect able to keep control of an increasingly complex colonial empire, and private business was, in turn, driven by the competitive spirit engendered by free trade. In this competitive arena, then, the most successful contestants were those able to use available resources most efficiently. In Kemper's words, 'Efficient outcomes, believed by intellectuals and capitalists alike to yield benefits to society as a whole, were the inevitable outcome of this encouraged competition' (Kemper 2). The Great Exhibition already hints at the close connection between the emergence of the machine and the emergence of the corporation with its demand for a free and potentially global market. In this context, it is undoubtedly significant that early opposition to the Exhibition in its planning stages came from commerce and industry, spearheaded by the protectionists who feared that the international constitution of exhibitors signalled the triumph of free trade. The defeat of the protectionists opened the door to global corporate capitalism.

The Great Exhibition of 1851 thus provides us with an early glimpse of efficiency as an emergent ideology. Although punctured by the great depression of 1873–95, the second half of the nineteenth century saw a veritable explosion of inventions that would carry on into the twentieth century. This period reinforced close connections between celebrations of devices making life more comfortable for ordinary people and competition among innovators and their financial supporters to supply markets with consumer goods. Among many other inventions in the late nineteenth and early twentieth centuries, we find the creation of the London Metro in 1860, the first transatlantic cable in 1866, the invention of the bicycle and typewriter highlighted at the Paris Exposition of 1867, the creation of Standard Oil by Rockefeller in 1870, Bell's invention of the telephone in 1877, Edison's invention of the microphone and phonograph in 1877, Daimler's internal combustion engine in 1886, Dunlop's rubber tire in 1888, Peugeot's first car in 1895, the Lumière Brothers' silent movie in 1895, the Diesel engine in 1897, Ford's Model T in 1908, the first transatlantic flight in 1919, and the talking movie and black-and-white TV in 1926. These inventions were accentuated by the construction of the first skyscraper in Chicago in 1884 and the Eiffel Tower in 1889.[5]

The splendours of the Great Exhibition of 1851 did not convince everybody that machines were changing the lives of people everywhere for the better. Opposition to the new Machine Age is famously entrenched in our cultural imagination by the Luddite revolt of 1811. Largely an agricultural nation not yet connected by a network of railroads, Britain at the beginning of the nineteenth century was characterized by goods produced in local workshops often located in people's homes. But by 1811 this localized and agriculturally based economy was plunged into crisis by declining exports as a result of Napoleon's protectionist measures and the enclosure movement forcing small landowners to migrate to the cities, where they were exposed to low-paying employment in factories with poor working conditions. The Luddite revolt started in the Midlands, where mass production in the knitting industry destroyed home-based manufacture. In addition to knitters in the Midlands, other workers in home industries found that their earning power and working conditions were rapidly deteriorating. Starting on 11 March 1811, frame knitters in Nottingham broke the frames of manufacturers who were exploiting them. In the ensuing weeks, over two hundred frames were broken throughout the region. In November of 1811 Ned Lud led a mob of disgruntled workers who broke several frames of a much-hated manufacturer; his act of rebellion spread rapidly and came to be associated with a desire to turn the wheel of time back to a time before modernity destroyed the old agricultural way of life. The Luddites are today associated with those resisting change. Confronted by the negative impact of modernization on wages and working conditions, the Luddites were indeed the first public dissenters to draw attention to the plight of the working classes, a plight later analysed in the writings of Karl Marx and Friedrich Engels, most notably in the latter's *Condition of the Working Class in England* (1845), published only a few years before the Great Exposition.

In spite of early enthusiasm for machines and gadgets, the Great Exhibition of 1851 lingers in the cultural memory less for what it contained than for the Crystal Palace building itself. Paxton's architectural extravaganza lives on as an ambiguous emblem of mid-Victorian attitudes and aspirations; it was both an expression of admirable human endeavour and also an ominous sign of the human spirit's degeneration into routine activities. The construction and fate of the Crystal Palace make the building itself a tribute to efficiency; using the 'most advanced modes of prefabrication,' it was built 'in six months in Hyde Park to house the Great International Exhibition of 1851' and then 'disassembled in three months' (Berman 237) at the end of the Exhibition, only to be reassembled on Sydenham Hill in 1854. This 'fairy palace' was a testament to the efficiency calculus later embodied at the Ford Motor Company; 'the rationalization of parts,' the use of 'standardized construction units,' and a 'large work force assembled for the job' allowed for 'the building to be rapidly assembled' (Miller 60) and later disassembled. In its dazzling elegance, the building fused aesthetic taste and industrial functionalism. But as a feat of engineering, it was also a model of exact planning and regimentation that presaged the nightmare vision of the modern cityscape. In his analysis of specifically literary reactions to the Crystal Palace, Berman highlights in *All That Is Solid Melts Into Air* the sociocultural implications of the 1851 Exhibition at issue in reactions to the commitment to efficiency that will mark the Ford Motor Company and its subsequent satirical treatment in Huxley's *Brave New World.*

Pointing out that the Crystal Palace had little impact on English literature,[6] Berman stresses that it became a prominent symbol in Dostoevsky's *Notes from Underground* (1864) and in Nikolai Chernyshevsky's *What Is to Be Done* (1863). Both Russian novelists had visited the Crystal Palace after it had been moved to London's Sydenham Hill.[7] Puzzled by the discrepancy between Dostoevsky's condemnation of the building as 'all ready-made and computed with mathematical exactitude' (in Berman, 236) and the more customary positive assessments stressing its 'lyrical' as well as 'visionary and adventurous' (237) dimensions, Berman surmises that Dostoevsky's rancour should be attributed to the envy experienced by a man from a backward country when confronted by this symbol of modern vision and development. However, by the end of his analysis, Berman concedes that Dostoevsky should be appreciated as an early 'prophet' who grasped the dark underside of this marvel of Enlightenment progress: 'His critical vision of the Crystal Palace suggests how even the most heroic expression of modernity as an adventure may be transformed into a dismal emblem of modernity as a routine' (248). As we will see, this transformation is particularly pronounced in the history of the Ford Motor Company.

In Berman's view, then, the Crystal Palace is both a monument and a threat to the creative human spirit. Far from being 'mechanically conceived and realized,' as Dostoevsky alleges, the Crystal Palace was, in many ways, the 'most visionary and adventurous building in the whole nineteenth century' (Berman 236, 237). Where many saw the Crystal Palace as the apotheosis of Victorian prosperity, Dostoevsky understands the building to signify not only a 'historic culmination' but also a 'cosmic totality and immutability' (236). In other words, Dostoevsky sensed that the dynamic energies and creative imagination that had gone into the engineering of the Crystal Palace had produced a static edifice hostile to the tumultuous social life resistant to the planned and regimented space of Paxton's achievement. Dostoevsky's diatribe against the Crystal Palace as an expression of 'arid Western rationalism, materialism, the mechanical view of the world' nevertheless acknowledges engineering as the 'primary symbol of human creativity' (241, 242). While Dostoevsky approves of the human desire to 'create and build roads,' he claims that 'man' is 'instinctively afraid of attaining his goal and completing the edifice he is constructing' (242). Endorsing the *process* of constructing a building, Dostoevsky deplores the completed edifice as a dead *product* antithetical to human creativity: 'The activity of engineering, so long as it remains an activity, can bring man's creativity to its highest pitch; but as soon as the builder stops building, and entrenches himself in the things he has made, the creative energies are frozen, and the palace becomes a tomb' (243). Or, in Dostoevsky's memorable words, man 'only likes to build [the edifice], and does not want to live in it' (242).[8] Berman concludes that there are two 'different modes of modernization: modernization as *adventure* and modernization as *routine*' (243). This differentiation also captures and anticipates the ambivalent attitude of the early twentieth century towards the promises and the perils associated with a new emphasis on the efficient execution of tasks in all areas of modern life.

Dostoevsky's perception of the symbolic significance of the Crystal Palace reveals itself in retrospect to have been prophetic in that it anticipates the death of public spaces in our cities and suburbs. The pleasantly planned Crystal Palace, with its elegant lines and its efficiently arranged flow of people through the exhibits, is the prototype of the modern mall and the gated community. Nature itself has not so much been subdued as incorporated. Paxton's building 'envelops rather than obliterates nature'; the 'great old trees' on its terrain are 'contained within the building,' giving it the look and feel of a 'greenhouse' (Berman 237). Viewed from a distance, 'the building closed itself off to outside scrutiny,' elevating the 'phantasmagoria of commodity culture' (Richards 23, 18) into a self-contained space of mobile consumers passively inspecting a proliferation of isolated and interchangeable objects.

Unlike Dostoevsky, the other Russian visitor to the Crystal Palace, Chernychevsky, applauded such inside/outside confusions to advance a utopian alternative to the messy cities of his homeland. For him, the 'new antithesis to the city,' explains Berman, 'is no longer the primitive countryside, but a highly developed, super-technological, self-contained exurban world, comprehensively planned and organized – because created *ex nihilo* on virgin soil – more thoroughly controlled and administered, and hence 'more pleasant and advantageous,' than any modern metropolis could ever be' (244). In his novel, he envisions 'a brave new world' (245) devoid of conflict and struggle.

Dostoevsky's reaction is filtered through Chernychevsky's social-engineering project, a utopian vision also satirized by Evgeny Zamyatin's *We,* a dystopian novel that heavily influenced Aldous Huxley's *Brave New World.* Zamyatin's novel conjures up 'a brilliantly realized visionary landscape of steel-and-glass skyscrapers and glassed-in arcades' (Berman 247); using ice as his dominant motif, Zamyatin condemns the 'freezing of modernism and modernization into solid, implacable, life-devouring forms' (247), an image that eerily announces the architecture of Henry Ford's River Rouge factory in Michigan. This image also relates to postmodern urban spaces that are, for Fredric Jameson, typified by John Portman's Westin Bonaventure Hotel in Los Angeles.

Jameson's analysis of the Bonaventure Hotel in his introduction to *Postmodernism or, the Cultural Logic of Late Capitalism* (1991) confirms

Berman's concluding claim that the 'real twentieth-century reincarnation of the Crystal Palace turned out to take place half a world away, in the U.S.A.' (248). The incorporation of outside spaces within the inside of the building that Paxton initiated finds its ultimate culmination in Portman's architectural style. With its 'great reflective glass skin,' the hotel 'repels the city outside' (Jameson 42), accentuating an emphasis on its interiority as hermetically sealed off from the outside world, a hyperspace of constant movement that violates 'the coordinates of an older space' (44). This hyperspace 'has finally succeeded in transcending the capacities of the individual human body to locate itself, to organize its immediate surroundings perceptually, and cognitively to map its position in a mappable external world' (44). Portman's architecture is symptomatic of the 'incapacity of our minds, at least at present, to map the great global multinational and decentered communicational network in which we find ourselves caught as individual subjects' (44). It is not insignificant that the Crystal Palace catered to the machine as a mutation in the mode of production while the Bonaventure Hotel gratifies the society of consumption having evolved from technological advancements. The trajectory from Paxton's modern building to Portman's postmodern one may well be symptomatic of the logic that drives efficiency from serving human aims to becoming an end in itself.

By the end of the century, when the Exposition of 1900 opened in Paris, the contradictory logic that had motivated the Great Exhibition of 1851 began to manifest itself more openly, showing a crack in the armour of nineteenth-century optimism. While there were some technical innovations,[9] the general emphasis was not so much on novelty as on improvements to already existing inventions: 'Railway engines, blast furnaces, cranes, and tractors were larger, faster, cheaper, and incredibly more efficient. In a march of material progress that shocked some observers, new machines rendered outmoded and, as if by magic, transformed into junk those that were the ultimates in efficiency just a few years earlier' (Mandel 68).[10] Moreover, as if to contradict Prince Albert's smug embrace of free trade as a pillar of a brotherhood of nations, the competitive spirit at the 1900 Exposition 'seemed to aggravate the already strained relations between nations' (Mandel 105) that were to explode in 1914.[11]

The 1900 Paris Exposition came to be seen as marking a 'profound psychological change' (Mandell 110), signalling the end of nineteenth-century confidence and optimism. 'In technology,' observes Mandell, 'there was not much that was new' (113), and the architecture of the exposition was deplored as confusing and overly conservative. In art, Paris 1900 was 'Europe's last great exhibition of allegory and the personification of virtues' (73). In other words, the many groups of colossal sculptures 'appeared strident and over-ripe' (73). Mandel sums up the 1900 Exposition in Paris as follows: 'Nothing the exposition might have produced could reach the heights that its enthusiasts had hoped for. But failure of the exposition went deeper. To nineteenth-century intellectuals raised on faith in science, reason, and progress, it seemed that the most complete and expensive demonstration to celebrate science, reason, and progress produced an impression of human uselessness, finiteness, and debility' (117). Indeed, a 'new, cold attitude toward world's fairs was most evident in France, but was also manifest elsewhere as the twentieth century began' (118). Although expositions and world's fairs continued to be mounted, the Paris 1900 Exposition had deflated the 'happy illusions that held the nineteenth-century intellectual consensus together' (118). It seems that in 1900 it was no longer possible to maintain the innocent enthusiasm of 1851. The Paris Exposition constitutes a turning point in the self-understanding of sophisticated Europe; 'after 1900 it seemed apparent to anyone who thought about it that material progress had far outstripped moral and social progress' (115). By the turn of the twentieth century, efficiency gains through technological advances gave rise to cultural anxieties that were to manifest themselves in various forms of resistance. This backlash focused primarily on the assembly line introduced by Henry Ford in 1913 and on the 'scientific principles' advanced by the 'efficiency expert' Frederick Winslow Taylor. While the celebration of efficiency at the Great Exhibition of 1851 was generally greeted with enormous enthusiasm, Dostoevsky's reservations and apprehensions proved prophetic, anticipating the contradictions and ambivalences that would surround the increasing commitment to efficiency both in the socioeconomic sphere and in its fictional representations.

2 Efficient Machines and Docile Bodies: Henry Ford and F.W. Taylor

Henry Ford's Model T remains to this day lodged in the popular imagination for being virtually synonymous with innovative engineering, a revolutionary production process, and an imaginative marketing strategy. The Ford Motor Company was contradictorily both a testament to the American entrepreneurial spirit and an emblem of the dehumanization of individuals and the homogenization of society. Although Ford applied for his first patent in 1898, it was not until 1908 that he produced the model T and not until 1913 that he introduced the assembly line. Although his roots were in the nineteenth century, Henry Ford is the quintessential 'modern' whose vision transformed the world. While the company he built up remained under his own control until shortly before his death, Ford Motor gradually evolved into a modern corporation or 'system' that compelled even its founder to adapt to its demands. Aside from the centrality of Ford in Huxley's *Brave New World,* Ford Motor imposes itself as an exemplary and well-documented case study of the role efficiency played in the socio-economic and cultural arenas in the early twentieth century. Henry Ford himself and the Ford Motor Company have been the subject of so many studies[1] that there is little controversy about the basic facts of their history. This analysis of Fordism is not meant to reinterpret this history but to cull from it a pattern illustrating not only the enormous influence of efficiency on the development of this American phenomenon but also the complexities and unintended consequences that flowed from it. To this end, I rely heavily on two excellent recent studies of Ford and his company: Douglas Brinkley's *Wheels for the World: Henry Ford, His Company, and a Century of Progress, 1903–2003* (2003) and Steven Watts's *The People's Tycoon: Henry Ford and the American Century* (2005). Brinkley's emphasis on the company and Watts's focus on the man complement each other without disputing each other's findings. Between them, they provide the cultural theorist with invaluable documentation and analyses of a phenomenon that is meant to inform and frame the literary responses to efficiency as an emergent ideology. In short, the intention here is neither to enhance nor to dispute the historical record that finds its current culmination mostly in Brinkley's rigorous and highly acclaimed study of the Ford Motor Company; the close reading of this historical record is meant to alert the literary-cultural establishment to the complexities of such apparently 'obvious' terms as 'Fordism,' 'Taylorism,' or 'Americanization.'

The chapter opens with a narrative of the dehumanization that occurred at the Ford Motor Company as an unintended by-product of Henry Ford's obsession with technical efficiency in the design and production of machines. Although the assembly line materialized as a logical solution to practical problems, it subordinated human beings to the demands of the production process. Yet it was not Henry Ford but F.W. Taylor who turned his attention to the techniques best suited to manage human beings. His book *The Principles of Scientific Management* is mobilized by a utopian desire to find a solution to the detrimental effect of class antagonisms on productivity. As events at Ford Motor demonstrate, Taylor's utopian ideal almost immediately resulted in the reduction of workers to mere cogs exposed to the disciplinary mechanisms of an indifferent social 'machine.' Departing from Ford's total-output model of efficiency, Taylor meant to institute a system premised on the optimal distribution of efforts contributed and rewards reaped for all participants. Central to his utopian ideal is the assumption that rational agents will choose to cooperate so as to optimize social efficiency. However, as game theory suggests, this reliance on a rational-choice model ignores the influence of intersubjective expectations on how individuals act. Although human behaviour as such could be said to be resistant to all social-engineering projects, game theory alerts us to the specific *systemic* features that both enable and undermine Taylor's utopian aspirations. Whereas Ford's assembly line conjures up images of workers oppressed by external forces, Taylor's principles of management compel human beings to reify their consciousness by internalizing the ideology of efficiency.

The Model T and the Assembly Line

The Model T initiated both freedom of movement for the masses and oppressive conditions for workers. Bucking the trend of competitors,

Ford turned his back on luxury cars, putting his efforts into manufacturing an affordable car for the masses. Intending the Model T to serve the needs of ordinary people, he agreed to few adjustments to its original design, making few concessions to aesthetic taste or drivers' comfort. Its most remarkable feature was precisely that it had none. The Model T came with few options; the joke of the day was Henry Ford's 'slyly generous offer that "you can have a Model T in any color you want, as long as it's black"' (Brinkley 181–2). The key to keeping the price down and production up was *standardization.* Until the Model T was discontinued in 1927, Ford used the same machine parts to eliminate costly recalibrations. Instead of concentrating on designing a better car, he became obsessed with streamlining the production process: 'Not much of importance may have changed on the T to make it new, improved, or different during its nineteen-year model run, but nothing remained the same about the methods used to produce it' (151). An unabashed proponent of efficiency, Ford doggedly chased the always receding promise of perfecting the manufacturing process. The appeal of the Model T, therefore, resided not in its speed, design, or drive for novelty but in its affordability. Ford was so committed to cost efficiency that he often risked losing customers by refusing to offer convenient options that had become standard for competing car manufacturers.

It was Ford's anti-elitist desire to create a 'universal car' for the people that necessitated the rationalized production process of the assembly line. Although Andrew Carnegie was the first industrialist to introduce 'the "line production" system,' it was Henry Ford who perfected it by designing 'conveyors, rollways, and gravity slides' to improve the 'flow of materials' (Micklethwait and Wooldridge 64). Ford's real 'stroke of genius was to introduce conveyor belts to move parts past the workers on the assembly line. This reduced the time it took to make a Model T from twelve hours to two and a half hours' (65). In Ford's early days at the Piquette Avenue factory, the realignment of 'machine placement for smoother flow' and the training of 'workers in more efficient production methods' (Watts 103) were already in place. But it was when Norval A. Hawkins was hired in 1907 that the idea of the assembly line took actual shape: 'Obsessed with logistical efficiencies, particularly those achieved by the ways of loading cars and parts into railroad boxcars,' Hawkins was able to 'fine-tune the sales organization' (Brinkley 99). Introducing an ominous comparison, Brinkley indicates that the Piquette Avenue factory 'hummed like a beehive, comparable in intensity of activity to the Manhattan Project that developed the atomic bomb during World War II. Both efforts were multifaceted, yet well coordinated in order to foster the creation of devastatingly practical products. That sort of outcome results only from an insistence on both efficiency and efficacy in every single task' (105). While 'conflicting stories about the origins of the assembly line circulated widely' (Watts 141), there seems little doubt that Hawkins was an early influence; however, it is Clarence W. Avery, an engineer said to be familiar with Frederick Winslow Taylor's principles of scientific management, who is credited with the 'mature form of the assembly line' (143), generally dated from 1913.

Once Piquette Avenue proved too small to accommodate the new production process, Ford planned his new factory at Highland Park, near Detroit. With an emphasis on efficiency rather than expense, Ford planned the arrangement of buildings so as to ensure the efficient flow of the production process. Growing out of the desire to 'improve the product at less cost' (Brinkley 151), the first assembly line was implemented at Highland Park in 1913. Ford's 'marvelous new factory' has been described as 'a monument to American productivity' that, 'with its class-encased roof,' was interestingly enough 'dubbed the "Crystal Palace"' (Watts 135). In an atmosphere in which 'improvement was the real product,' efficiency became 'the company's most important concern' (Brinkley 151, 156). The introduction of the assembly line resulted in impressive productivity gains for Ford Motor; 'production nearly doubled every year for a decade after 1913' at the same time as the 'price of a Model T dropped by two thirds' (155). But the assembly line at Highland Park came to symbolize the subordination of workers to the machines they operated. In *Abroad at Home* (1914), Julian Street was one of the first to record his impressions after a visit to the factory in 1914. Feeling both 'overwhelmed' and frightened by the 'delirium' caused by the hellish scene of movement and noise, he comments: 'Of course there was order in that place, of course there was system – relentless system – terrible "efficiency"' (quoted in Brinkley, 155). Admiration for Ford's achievement was already tainted by sympathy for the plight of the alienated workers who were dwarfed by the machines controlling them.

Once the factory at Highland Park had maximized its efficient functioning, Ford had to build an even newer plant at River Rouge. Engineered 'from a blank sheet of paper,' this new monument to efficiency was 'intended to allow the maximum number of operations without wasted time, effort, or cost' (Brinkley 282). River Rouge so impressed through its careful planning and skilled engineering that Henry Ford's son Edsel commissioned the renowned photographer Charles Sheeler to capture 'the essence of "the great American machine"' (290). One of the first to recognize the aesthetic potential of the new industrial design pioneered at the Rouge, Sheeler took about thirty-two photographs of 'smokestacks and conveyor belts,' including *Criss Cross Conveyor No. 6*, which has become a classic example of 'industrial design becoming modern art' (291). Instead of focusing on suffering human subjects, the photographer elevated industrial objects to the status of art. 'Overwhelmed by the modern efficiency, cleanliness, and size of the Rouge,' Sheeler concentrated on 'the "functionalism" of the factory, the geometric perfection of the steel beams and iron pillars and coal heaps' (291). In her discussion of scientific management and Precisionist art, Corwin singles out Sheeler for his tendency to reinforce the depersonalization arising with the privileging of Taylor's system, the painter/photographer's 'inclination toward artistic self-effacement,' and the concomitant erasure of 'visible labor' (Corwin 152, 154). This aestheticization of the modern industrial era reflects the 'terrible beauty' that was being born at Ford Motor from 1913 onwards. In 1851 Prince Albert had conceived of art as an ornamental addition to the industrial might celebrated at Crystal Palace; by 1913 Ford Motor was in the process of aestheticizing the machine.

Highland Park and River Rouge exuded a mesmerizing atmosphere of irrepressible human creativity and ingenuity at the same time as they stood for an exponential increase in the social cost incurred as all aspects of the production process became integrated into a totally efficient system. While the assembly line is the privileged metaphor for what Dostoevsky called 'modernization as *routine*,' Henry Ford himself can, perhaps, be said to epitomize 'modernization as *adventure*' (Berman 243). As long as he applied his ingenuity to squeezing efficiency gains out of machinery and buildings, his revolutionary makeover of the factory system remained in keeping with the self-understanding of America as a land of self-made men who tamed nature for human consumption. But the assembly line was also held directly responsible for the devaluation of the workers' individual uniqueness. The standardization of the production process meant that the tasks performed by workers were also being standardized. By breaking down the construction of cars into discreet fragments, Ford Motor asked workers to perform automatic and routinized operations that turned them into extensions of the machines they handled. The machines dictated their tasks, denying them the exercise of individual initiative and creative input. The simplicity of these tasks meant that they could be carried out after minimal training;

there was no longer any demand for skilled workers. It was as if, in 1913, we had, as a society, irreversibly crossed some line. 'Until 1913,' specifies Brinkley, the Ford Motor Company assemblers had been skilled mechanics. The assembly line, however, replaced craftsmanship with sheer systematic toil' (154). From the beginning, workers objected to the assembly line for 'the demeaning effect of the conveyor and the violation of the workman's integrity by the division of labor' (Brinkley 154). While the 'assembly line, much like the Model T it produced, became a symbol of modern America and its prosperity,' its advent also 'created the quintessential system for regulating and rationalizing the labor process,' thereby eroding the Protestant work ethic which 'insisted that work must be morally meaningful and spiritually fulfilling as well as economically sustaining' (Watts 135, 154–5).

In his effort to make an affordable car available to the masses, Ford thus rewarded a class of creative employees with a great deal of individual freedom while simultaneously tying a great mass of labuorers to a machine system that denied them all individuality. The 'moving assembly line changed Ford Motor Company – for good and ill – from a fine, successful car company into the greatest industrial enterprise in the world, and from a high-quality workshop into an unskilled-labor mill' (Brinkley 155). In the popular imagination, then, Ford is guilty of subordinating human beings to machines; with the arrival of the assembly line, workers began to think of themselves as mere 'cogs' in an indifferent machine. It is not entirely surprising that the assembly line remains lodged in our cultural imagination as a symbol for the pursuit of efficiency gains for their own sake. What motivated Ford's decisions was the ambition to control every last aspect of the production process, that is, his desire for an always elusive perfection of the system. Unlike other tycoons, Ford was decidedly not driven by the lure of profit; he disliked high society (especially financiers) and disdained the lavish luxuries of the affluent life, which were an affront to his ascetic temperament. As a Ford engineer stressed, 'They weren't interested in anything except efficiency of production. They wouldn't talk dollars and cents at all. They talked in terms of the minutes that the thing cost' (Watts 154).

Efficient Management of Workers (Frederick Winslow Taylor)

As long as we conceive of an efficient system as a machine designed to maximize the total output of energy, the assembly line offers itself as the appropriate metaphor for an unassailable objective reality that brooks no resistance. But the assembly line at the Ford Motor Company is not equivalent to the car it produces; the factory floor is in reality not a machine but a social space. The metaphorical compression of the end product (the car) and the production process (the assembly line) suggests that the social relations between workers and management took on the form of the objective relations between the parts of a machine. In actuality, though, the apparently rational organization of the production process rests on a hierarchical social order, which subordinates workers to managers. There is thus a subjective aspect to the apparently objective totality of the factory floor. It appears that the machine metaphor applies primarily to the oppressed workers who see themselves as cogs in an indifferent system. This perception ignores the equally reifying impact of the system on the managers, who seem to benefit from the results of the efficient totality. The Ford Motor Company represents a microcosm of the opposition between the bourgeoisie and the proletariat in Marx's analysis of capitalism. In Marx's view, the material privileges enjoyed by the bourgeoisie conceal from them the dehumanizing impact that the capitalist mode of production has on them. In contrast, oppressed by the bourgeoisie, the suffering proletariat proves to be alert to the alienating conditions that afflict capitalist society as a whole. At the bottom of the social order at the Ford Motor Company, the workers felt treated as dehumanized drudges roughly on the same level of importance as steel rods or pumps. Performing repetitive actions, they were forced to think of themselves as an extension of machines rather than their masters.

However, whether in the form of the car or the assembly line, the concept of the machine had not emerged spontaneously to inflict its regimen on hapless workers; machines had been designed and constructed by social agents who also determined their place and function in the production process. The animosity against the machine conceals the part played by social conflict and antagonisms. While working for U.S. steel companies, Frederick Winslow Taylor realized that the class struggle was highly detrimental to a factory's overall productivity. In his *Principles of Scientific Management* (1911), he proposed a system, embodied in the 'efficiency expert,' that would act as an arbiter empowered to resolve class conflicts. The system he devised sought to maximize efficiency gains not through the elimination of waste typical of the combustion engine's focus on total output, but through the optimal equilibrium between advantages and disadvantages typical of social-contract theories from Plato to Marx. Where the total-output model worked primarily through violently repressive means, the equilibrium approach depended more heavily on ideological manipulation to do its work. Although Taylor's equilibrium model became subservient to Ford's output model, his principles of scientific management insist on being analysed in this study of efficiency not so much for their association with the Ford Motor Company as for their own sake. Although the main emphasis is on the nature and failure of Taylor's utopian ideal, the much-debated question of whether Taylor influenced Ford or the other way around deserves some comment. The chronology of events seems to suggest that the principles Taylor described and publicly defended in 1911 had been a source for the introduction of the assembly line at Ford Motor in 1913. However, Douglas Brinkley contends that Ford 'learned nothing from Taylor; he never read his books nor mentioned him' (Brinkley 140). In defence of Taylor, Robert Kanigel avers that the Ford team developing the assembly line included men like Clarence Avery, who 'kept in touch with the ideas of a man like Frederick W. Taylor' (Kanigel 497). There is in effect historical evidence that Ford and Taylor anticipated aspects of each other's ideas and practices.[2] Moreover, as Alexander points out, it is often forgotten that Taylor's theories of management were influenced by his 'metal-cutting work,' by the way 'things should move and in what relation to each other' (Alexander 11). Not unlike Ford, his concerns with organization were never far removed from 'the fundamental technological activities through which efficiency was defined' (11–12). No matter what conclusion one may want to reach, these two figures were individually and jointly responsible for the public prominence accorded to efficiency as both a benefit and a curse.

As a mark of the difficulty of establishing who contributed what to society's obsession with efficiency, Taylor's efficiency expert tends to be remembered as a figure standing over workers with stopwatch in hand,[3] inviting the by now customary conflation of his ideals with the rationalizing logic of Ford's assembly line. Typical of a propensity to see the *Principles of Scientific Management* merely as a theoretical justification for the oppressive practices at the Ford Motor Company is *The Big Money* (1936), the third volume of John Dos Passos's *U.S.A.* trilogy, in which he condemns both Ford and Taylor for the dehumanization he dramatizes in the portrayal of his fictional characters. The mini-biography on Ford ('Tin Lizzy') singles out the homogenizing impact of increases in efficiency made possible through 'cheap interchangeable easily replaced standardized parts' (Dos Passos 49). But the assembly line is for Dos Passos merely the practical application of the 'Taylor Plan that was stirring up plantmanagers and manufacturers all over the country' (49). Once 'efficiency was the word,' human beings became the subject of social-engineering strategies: 'The same ingenuity that went into improving the performance of a machine could go into improving the performance of the workmen producing the machine' (49). Where the taming of workers was but an incidental aspect of Ford's obsession with efficiency, it was the central feature of Taylor's preoccupation.

What is usually overlooked in the rush to condemn the efficiency expert is a utopian dimension in Taylor's thinking that clearly distinguishes him from Ford's search for practical solutions. Although Taylor is thought to have provided the theoretical justification for the deskilling of workers and the practices of surveillance at Ford Motor, he himself was genuinely baffled to find himself vilified as an oppressor of workers. In his view, he had devised a system intended to alleviate the class conflict pitting management against workers; in effect, he saw himself as the workingman's best friend. On closer inspection, his *Principles of Scientific Management* are indeed motivated by a utopian desire to replace exploitation with cooperation; however, as with all social utopias, he was unable to imagine a social-engineering project that was not open to infiltration by totalitarian social-control mechanisms. The deterioration of Taylor's utopian workplace into the nightmare reality of the assembly line at the Ford Motor Company does not necessarily constitute a misapplication of his principles but may be implicit in the very concept of social-contract theories designed to ensure the smooth functioning of social spaces. Yet Taylor's emphasis on cooperation introduces a model of efficiency that is distinctly different from the assembly line Huxley satirizes in the opening pages of *Brave New World.* Where the machine metaphor implies that every last ounce of energy had to be extracted from each worker, Taylor's system promoted a social-engineering scheme premised not on the notion of *total output* but on the *optimum distribution* of effort and reward. The conflation of Ford and Taylor thus conceals a model of efficiency that today's economists and social theorists trace to the Italian mathematician Vilfredo Pareto (1848–1923), who is retrospectively credited with the development of 'Pareto efficiency' or 'Pareto optimality,' a preference-based approach to equilibrium theory.

With Taylor's emphasis on 'managing' both workers and bosses, the efficiency expert emerged as a new social figure whose skill lay not in the production of goods but in the management of human bodies. These figures differ from the worker in that they decide on how work is to be done that they themselves do not perform. But they also differ from the boss in that they were 'people who didn't own the organizations they worked for but nevertheless devoted their entire careers to them' (Micklethwait and Wooldridge 60). On the one hand, as disciplinary managers of men and women, efficiency experts were deeply resented and vilified by workers. On the other, they freely consented to serve a master who used them to enslave the workers for gains in which they did not directly share. However, it is only by situating this controversial figure within the utopian features of Taylor's 'American plan' that we can assess the 'dystopian' consequences Huxley foregrounds in his satire of the Ford Motor Company. As we will see, what emerges from this 'narrative' of the Taylor and Ford interplay is the uncomfortable realization that the good intentions of both men produced consequences they had neither desired nor anticipated. Yet, as game theory clarifies, such unintended consequences are not the outcome of some mechanistic principle beyond human control, but of the deliberate rational choices made by individual social agents.

Taylor's lifespan, from 1856 to 1915, 'almost exactly coincided with the Industrial Revolution at its height' (Kanigel 7).[4] His ideas can easily be seen as the culmination of the nineteenth century's optimistic faith in technological progress. The experiments described in *The Principles of Scientific Management* reach back thirty years to the days when he worked at Bethlehem Steel and the Midvale Steel Company. The work is an amalgam of earlier speeches and publications by Taylor himself and of experiments carried out by other efficiency experts (Carl G. Barth, Frank B. Gilbreth, Henry L. Gantt). But it was not until Louis Brandeis, the 'people's lawyer' (Kanigel 430), invoked Taylor to block rate increases demanded by the powerful railroads that his principles were catapulted into the limelight. Arguing that consumers should not be asked to subsidize a poorly managed public company, Brandeis appealed to efficiency to defend the 'little man' from 'big business.' Taylorism was thus initially hailed as a progressive theory destined to 'secure prosperity for all and rid the industrial world of its inequities' (Kanigel 443). However, labour immediately condemned Taylorism as an enemy of workers who were 'treated as a machine' and 'robbed of independence, judgment, and thought' (444). The publicity generated by the Eastern Rate Case and also by workers' protests at Rock Island and Watertown Arsenals convinced government to strike the 'Special

Committee to Investigate Taylor and Other Systems of Shop management,' chaired by William B. Wilson; the committee started hearings on 4 October 1911 that were to last five months and generate 1935 pages of printed transcript. Efficiency had not only become a hot issue but was in effect put on trial.

Taylor's experiences labouring in steel mills in the late nineteenth century had taught him that human beings tend to make decisions serving their short-term self-interest without taking into account the benefits of cooperation in the longer term. He noticed that workers would engage in 'soldiering' (or 'slacking') either because they saw no reason to exert themselves or, even more scandalously, because they wanted to spite their employer. In the absence of incentives from management, workers had every reason to do as little as the least productive among them: 'Why should I work hard when that lazy fellow gets the same pay that I do and does only half as much work?' (Taylor 20). It seemed collectively to be more advantageous for every worker to do as little as possible. As new workers often discovered, those who worked hard incurred the wrath of their co-workers who had agreed on a slower pace. Class antagonism, Taylor realized, resulted in slacking as a 'fixed habit,' to the point where 'men will frequently take pains to restrict the product of machines which they are running when even a large increase in output would involve no more work on their part' (24). In an effort to avoid being suckered by a freeloader, each worker will feel compelled to keep reducing his own productivity.

Management engaged in its own race to the bottom. Blind to counterproductive consequences, bosses considered it in their self-interest to enrich themselves at the expense of workers. The result was that inefficiently run factories kept wages low to keep what little profit there was in the hands of management and to punish a non-cooperative workforce. If the workers could be forced to optimize their productivity without being compensated, then management would gain an advantage at the expense of the workers. However, since workers have no incentive to enrich their masters, they protect their own self-interest by slacking on the job. This antagonistic situation has a tendency to become so collectively self-defeating that the factory may be forced to close. Moreover, as Taylor saw it, financial incentives alone would not be enough to increase a worker's productivity. Management expected workers to perform their tasks efficiently without providing them with the necessary equipment and training to achieve this goal. According to his observations, workers were left to struggle with tasks that neither they nor their ignorant managers understood. When management blames workers and workers blame management for the inefficiencies they have jointly created, then the system's overall efficiency is impeded and everybody is worse off.

In an effort to counteract this collectively self-defeating situation, Taylor opens *Principles of Scientific Management* with the following often quoted statement: 'The principal object of management should be to secure the maximum prosperity for the employer, coupled with the maximum prosperity for each employee' (Taylor 9). This formula was intended to create an optimum equilibrium between the demands and responsibilities of both management and workers. Taylor's observations of human behaviour convinced him that this ideal equilibrium could not be achieved if human beings were left to their own devices. Collectively self-defeating antagonisms could only be resolved if management and workers could be convinced to accept the decisions of an external arbiter. Filling this new role, the efficiency expert was to devise a system outlining a rational process designed to increase overall productivity to the benefit of everybody. It was not that the expert was endowed with superior human qualities or knowledge; on the contrary, he was a narrow specialist in the study of speed and motion. What counted was the *system* rather than the human agent. Bosses and workers alike would act on the advice of efficiency experts who had developed the most efficient operational plan. In theory, at least, happy managers would be profiting from higher profits and happy workers would appreciate not only higher wages but also a more efficiently organized workplace. In spite of his severe critique of Taylor, even Dos Passos acknowledges the utopian aspirations implicit in the principles of scientific management: '*There's the right way of doing a thing and the wrong way of doing it; the right way means increased production, lower costs, higher wages, bigger profits:* the American plan' (Dos Passos 20). Although tinged by the ironic knowledge of the failure of Taylor's ideal, Dos Passos nevertheless acknowledges the intention behind Taylor's plan as an updated version of the American dream for the twentieth century.

Taylor was one of the first to understand the importance of *managing* the operations of a company; where some of the practices introduced at the Ford Motor Company touched on aspects of management, it was Taylor who theorized the revolutionary notion of a planning department acting on the advice of experts in *management* rather than in engineering or manufacturing. Disputes between class antagonists were to be arbitrated through appeals to the principles of a rationally constructed system. As Kanigel stresses, it was 'Taylor, after all, who first said in 1911, explicitly and without apology: "In the past the man has been first. In the future the system must be first"' (Kanigel 19). Where detractors read this privileging of the system as evidence of the dehumanizing repercussions of Taylorism, Taylor himself understood himself to be curbing the abusive practices of wilful individuals through the application of objective standards. The scientific-management model was above all designed to replace the old 'rule-of-thumb method' (Taylor 36) with a new rational approach. Instead of allowing management and workers to indulge individual or self-interested whims, Taylor's system championed a communal, cooperative method based on rational planning.[5] Taylor concluded optimistically that 'it is this combination of the initiative of the workmen, coupled with the new types of work done by the management, that makes scientific management so much more efficient than the old plan' (37). In the first instance, workers would now be happier performing clearly explained duties for which they were well suited. They would be able to increase their individual efficiency, thereby improving the productivity of the factory so that there would be enough money to reward management with profit and workers with higher wages. Freed from 'discord and dissension,' both management and workers would be 'far more prosperous, far happier,' and ultimately would be 'suffering less' (29). This equilibrium theory of efficiency was predicated on a trade-off in which each antagonist in the class conflict sacrifices some short-term interest in order to achieve a result in the longer term that would make both sides better off than if either had acted strictly according to self-interest.

Taylor's equilibrium model of efficiency proved an elusive ideal. It is not coincidental that Taylor is primarily remembered for the time and motion studies that he introduced to assist the very workers who vehemently objected to their implementation. What Taylor evidently failed to take into consideration was the persistence of a power asymmetry between management and workers. Management immediately appropriated the time and motion studies meant to assist workers in the execution of their tasks to exploit and oppress labour. In the popular imagination, Taylor's stopwatch took its place next to Ford's assembly line. The experience of the worker was denigrated in favour of a theoretical knowledge often in the hands of college students hired to measure and time how man and machine were best designed to work effectively together. With the introduction of the efficiency expert – 'the college students with stopwatches and diagrams, tabulating, standardizing' (Dos Passos 20) – the workers lost not only their autonomy but also their human dignity. The stopwatch thus became 'the very symbol of Taylorism' (Kanigel 466). Where Taylor saw the stopwatch as a neutral tool in the fight against waste, to the workers it was 'a hideous invasion of privacy, an oppressive all-seeing eye that peered into their work lives, ripping at their dignity' (466). Instead of being encouraged to use their own judgment as skilled craftsmen, the workers felt reduced to mere automatons asked to carry out repetitive actions in a standardized workplace. In facilitating the most efficient flow of materials and workers' tasks, Taylor was accused of making redundant the '*great mass of traditional knowledge which in the past has been in the heads of the workmen and in the physical skill and knack of the workman*' (Dos Passos 19). The problem for Dos Passos was that Taylorism extended the logic of standardized 'tools and equipment' (20) to standardized men and women. Exacerbating the division of labour and the power asymmetry that Taylor sought to alleviate, his system 'broke up the foreman's job into separate functions, speedbosses, gangbosses, timestudy men, orderofwork men' (20). The efficiently arranged workplace reproduces the hierarchical social relations that subordinate workers not only to bosses but also to college boys. Transforming the worker from 'a skilled mechanic into that of *factotum* or machine' (Kanigel 445), the *Principles of Scientific Management* supplied the theoretical justification for the assembly line at Highland Park.

What bothers Dos Passos above all is the insidious way in which the 'American plan' encourages human beings to internalize the ideology of efficiency. Referring to one of Taylor's 'famous experiments,' Dos Passos tells us how Taylor triumphantly recalls having taught a 'Dutchman named Schmidt' to 'handle fortyseven tons instead of twelve and a half tons of pigiron a day.' Aside from reducing Schmidt to an automaton or 'gorilla,' Taylor got him to 'admit that he was as good as ever at the end of the day' (Dos Passos 21). The victim of Taylor's celebration of efficiency as the path to overall prosperity is here forced to internalize the ideology that in fact dehumanizes him. In *The Big Money*, Taylor himself is accused of having succumbed to the power asymmetry he had tried to avert. Once Taylor 'got to be foreman,' claims Dos Passos, he moved to 'the management's side of the fence,' contradicting his earlier argument that one 'mustn't give the boss more than his money's worth' (19). In an ironic twist, though, Taylor himself became the victim of his own investment in the 'American plan.' When he insisted that a 'firstclass man' who did 'firstclass work' should receive 'firstclass pay,' he 'began to get into trouble with the owners' (20). Dos Passos reports with considerable ironic glee what happened after Charles Schwab took over Bethlehem Steel in 1901:

> Fred Taylor
> inventor of efficiency
> who had doubled the production of the stamping-mill by speeding up the main lines of shafting from ninetysix to twohundred and twenty-five revolutions a minute
> was unceremoniously fired. (21)

In the final analysis, Taylor's objective system is hijacked by the corporation it had empowered, a monster that now mocks Taylor's plan by revealing itself as a coldly efficient machine bent on perpetuating itself without regard for the human dignity of either Schmidt or Taylor. For Dos Passos, the efficient equilibrium between effort and reward cannot maintain itself in the face of power asymmetries. The concept of the whole human engaged in a meaningful creation of goods was being supplanted by the isolated worker confronted by a process of production broken up into standardized fragments to be arranged and integrated into an expert's total plan.

To the end of his days, Taylor remained puzzled by the hostile reception of his equilibrium-efficient principles. Shortly before his death, he wrote to a Harvard engineering professor that his system 'prevents arbitrary and tyrannical action on the part of the foremen and superintendents quite as much as it prevents "soldiering" or loafing or inefficiency on the part of the workmen' (Kanigel 510). But to his detractors he was responsible for the standardization of the workplace, the subordination of the worker to the machine, the reliance on methods of surveillance, and the introduction of experts. Kanigel best sums up the contradictory reception of Taylor's method when he says: 'To organized labor, he was a soulless slavedriver, out to destroy the workingman's health and rob him of his manhood. To the bosses, he was an eccentric and a radical, raising the wages of common laborers by a third, paying college boys to click stopwatches. To him and his friends, he was a misunderstood visionary, possessor of the one best way that, under the banner of science, would confer prosperity on worker and boss alike abolishing the ancient class hatreds' (1).

The conflicted reactions to Taylor's *Principles of Scientific Management* allow us to gain some understanding into the process by which efficient social-engineering projects tend to undermine themselves. Taylor started out with the laudable intention of improving the conditions for all players in the socio-economic system. Realizing that the human being could not be taken out of the worker, he devised a model of rational utility or Pareto efficiency: 'Maximum efficiency has been achieved when it is impossible to improve anyone's condition without harming someone else' (Heath 23). Taylor's theory of a Pareto-efficient distribution of responsibilities and benefits proved in practice to be so harmful to the workers that they immediately protested against its implementation. Leaving the power asymmetry between workers and bosses intact, Taylor's system seemed merely to replace the machine with the efficiency expert in order to subordinate workers to an *external* force beyond their power. However, a closer look at the history of the Ford Motor Company tells a more complex story. Although the power asymmetry was certainly a constant factor in the dehumanization of workers, social-engineering projects influenced by Taylor's principles surreptitiously exacerbated the lot of workers by compelling them to *internalize* their victimization. Since Ford and Taylor tend to be conflated in the cultural imagination, the Ford Motor Company serves as a particularly 'good example' to trace this psychologically damaging management of the human body.

Social Engineering at the Ford Motor Company

The history of the Ford Motor Company illustrates that any commitment to efficiency in the social sphere conjures up the spectre of surveillance and social control. Although the actions of individuals affect events at the Ford factory, these actions are made possible by Taylor's 'mental revolution' as it gradually influenced the decisions of the man in power, Henry Ford himself. Once Taylor privileged the system over the human being, he implicitly suggested that the conditions created by the drive for efficiency could no longer be simply dismissed as the outcome of either good or evil intentions by social agents like Henry Ford. As long as Ford was happily solving technical problems to increase the efficiency of his cars and production process, his obsession with the elimination of waste testified to an admirable sort of ingenuity. However, when he discovered that the most perfect technical blueprints were open to disruption by resistant human beings, the drive to maximize the output of cars compelled him to resort to manipulative and coercive measures that were in many ways contrary to his social conscience and Puritan temperament. Ford could be said to have started out as an active engineer representative of 'modernization as *adventure,*' who initiated a socio-economic revolution responsible for modernity's degeneration into 'modernization as *routine*' (Berman 243). In Dostoevsky's terms, Ford is typical of the man who 'likes to build [the edifice]' but 'does not want to live in it' (in Berman, 242). The history of the Ford Motor Company allows us to draw out at least major aspects of the complex systemic logic that transformed the company from a relatively benign paternalistic enterprise into a malignantly oppressive behemoth. Whether consciously or not, Ford Motor underwent this transformation by exploiting Taylor's principles of scientific management to social ends the efficiency expert had specifically proscribed.

Having rationalized the production process of the Model T through the introduction of the assembly line, Henry Ford discovered that productivity did not rise according to his orderly plans; to his great dismay, he discovered that workers could not be relied upon to create the efficiencies he had calculated. It soon became apparent that individuals were not as predictable as rods of steel; the human element confronted Ford as a potential residue of useless energy that threatened the efficient system as a site of resistance. Conditions at Highland Park and later at River Rouge were such that worker turnover was high, and those who stayed saw no reason for exerting themselves. Recognizing that workers did not necessarily share his commitment to the company, Ford set out to correct the weakness that lack of human motivation introduced into his organization. Having imposed his will on the material world, he now turned his attention to the taming of human bodies.

As the Ford Motor Company grew in scale and complexity, Henry Ford could no longer keep an eye on all of its operations. Once the system took over from Ford's personal control, even his best intentions shifted the ground from a paternalistic form of social engineering to a totalitarian form of social control. In the early days, Henry Ford established himself as a paternalistic manager who was convinced that overall productivity could be increased if he encouraged employees to work at their own pace and to use their own creative initiative. While located at Piquette Avenue, the company was a small family enterprise, with Ford personally knowing each of his workers. His personal leadership ensured the loyalty of his workforce. However, as the company expanded to occupy ever larger physical locations, Ford could no longer depend on personal ties to motivate his workers. In line with his paternalistic principles, he decided to encourage productivity and loyalty at River Rouge by offering his workers the incentive of five dollars a day, a wage three times the going rate. This incentive established Ford Motor as a desirable, enlightened employer. With the Five-Dollar Day, 'Ford overturned the older robber-baron image of the American big businessman,' introducing 'an unfamiliar yet inspiring social type: the businessman as reformer' (Watts 178). On the surface it appeared as if this incentive actualized and confirmed Taylor's ideal plan; the workers would be paid well and Ford would reap higher profits.

Ford's apparent munificence was from the start a shrewd business move, which brought with it the temptation to control the workforce. To remedy the gap between 'human efficiency' and the 'technological and organizational efficiency in the production of the Model T,' Ford introduced broad policies and practices to reduce worker dissatisfaction arising from 'excessively long hours, low wages, poor housing and home conditions, undesirable shop conditions, and arbitrary handling of workers by foremen' (Watts 181, 180). Having secured a more motivated and stable workforce through the incentive of the Five-Dollar Day, Ford thus turned his attention to factors outside the immediate workplace that might affect a worker's ability to perform at optimum capacity. In an apparently progressive move, Ford Motor started to take care of its employees by gradually making available such incentives as savings options through a company bank, health care through a company hospital, a profit-sharing scheme, and English classes for immigrants. However, offering high wages as well as favourable working conditions, Ford felt empowered to be selective in hiring and firing practices. He set out to attract and retain the kind of person best suited to help the company achieve its productivity goals. Combating absenteeism, high turnover, and indolence, Henry Ford was looking for employees who were like him – sober, clean living, and thrifty. The criteria the company developed included married men who took 'good care of their families,' single men over the age of 22 of 'proven thrifty habits,' and men 'under 22 and women of any age, who provided the sole support to some next of kin or blood relative' (Brinkley 172). In his desire to pick 'the cream of the crop of employees,' Ford thus focused on 'the *whole* employee, not merely his or her specific job skills' (174). At first glance, the incentive structure at River Rouge seemed to reproduce the early paternalism at Piquette Avenue: Henry Ford took care of his employees both at work and at home. The Five-Dollar Day came out of a genuine 'populist strain' on the part of Ford, whose career had been marked by 'loyalty to working people and suspicion of financial power' (Watts 183). But, by paying his workers well, he also happened to create a 'lot of customers' (183). In Ford's mind, he had produced a Pareto-efficient system securing the 'greatest happiness for the greatest number' advocated by Jeremy Bentham.

On closer inspection, though, Ford's paternalism concealed a desire for social control, which became increasingly visible. Before paying five dollars a day, Ford felt justified in checking up on employees and job applicants by enlisting 'a team of thirty investigators charged with ascertaining which employees and applicants fully qualified' (Brinkley 172) for the privilege of working for Ford. Ford Motor's 'clean-living rules and monitoring of compliance' led to the institution of the company's 'full-fledged sociological department' (173). While this department could be accused of interfering in the private lives of Ford's employees by 'keeping records on the lifestyles and spending habits of each of Ford's hourly employees,' Ford initially created it to 'take care of its own people in the midst of an urban-industrial environment that was brutal on the uninformed employees' (276). It 'helped people in trouble,' especially in the areas of 'broken families and confused immigrants' (276). However, the more actively Ford Motor sought to assist its employees, the more it arrogated to itself the right to exert undue influence. As Brinkley points out, Ford Motor 'intervened in fundamental ways: by moving families into satisfactory housing or by insisting that errant husbands break habits such as drinking whiskey and chasing women (at the risk of losing their jobs). It was also empowered to forward the paychecks of irresponsible workers to other members of the household, and it extended loans against future pay' (276). In other words, the 'sociological department succeeded in its effort to promote thrift, at least among long-term employees' (278). But, 'although these early expectations and investigations of its employees were well meant,' Brinkley writes, 'their invasions of workers' privacy stood unprecedented among automobile manufacturers and listed into the troubling realm of social engineering' (173). Paternalistic concern for the prosperity of employees concealed Ford Motor's desire to exert control over their whole lives. In a totalizing gesture, the company compelled workers to fit into pre-established categories and weeded out those who refused to conform.

Where the machine metaphor for efficiency implied an oppressive force descending on passive victims from outside, the actual practices at Ford Motor relied on the ideological complicity of the so-called cogs to serve the needs of the system. Increasingly, Ford sought to create a predictable and productive labour force through a combination of incentive structures and intimidation. As we have already seen, the lure of the Five-Dollar Day enticed employees to submit voluntarily to a regime of surveillance designed to deprive them of their claim to privacy and later their right to unionize. Ford Motor proved that 'while it may indeed be impossible to legislate morality, it is quite easy to buy it' (Brinkley 174). The company was able to 'buy' the morality conducive to its economic interests only because employees agreed that it was reasonable for the employer to expect optimum efficiency from them in exchange for five dollars a day. Efficiency worked as an unacknowledged ideology that legitimated the often dehumanizing expectations the company imposed on its workers. Where money was the economic currency of exchange, efficiency was its ideological support. In the name of efficiency, people accepted a degree of social manipulation entirely at odds with the self-understanding of the autonomous bourgeois individual of liberal humanism.

So far, then, the story of the Ford Motor Company illustrates how a well-intended paternalism resulted in conditions that reduced workers to functioning like the machines they operated: 'By making the productivity of labor as predictable as that of the machines that set its pace, Ford Motor Company had done no less than redefine industrial capitalism' (Brinkley 174). During this revolution in efficiency, the 'intrusive employee investigations skated right up to the line between social engineering and social control' (174), and Ford, the creative industrial entrepreneur of Piquette Avenue, had transformed himself into the routinized industrial manager of Highland Park.

Although Henry Ford's success was from the start predicated on the pursuit of efficiency gains, he became more and more obsessed with efficiency for its own sake. Once efficiency had become an end in itself, Ford lost sight of the social costs his strategies incurred. Once he treated employees as mere cogs in the system, he cut himself off from the camaraderie he had enjoyed in his early days. Committed only to gains in productivity, he distanced himself not only from the workers with whom he used to identify but also from loyal managers whose advice he had sought and often heeded in the past. He had allowed himself to become dehumanized by the very system over which he seemingly exercised total control. The story of the Ford Motor Company in the 1920s reveals a mutually reinforcing connection between Ford's reified consciousness and the increasingly oppressive conditions under which employees laboured, not only on the factory floor but also in the offices of middle management.

After the First World War, the Ford Motor Company discovered that the paternalistic approach symbolized by the sociological department was no longer sufficiently cost-effective. A surplus of workers and the deskilling of tasks in the factory meant that incentives to attract highly skilled and motivated workers were no longer needed. In his 'zeal for efficiency' (Brinkley 278), Henry Ford decided that treating workers as individuals was a waste of resources; in 1921 the sociological department was closed down, soon to be replaced by the Ford Service Department run by Harry Bennett, a man who 'followed orders without question, fired workers without asking the reason, and gathered information within the factory through spying or brute force' (344). In the new atmosphere, 'the fact that well-adjusted workers were more productive than troubled ones no longer seemed to be worth the bother of hovering over them with a battalion of investigators' (279). But the surveillance apparatus did not disappear; it simply took on a new and more sinister function. Fearing worker unrest, Bennett employed a large private police force consisting of thugs who 'paraded along the assembly lines enforcing discipline and looking for slackers' (449) in order to control workers and prevent unionization. Exemplifying the 'tough school' of management that operated through discipline and spying, Bennett controlled an elaborate spy network extending 'from the factory floor to executive offices' (Watts 449). Planting men 'throughout the workforce' (Brinkley 383), Bennett was allowed to intimidate workers at will. Heading a 'repressive organization,' Bennett presided over the infamous 'speed-up' at Ford Motor that 'demanded a faster pace of work' so intense that 'men were forbidden to talk with one another,' were given few opportunities to go to the toilet, and were allowed only 'fifteen minutes for lunch' (Watts 455).[6] Directing 'an army of henchmen,' Bennett systematically created an atmosphere of surveillance and 'physical intimidation' (449). Developing 'files on the personal transgressions – drinking, gambling, womanizing – of even the highest-placed executives,' Bennett 'carefully cultivated an ethos of violence' (449, 450) that eventually established him as the most powerful figure in the Ford organization. Although there is disagreement as to how much Henry Ford knew about Bennett's tactics of intimidation, it is clear that the head of the Service Department traded on a close personal friendship with the company's founder. At the very least, Ford must have turned a blind eye to Bennett's repressive measures. Ford allowed his initial paternalism to deteriorate into Bennett's explicit aim to tame men so as to turn them into docile bodies.

Workers at Ford Motor felt dehumanized not only because they considered themselves to be extensions of the machines they operated but also because they were treated as expendable and infinitely exchangeable parts within a totalized and objectified social system. On the one hand, management and machines confronted employees as hostile forces beyond their control. On the other, employees consented to their victimization by competing with each other for privileges bestowed on those complying with the demands of management. The system thus worked by a combination of Bennett's repressive measures (Althusser's RSA) and the ideological complicity of workers/consumers (Althusser's ISAs).[7] Being able to draw on a surplus army of workers, the Ford Motor Company eventually dispensed to a large extent with ideological strategies to concentrate on tactics of intimidation. Inside the factory, 'service men marched through the factory, displaying their guns, sticks, and other weapons and enforcing obscure rules at their whim' (Brinkley 383). But Bennett's long arm reached even beyond the factory gates; his thugs were known to beat up suspected political activists who sought to unionized the Ford Motor Company. Ironically, it was Ford's 'traditionalist ideology of individualism' (Watts 458) that accounts for his unremitting hostility to unions. In agreement with Taylor's rational-choice model, Ford was determined to deny workers a collective voice because he 'remained convinced that the employer-employee relationship was an agreement between consenting individuals' (458). Disavowing what he tacitly authorized, Ford thus exposed workers not only to exploitation and abuse but also to being terrorized into submission or summarily fired by Bennett and his spies.

Conditions at River Rouge deteriorated to the point where workers were galvanized into open rebellion. The combination of resentment at layoffs in the magnitude of 50,000 and 60,000 workers and outrage at 'the abuses of the company's obsession with surveillance' (Brinkley 391) finally resulted in the 'Ford Hunger March' on Sunday 6 March 1932. In stark contrast with Henry Ford's earlier empathy for his workers, he now allowed Bennett to put down this spontaneous uprising with brute force: 'Harry Bennett ordered two high-pressure fire hoses to blast the marchers with freezing water. This only infuriated them. Suddenly, as the marchers approached, Bennett's forces opened fire, killing one man' (391). After a later incident in 1937, a bloody assault on labour leaders Walter Reuther and Richard T. Frankensteen, the press indicted the Ford Motor Company as fascist and called Harry Bennett an 'embryonic Hitler' in charge of his own 'blackshirts' (Watts 454). From being called a 'pioneer of humane industrial reform,' Henry Ford now 'appeared as a despot who abused his workers' (454–5). The 'Eden of harmony and high wages' at Highland Park in the 1910s had been replaced in the 1930s by the River Rouge factory, which 'resembled an armed camp' (454). In his relentless drive for efficiency gains, Ford was caught up in a system he had founded but which now controlled him more than he controlled it.

Initially designed as a paternalistic support system for Ford employees, the social-engineering project conceived along Taylor's equilibrium model had gradually transformed itself into the totalitarian system presided over by the autocratic Henry Ford and the sadistic Harry Bennett. Ford Motor clearly increased its productivity by dividing the production process into isolated segments and strengthened its power position by treating workers as so many interchangeable 'cogs.' The fragmentation that began as a by-product of Henry Ford's obsession with efficiency gains eventually became a central strategy of social control. As Brinkley emphasizes, 'Bennett's job was to keep the Rouge workforce of 70,000 as a group of isolated individuals, and not let them create community' (Brinkley 384). Fragmentation was an integral aspect of the totalitarian regime of terror that came to be entrenched with the tactics of Harry Bennett. However, Bennett's open violence was ultimately less corrosive than the investment in efficiency that worked ideologically to create docile bodies. Ford's contributions to corporate capitalism were the assembly line to make production more efficient and the Five-Dollar Day to increase the consumer base necessary for the efficient functioning of the market.

Unintended Consequences (Game Theory)

What accounts for the deterioration of Taylor's utopian ideal into the horrors of Bennett's carceral regime at the Ford Motor Company? As with any other development, the causes of these events are undoubtedly overdetermined. One could presumably delve into the psychological predispositions of the individual social agents involved or one could speculate on the moral weaknesses of human nature in general. But the obsession with efficiency in Fordism and Taylorism is also beset by systemic problems endemic to rational-choice models of which Taylor's utopian ideal is a prime example. His solution to the elimination of waste in the production process presumes that rational human beings will act rationally when they are presented with a rational plan outlining the benefits accruing to everybody when the system's efficiency is increased through cooperation. Although the Pareto-efficient equilibrium he envisaged was utilitarian in its aims, he nevertheless prided himself on the social justice implied in the alleviation of the class conflict. His genuine dismay at being the target of worker hostility indicates how deeply he felt about improving the conditions of existence for workers. The instrumental emphasis on increasing an industrial concern's output points to broader ethical implications for the organization not only of the factory floor but also of modern society as a whole.

Through its economic orientation, Taylorism raises issues of particular interest to socio-economic theorists, sharing with them assumptions most prominently discussed in game theory. Beginning with the publication of *The Theory of Games and Economic Behavior* by John von Neumann and Oskar Morgenstern in 1944, game theory was, by the 1970s, 'spreading like a bushfire through the social sciences' (Hargreaves and Varoufakis 1). The analyses of game theorists were thus not available to Ford and Taylor, but from a retrospective position game theory allows us to gauge the theoretical impasses faced by Taylor himself and by those affected by his mental revolution. Although game theory remains popular with social economists, its narrow focus on intersubjective relations within primarily utilitarian contexts makes it far less attractive to literary and cultural critics, for whom literary texts and cultural practices owe their social currency to claims of resistance to the rational-choice models at stake in game theory. Nevertheless, game theory provides us with insights into systemic conditions that retrospectively explain aspects of Taylorism which also surface in the novels to be treated later in this study.

What, then, is game theory? In an attempt to predict probabilities, game theorists construct mathematical models that analyse probable outcomes when competing social agents consider which strategies best serve their interests under given circumstances. If rational players adopt rational methods to achieve rational goals, then competition will spontaneously create an optimal distribution of interests or an ideal state of equilibrium. It is immediately obvious that such rational-choice models are insufficiently alert to the psychological depths of human motivation; they reduce complex human beings to their instrumental utility. Despite its obvious limitations, game theory provides insights into the systemic tendencies of conscious decisions to result in outcomes that were neither intended nor anticipated. Although our rational choices are undoubtedly driven by deeply unconscious motivations, they are also conditioned by the expectations we have of others and by the expectations we assume they have of us. If game theory has something to tell us about social interactions, it is not in its ability to predict behaviour, but in its predominantly descriptive analyses of an intersubjective dynamic manifesting itself in a broad range of social activities.

Since game theory is generally taxed with a reductive view of human motivations, it is at least interesting to note that the most prominent psychoanalytic theorist of our own day, Jacques Lacan, expressed considerable appreciation for game theory as a complement to his own concerns. Intersubjective expectations play a role in Lacan's theoretical emphasis on the *social* constitution of the unconscious; if we construct our identity through the discourse of the Other, then we are formed through our intersubjective encounters. In 'The Function and Field of Speech and Language in Psychoanalysis,' Lacan refers to game theory precisely in the context of his contention that a subject's speech necessarily 'includes the other's discourse in the secret of its cipher' (Lacan, 'Function' 69). The source of the subject's 'most profound alienation' is to be located in 'our scientific civilization' in which the subject 'loses his meaning in the objectification of discourse' (69). Taking this 'presence of intersubjectivity' (Lacan, 'Instance' 163) in our unconscious from Alexandre Kojève's reading of Hegel's master-slave dialectic in his *Introduction à la lecture de Hegel* (1947), Lacan incorporates into his analysis of the unconscious what was for Hegel an objective condition of intersubjective recognition and acknowledgment. It could be argued that game theory returns us to Hegel, constituting an elaboration of the objective conditions that organize his dialectical process of subjectification. As we will see, game theorists schematize how a player's strategies are determined by the expectations that she harbours of the expectations she attributes to other players. What is of interest to both Lacan and game theorists is the intersubjective dynamic operative in the decisions reached by apparently autonomous and rational subjects. But game theory clarifies above all the logic that doomed the ideal state of equilibrium between antagonistic forces at stake in Taylor's utopian aspirations.

It is certainly the case that game theory treats human beings as rational agents intent on rationally satisfying rational preferences. This emphasis on social agents as preference-satisfiers foregrounds concealed motivations that are not so much mobilized by the desires Lacan locates in the unconscious as by conscious decisions we make as we interact with others. In the final analysis, Lacanian subjects are, of course, also preference-satisfiers; they seek ways to gratify desires they do not consciously recognize as their own. But the preferences at issue in game theory are typically examined in often trivial, everyday situations that rarely demand sophisticated psychological analysis. Although our attitude towards stock-market risk may be tied to deep-seated cathexes, game theorists assume that our risk tolerance can be calculated by assessing it a utility number or 'util.' The probability of an outcome is thought to be calculable on the basis of the most complete set of pertinent information. In *Game Theory: A Critical Text* (1995; 2004), Shaun P. Hargreaves Heap and Yanis Varoufakis provide us with a simple example of a typical game situation: 'Imagine for instance that you are about to leave the house and must decide on whether to drive to your destination or to walk. You would clearly like to walk but there is a chance of rain which would make walking awfully unpleasant' (9). If the weather bureau predicts a 50–50 per cent chance of rain, then the person's preference to walk will tilt the decision towards walking. Under such uncertain conditions, 'utility maximization' describes an action that best satisfies a person's 'preferences' (11). Walking would then be given a higher 'util' number than driving. 'Utils' allow game theorists to construct and test social interactions in what appear to be mathematically precise formulations. Such precision is, of course, always undermined by indeterminacies that even ardent game theorists tend to acknowledge.

Indeterminacies do indeed severely undermine game theory's emphasis on calculating and hence predicting probabilistic outcomes. Treating social agents as preference-satisfiers, game theory tries to quantify what are ultimately subjective criteria. Since '"preferences" are not given independently of beliefs and the indeterminacy of belief yields indeterminate preferences, so talk of acting on preferences becomes difficult to sustain,' making indeterminacy '*the* weakness of game theory' (Hargreaves and Varoufakis 285, 206). The subjective dimension in preference satisfaction compromises the precision of game theory down to its 'four key assumptions': '*Agents are instrumentally rational;* they have *common knowledge* of this rationality; they hold *common priors;* and they *know the rules* of the game' (6).[8] It is immediately obvious that not all actions are informed by instrumentally rational considerations, that not all individuals interpret rationality the same way, that they bring different assumptions (priors) to the game, and that their knowledge and interpretation of the rules may vary significantly. Given that 'it seems quite unlikely that a person's ethical concerns will be captured by a well-behaved set of preferences' (Hargreaves 188), game theory's 'model of rational agency' (205) is clearly not sufficiently sophisticated to cope with moral and psychological motivations exceeding the limits of instrumental rationality. In spite of such limitations, game theory helps us to understand why (Pareto-efficient) equilibrium models fail to achieve their theoretical intentions.

No matter how heuristic a fiction, game theory draws attention to two competing forms for achieving a state of equilibrium. Social agents can either compete or cooperate to balance what they put into a system and what they take out of it. Competition is the driving force behind the *Nash equilibrium,* a model developed in the 1950s by John Nash, on whose life the film *A Beautiful Mind* is based. This model is predicated on the 'idea that, in equilibrium, players' strategies must be best replies to one another' (Hargreaves and Varoufakis 4). This formula assumes that a rational player will always try to maximize his payoff, thereby creating a gain for himself at the expense of his opponent. But if each player is apprised of all aspects of a situation and acts strictly according to his self-interest, then neither will be able to take advantage of the other, thereby creating a perfect state of equilibrium. Nash equilibriums posit quintessentially competitive strategies within a finite or static totality. If one player scores +1, then the other one must score –1. In strictly mathematical terms, this zero-sum game ensures that all moves reinforce the game's homeostatic equilibrium. When applied to the broader social arena, the 'best' or most efficiently organized society would be the one maximizing its utility; the sum total of advantaged and disadvantaged members of a society ought to be averaged out so as to form an exact equilibrium. Unlike Taylor, who based his equilibrium model on cooperation, Nash trusts that a state of equilibrium emerges spontaneously when competitors calculate their 'best replies to one another.' In spite of the many permutations and refinements that the Nash equilibrium formula has undergone, game theory remains committed to the idea that rational agents will seek to maximize the outcome of a situation by pursuing their self-interest or preference satisfaction.

In his reliance on 'best replies,' Nash provides a model justifying the competitive basis of free-market capitalism. It could be argued that Adam Smith anticipated Nash's equilibrium theory when he advanced his famous doctrine of the 'natural identity of interests,' which posits that each individual's free pursuit of self-interest will result not only in economic efficiency beneficial to all but also in social cooperation conducive to peace and harmony. Once alerted to the benefits of the division of labour, rational social agents would immediately accept that it was more efficient for a vintner to produce wine and exchange his surplus for the bread, vegetables, or meat constituting the surplus of the baker, farmer, or butcher. In his *The Wealth of Nations* (1776), Smith contended that 'the greatest improvement in the productive powers of labor, and the greater part of the skill, dexterity, and judgment with which it is anywhere directed, or applied, seem to have been the effects of the division of labor' (1). The division and subdivision of labour was for Smith a great advance in civilization. He presumed that free individuals would spontaneously choose to organize themselves in such a way as to increase the public good through the maximization of economic efficiency. *The Wealth of Nations* thus links the way we produce our material means of production to an ideal social order. In a 'well-governed society,' he writes, the division of labour will lead to 'universal opulence which extends itself to the lowest ranks of the people' (8). In the pursuit of self-interest, then, social agents make 'best replies' to each other, thereby ensuring the public good through the 'invisible hand' of the market. The spontaneously emerging economic equilibrium is for Smith unproblematically duplicated in the political arena of the 'well governed society.' Like Taylor after him, he assumes that economic prosperity underpins social justice. The best replies that competitors in a free market make to each other are thus the cornerstones of a stable but not necessarily just society.[9]

Game theory illustrates why competition may increase efficiency but surreptitiously does so at the expense of social justice. The logic of Nash equilibriums is not unlike the argument that the free market is the most efficient provider of the public good. In his explanation of economic theory, Tim Harford condenses David Ricardo's *On the Principles of Political Economy and Taxation* (1817) by resorting to an example that incidentally dramatizes Nash's 'best reply' scenario. Using one of Ricardo's own illustrations, he asks us to imagine 'a wild frontier with few settlers but plenty of fertile meadow available for growing crops.' A newly arrived young farmer 'offers to pay rent for the right to grow crops on an acre of good meadow' for which the rent will be low because the landlord is competing with other landlords at a time when there is a lot of fallow land on offer. In short, 'settlers are scarce and meadows are not, so landlords have no bargaining power.' But, as fertile land becomes scarce, newcomers will 'offer to pay good money to any landlord who will evict [established farmers]' to let them 'farm there instead.' Now 'the landlords have acquired real bargaining power, because suddenly farmers are relatively common and meadows are relatively scarce' (Harford 7). The amount by which 'the landowners will be able to raise their rents' is determined by the 'best replies' competitors make to each other. The increase in rent 'will have to be enough that farmers earn the same farming on meadows and paying rent, or farming on inferior scrubland rent free. If the difference in productiveness of the two types of land is five bushels of grain a year, then the rent will also be five bushels a year. If a landlord tries to charge more, his tenant will leave to farm scrubland. If the rent is any less, the scrub farmer would be willing to offer more' (8). But this equilibrium will only produce and reproduce itself as long as there is no political interference to regulate market forces and as long as moral and psychological factors are left out of account.

Aside from their limited applicability in the real world, Nash equilibriums prove most problematical, Taylor noticed, in that, contrary to the reliance of Smith and Ricardo on the 'best reply' logic, the pursuit of self-interest did not necessarily lead to healthy competition among free individuals but to sub-optimal 'free rider' problems. We may recall that Taylor devised his system of management to counteract the tendency of 'slacking' workers to catch a free ride at the expense of others. Where Smith considered the market to be the most efficient provider of the public good, Taylor realized that competitors rather consistently selected options that were distinctly not the 'best replies' to each other posited by Nash. Although a company's productivity was clearly spurred on by competition with other companies, it was equally clearly hampered by internal conflicts between competing class interests. If social agents could be convinced of the benefits of cooperation, the system's overall effectiveness would be greatly enhanced. In its utopian conception, Taylor's system assumed that managers, workers, and efficiency experts would spontaneously cooperate once they rationally understood the 'common priors' and the 'rules of the game.' What he discovered, of course, is that rational people will rationally choose a course of action that demonstrably makes their situation worse than it has to be.

Game theorists have debated this puzzling situation since the early 1950s, when Albert Tucker first modelled its logic in the scenario of the Prisoner's Dilemma. This dilemma 'fascinates social scientists because it is an interaction where the individual pursuit of what seems rational produces a collectively self-defeating result. Each person does what appears best (and there is nothing obviously faulty with their logic) and yet the outcome is painfully sub-optimal for all' (Hargreaves and

Varoufakis 172). Although one might be tempted to attribute the deterioration of Taylor's utopian model into the assembly-line nightmare at the Ford Motor Company to a faulty implementation of the principles of scientific management, this explanation is clearly not sufficient. The unintended consequences of Taylor's cooperative model exemplify the 'free-rider problem' that the Prisoner's Dilemma shows to be endemic to the intersubjective dynamic of social interactions.

The well-known Prisoner's Dilemma has seen many paraphrases. Offering an 'embellished version of the tale,' Hargreaves and Varoufakis begin by reminding us that 'Tucker's original illustration has two people picked up by the police for a robbery and placed in separate cells. The police know that they are the culprits but have no hard evidence on which to found a prosecution.' Their version then captures what the District Attorney must have said to each of the prisoners as he languished in his cell:

> If you both 'confess' then the judge, being in no doubt over your guilt, will sentence you, give or take, to 3 years imprisonment. Of course, you know that our evidence against you is insufficient to convict and, hence, if you both deny the charge, I shall have to set you free. However, if you deny the charges *but your friend in the next cell confesses,* I shall make sure that the judge will take a dim view of your recalcitrance and that an example be made of you; let's say an exemplary punishment of at least 5 years. On the other hand, if the situation is the reverse (with you confessing and your next door neighbour denying the charge), I am sure I can intercede with the judge to give you a suspended sentence, on account of your assistance in bringing about a conviction. Not only that but, do you recall that alcohol license which you requested last month? The one that was turned down? I am sure I could swing it for you. (173)

As Hargreaves and Varoufakis put it, 'The analysis of the game is startling. Each selects their dominant strategy, confess (or defect), and they both go to prison for 3 years when they could have both walked by both denying the crime' (174). In other words, instead of choosing the win-win scenario, they invariably opted for the lose-lose one. Even if the two prisoners could have communicated with each other, they would still have found themselves caught in the Prisoner's Dilemma. While they might agree that the best outcome for both of them was to deny the charge, they could not be certain that the accomplice in crime would honor the agreement. Communication alone is not sufficient: 'Promises as well as threats must be *credible*' (174). The game's structure thus 'remains intact: the best action in terms of pay-offs is still to confess' (174). Agreements are consequently not worth the paper they are written on unless they can be enforced: 'As Hobbes remarked when studying a similar problem "covenants struck without the sword are but words"' (174). In short, cooperation involves a complex set of assumptions that involve what we expect of others, what they expect of us, and what we expect others are expecting of us. Taylor's solution was to instal an efficiency expert capable of objectively assessing intersubjective tensions in order to persuade antagonists that cooperation is the rationally superior option.

In slightly more elaborate terms, the Prisoner's Dilemma illustrates that individuals tend to choose a suboptimal outcome either because they themselves want to enjoy a free ride or because they want to prevent an opponent from enjoying a free-ride at their expense. Essentially a matter of trust, the free rider problem is a 'collective action problem' (Hargreaves and Varoufakis 176) that, according to the examples Hargreaves and Varoufakis analyse in *Game Theory*, manifests itself in such topical areas as global warming, domestic labour, public goods, disarmament, Adam Smith's invisible hand, trade unionism, Marx's theory of class conflict and capitalist crises, corruption, the diminished effectiveness of antibiotics, and the 'collective irrationality' that has us all stand at sporting events when we would be much better off sitting down (176–80). In all these cases, rational decisions by rational actors intent on satisfying individual preferences produce collective outcomes that defeat the initial intention. Given the broad applicability of scenarios that demonstrate the Prisoner's Dilemma, it is not surprising that the 'problems of aggregating individual preferences into collective decisions' has generated a 'considerable literature' (Hindess 11–12). In his critique of rational-choice models, *Choice, Rationality, and Social Theory* (1988), Barry Hindess refers repeatedly to the game of the Prisoner's Dilemma to show that 'it is a mistake to suppose that rational individuals sharing an interest in a collective outcome can normally be expected to act so as to produce that outcome' (12). If individuals cannot see how their contribution to a public good serves their rational self-interest, they will have 'no incentive to contribute towards the provision of that good' (14). More importantly, in complex societies, the connections between our contributions to a public good (usually in the abstract form of taxes) and our enjoyment of it (as in our use of roads, electricity, water, etc.) are obscured, if not rendered entirely invisible to us.

In *The Efficient Society,* Heath takes great delight in listing numerous examples in our daily lives illustrating the 'collective irrationality' typical of the Prisoners' Dilemma. One of his most telling cases of a 'perverse outcome' (Heath 42) has to do with our habit of passing slower cars on our freeways. We tend to assume that the most efficient way for reaching our destination as fast as possible is to overtake slower cars whenever an opportunity presents itself. However, since the passing driver often has to squeeze in between cars, she causes drivers behind her to slow down. Although it seems in your interest to reduce your commuting time even though it increases that of the drivers behind you, on closer inspection it turns out that weaving in and out of traffic slows down everybody because 'just as your passing slows down everyone, behind you, the people passing in front of you slow you down' (44). As Heath points out, 'Commuters all slow each other down, even though not one of them slows him or herself down' (44). From a rational perspective, then, our self-interested tendency to gain an advantage over other drivers by passing them is collectively self-defeating. This conclusion was borne out when the Dutch 'enacted a law that prevented trucks from entering the *second* lane' of freeways, thereby making it 'illegal for trucks to pass on more than 70 per cent of highways in the country' (43). While the truckers at first went 'ballistic,' they soon 'discovered that it was starting to take them *less* time, not more, to reach their destinations' (43). Before reaping the unexpected benefits of a law they considered not to be in their self-interest, the Dutch truckers had been no more aware than other drivers of the way their behaviour had been collectively self-defeating. Heath calls our tendency to drive as fast as possible on our freeways 'a paradigm case of individual efficiency generating collective *inefficiency*' (45). But cooperation does not come easily to us; the Dutch truck drivers did not voluntarily stay in the slow lane, but only after they had been legislated to do so. If the legislation were to be rescinded, it is highly likely that they would revert to their previous sub-optimal behaviour.

We may take some comfort from the discovery that 'people cooperate ... much more than game theory leads one to expect' (Hargreaves and Varoufakis 181). Game theorists show that the Prisoner's Dilemma is often overcome when we move from synchronic to diachronic analyses. Once games are repeated over time, players are more likely to opt for cooperative rather than competitive strategies. High on the list of influences counteracting the free-rider problem is 'the positive influence of pre-play discussion on co-operation' (181), a discussion constituting a 'rhetorical effort' that 'encourages appeals to general (i.e. impersonal) principles regarding action rather than the ones (like game theoretical instrumental reasoning) which are simply based on the individual pursuit of narrow self-interest' (181). In other words, moral consideration of the common good affects the degree of cooperation players are willing to select. But unselfishness need not be the outcome of a struggle against egotistical self-interest; it is also possible that 'actors are not always sufficiently sophisticated to realize that selfishness is the only rational approach' (Hindess 33). Most crucially, perhaps, game theory illustrates that 'co-operation is likely to be highest when it is expected to be reciprocated' (Hargreaves and Varoufakis 183). Once 'a pattern of costs and incentives different to once-only interaction' (Hindess 33) is at stake, we discover that a race to the bottom is no longer necessarily the dominant strategy. An anticipation of reciprocation arises when a Prisoner's Dilemma is repeated over time; the success or failure of a collectively desirable outcome then depends on how much trust we have in an opponent's promise to perform as expected. Kant's categorical imperative, for instance, shows that co-operation is conditional on the level of trust created when an indefinitely repeated Prisoner's Dilemma game convinces us to abandon short-term self-interest for the benefit implicit in long-term reciprocation. If I want my street to be clean, I will refrain from littering because I am reasonably certain that my neighbour will reciprocate by similarly refraining. Experiments have shown that people are more likely to litter if there is already litter on the street. Our immediate cooperation is thus conditional on future pay-offs.

The logic of a 'simple, reciprocal co-operative strategy that punishes defectors with defection in the next round' is illustrated in Anatol Rapport's game of Tit-for-tat. This game gives players '*the opportunity to condition their behaviour on what their opponent did earlier*' (Hargreaves and Varoufakis 192). Moves are predicated on expectations formed by previous experience. A 'collective fund for "health care"' (195) is an obvious example of this Tit-for-tat logic. We pay into the fund on the expectation of benefiting from it should we fall ill. It is only because we know from experience that this health-care scheme has worked that we trust it enough to pay into the fund. What is at stake for us is people's reputation as trustworthy players, moral individuals, or upstanding citizens. As Hargreaves and Varoufakis will conclude, once 'the *Prisoner's Dilemma* (or free rider problem) is repeated, everything seems to change. Co-operation gets a chance and defection is suddenly no longer a dominant strategy' (196). In other words, 'what sustains co-operation in the indefinitely repeated version with a strategy like *Tit-for-tat* is the desire to remain in "good standing" in future periods so that the co-operative outcome is obtained in those periods and it is this desire which can outweigh the immediate gain from defection now' (203). The 'normative and reciprocal nature of co-operation' (205) introduces aspects of human reasoning that exceed the narrow boundaries of instrumental rationality. Cultural norms and alternative forms of reason thus alleviate the tendency of racing to the bottom in the Prisoner's Dilemma, but reinforce the problem of indeterminacy so inimical to game theory.

Heath's example of freeway passing suggests that we are often not able to work out what is rationally in our best interest. And if we are able to work it out, we may perversely persist in behaviour that we know to be counterproductive or collectively inefficient. It is not only because we are motivated by unconscious desires that we are unable to understand why we act as we do; we also do not wish to recognize the collectively self-defeating decisions we often make without taking into account the unintended consequences that could flow from them. In short, we disavow what we know. The logic of disavowal has been the subject of Slavoj Zizek's radicalization of Lacan's psychoanalytic theory. His most frequently cited articulation of this logic is the formula of fetishism that is '"I know, but nevertheless ..." ("I know that Mother doesn't have a penis, but nevertheless ... [I believe that she has]")' (Zizek 245). The 'paradox' of disavowal is for Zizek at the 'core of so-called doublethink' in Orwell's *1984:* '"We must consciously manipulate the whole time, change the past, fabricate "objective reality," at the same time sincerely believing in the results of this manipulation.' It follows, then, that the '"totalitarian" universe is a universe of psychotic split, disavowal of the obvious evidence, not a universe of "repressed secrets": the knowledge that we "deceive" in no way prevents us from believing in the result-effect of the deception' (244). What game theory illustrates is that even in our most instrumentally rational interactions, we subject ourselves to the logic of disavowal in order to persist in behaviour we know to be collectively self-defeating.

We can now see that Taylor's response to the free-rider problems he had observed on the factory floor encouraged him to develop an equilibrium model predicated on increasing a system's efficiency through spontaneous cooperation. Yet, by putting his trust in cooperation to achieve a Pareto-efficient equilibrium, he failed to consider the tendency of players to make short-term choices that prove in the long term to be collectively self-defeating. The Prisoner's Dilemma illustrates that spontaneous cooperation is an unlikely outcome even if it is theoretically acknowledged as a desirable strategy. The Ford Motor Company shared Taylor's concern that class and other social antagonisms were serious impediments to the efficient output of cars. The initial solution of higher wages and social assistance to increase efficiency was a winning formula; it lowered the price of cars to the point where workers could be turned into consumers so that the company was able to profit from the massive scale of sales. Although this essentially cooperative model was already inhabited by a paternalistic desire for control, it produced social benefits for workers that were at least approaching Taylor's ideal notion of a system in which input and output were to be evenly distributed. In spite of benefiting less than the managers, the workers enjoyed an increase in prosperity that included aspects of social justice. But the primarily ideological means by which bodies were rendered docile soon proved too cumbersome and expensive. Once the service department under Bennett came into full force, workers were exposed to strategies of undisguised coercion and oppression. The Ford Motor Company acted on the realization that cooperation could only be achieved through coercion, thereby illustrating the failure of Taylor's assumption that an efficient plan would spontaneously encourage voluntary cooperation. Yet the Ford Motor Company is not only a distortion of Taylor's utopian vision but also its most logical extension. Ford's reduction of the human being to a 'cog in the machine' is the paradigmatic consequence of Taylor's central contention that from now on the system is to take priority over the individual.

When Taylor called his reconfiguration of the relationship between the individual and the social collectivity a 'mental revolution,' he situated his principles of scientific management within a philosophical tradition of social-contract theories that, not coincidentally, also receive attention in discussions of game theory. Illustrating the broader social and philosophical ramifications of game theory, Hargreaves Heap and Varoufakis comment most prominently on philosophers like Bentham, Hegel, Hobbes, Hume, Kant, Marx and Engels, Rousseau, Adam Smith, and Wittgenstein. Although there were undoubtedly many reasons for Ford and Taylor becoming flashpoints for cultural anxieties, they inadvertently demystified the comforting illusion that the conflict between subject and object is open to a sublation Hegel called an 'identity of non-identity.' Most significantly, perhaps, the emphasis on efficiency contributed to a shift from old liberal values understood to liberate individuals from external constraints to a new liberalism advocating the primacy of the public good over individual self-interest. The older liberal conception assumed that individuals endowed with civic, economic, and political freedom would spontaneously organize themselves into harmonious communities. They would voluntarily cooperate just as Taylor had envisaged when he trusted rational social agents to work together to achieve a mutually beneficial collective outcome. In contrast, the more recent conception of liberalism sought to regulate the behaviour of individuals and empowered the state to intervene both in the market and in the social arena. Instead of affirming 'negative freedom' from compulsion, the 'positive freedom' in new liberalism focuses on the benefits accruing to individuals from the collectivity. The new social figure of the efficiency expert or 'manager' situates Taylor on the side of paternalism and social control; he believes in the regulation of individual behaviour so as to ensure an optimal collective outcome. Contradictorily embracing two conflicting ideological investments, Taylor thus embodies a crisis in the cultural self-understanding of his day when utilitarian economic considerations were visibly eroding political appeals to traditional notions of individual freedom. As we will see in the second part of *Modernism and the Culture of Efficiency,* such conflicted strands manifest themselves in the everyday lives of characters depicted in modern novels published in the early twentieth century.

The most graphic recent representation of the disciplinary regime emerging at the Ford Motor Company is undoubtedly the metaphor of Jeremy Bentham's Panopticon that Michel Foucault makes central to his influential study of prisons in *Discipline and Punish.* The Panopticon is for Foucault an architectural configuration functioning metaphorically to explain the ways surveillance operates as an asymmetrical form of power. In *Discipline and Punish,* Foucault highlights the salient features of Bentham's Panopticon by describing it as a prison consisting of a central tower from which a guard is in a position to observe all the prisoners, who are backlit to increase their visibility to the guard and isolated from each other by walls. The prisoners cannot see the guard but know that he sees them; they are collectively and individually subjected to the gaze of a potentially violent locus of power. 'All that is needed,' writes Foucault, 'is to place a supervisor in a central tower and to shut up in each cell a madman, a patient, a condemned man, a worker or a schoolboy' (*Discipline* 200). Feeling at all times exposed to this surveillance, the prisoner internalizes the disciplinary gaze and submits to its power. The Panopticon 'reverses the principle of the dungeon; or rather of its three functions – to enclose, to deprive of light and to hide – it preserves only the first and eliminates the other two' (200). At the Ford Motor Company, the worker enters the factory voluntarily; however, once inside, he is tied to a place from which he is fully visible to the supervisor and prevented 'from coming into contact with his companions' (200). He is thus 'enclosed,' cruelly exposed to a light shining on all his moves, and thus unable to hide even aspects of his life outside the workplace from the company. The deskilled assembly-line worker is not unlike the prisoner: 'He is seen, but he does not see; he is the object of information, never a subject in communication' (200). Although the disciplinary mechanisms Foucault describes work 'without recourse, in principle at least, to excess, force, or violence' (177), in Highland Park and River Rouge this insidious form of domination was reinforced by the often spectacular production of terror symbolized by Harry Bennett's thugs. The docility of the prisoner in the Panopticon is, of course, ultimately secured by the knowledge that the guard's gaze is backed up by his gun. It is enough for workers to know that Bennett may see them and is prepared to harm them for the production process to function efficiently.

Although Foucault does not deal specifically with the factory, he references it as a carceral model by including the 'worker' in his list of possible prisoners. For Foucault, the 'ideal model' for the modern carceral society is 'the military camp,' an engineered society dedicated to violence in which 'all power would be exercised solely through exact observation' (*Discipline* 171). It is through the classification and regulation of the human body that administrative organizations train and discipline the bodies of soldiers, schoolboys, factory workers, and citizens so as to ensure 'their utility and their docility, their distribution and their submission' (25). Although submission in the military camp is mostly achieved through surveillance, this ideological form of domination is reinforced by displays of authority figures carrying guns. Backed up by Harry Bennett's repressive measures, Henry Ford's disciplinary gaze becomes so internalized by the workers that violence is rarely necessary. The Ford Motor Company emerges as a disciplinary apparatus exemplifying a mechanism of surveillance, which Foucault extends to modern society as a whole:

> By means of such surveillance, disciplinary power became an 'integrated' system, linked from the inside to the economy and to the aims of the

> mechanism in which it was practiced. It was also organized as a multiple, automatic, and anonymous power; for although surveillance rests on individuals, its functioning is that of a network of relations from top to bottom, but also to a certain extent from bottom to top and laterally; this network 'holds' the whole together and traverses it in its entirety with effects of power that derive from one another: supervisors, perpetually supervised. And, although it is true that its pyramidal organization gives it a 'head,' it is the apparatus as a whole that produces 'power' and distributes individuals in this permanent and continuous field. This enables the disciplinary power to be both absolutely indiscreet, since it is everywhere and always alert, since by its very principle it leaves no zone of shade and constantly supervises the very individuals who are entrusted with the task of supervising; and absolutely 'discreet,' for it functions permanently and largely in silence. (176–7)

Far from conforming to the integration symbolized by organic metaphors of the totality favoured by the Enlightenment tradition, this system is indifferent to the atomized subjects it distributes and regulates. What is most striking about the Ford Motor Company is the way it inflected power by uncannily reinforcing the terror of Harry Bennett's tactics with the disciplinary mechanisms of the administrative apparatus. In its correspondence to the Panopticon, the factory exercises a form of power that is experienced as both totalizing and individualizing; it pervades the whole system at the same time as it seems designed to oppress each individual worker. Workers at the Ford Motor Company were all the more easily persuaded to internalize the disciplinary gaze as they accepted as reasonable the expectations imposed on them by appeals to efficiency.

Why did people accept the increasingly demeaning and dehumanizing conditions at Ford Motor? While they might have objected to repressive measures instituted by management, they initially accepted the argument that it was reasonable for Ford to expect efficiency from employees who were well remunerated. They were willing to sell their souls for $5 a day and for company benefits. They did not feel 'owned' by the company but understood themselves to be contributing their part to the overall efficiency gains that made Ford Motor competitive in the market. Americans, who prided themselves on having escaped from oppressive regimes in Europe, willingly accepted the company's right to interfere deeply into their personal lives. The company was allowed to own not only their labour but also their whole body and mind. This complicity with their victimization can only be explained by an overall agreement that efficiency is an unquestioned 'good.' Infected by the desire for the ultimately efficient totality, they were prepared to endure hardships that extended beyond hard work to loss of control over their personal lives.

3 An Experiment in (In)Efficient Organization and Social Engineering: Auschwitz

There were, of course, many repercussions of the sociocultural 'revolution' that crystallized around Fordism and Taylorism in the early twentieth century. A commitment to efficiency manifested itself in various forms, ranging from the extreme case of the concentration camp in Nazi Germany to the apparently innocuous pleasures afforded by the consumer society. The openly repressive tactics employed by Harry Bennett to ensure an efficiently functioning workforce at the Ford Motor Company finds one particularly troubling logical extension in the social-engineering experiments carried out most notoriously at Auschwitz-Birkenau, while the ideological measure of Henry Ford's Five-Dollar Day to incite conformity to socio-economic norms saw its most prominent and apparently innocuous exemplification in the American suburb. Whether efficiency is imposed or inculcated, repressive violence and ideological manipulation could be said to reinforce each other to varying degrees throughout modern society. Although Auschwitz worked through brutal violence, it nevertheless relied to a considerable extent on ideological commitments on the part of both victims and henchmen. At the other end of the scale, the eager embrace of the good life in suburbia by the increasingly dominant middle classes was not without its compulsions and coercions.

Both Auschwitz and suburbia constitute social-engineering projects driven by the utopian dream of creating the most perfectly efficient society. The racial targeting in the Nazi persecution of the Jews was justified on eugenic grounds; the perfect society could only be constructed after the 'weeds' in the 'garden' had been eliminated. The Jews were equivalent to the 'waste' impeding the optimal performance of the combustion engine. But the concentration camps were also used as laboratories to test the most efficient strategies for converting individuals into docile bodies. The domination of the Jews was to be a model for the management also of the German population and, ultimately, for world control. If Auschwitz represents the most sinister side of our investment in efficiency, the suburb is its most benign one. As Taylorization infiltrated the social arena from the factory to the educational system, the benefits of the system were generally thought to outweigh its disadvantages. The suburb is the image not only of the most efficient but also of the happiest society. However, the happy suburban consumer is in fact subject to a disciplinary regime that encourages submission to the requirements of corporate capitalism.

Although the efficient organization of factory or suburb is not to be confused with efficient technologies designed to kill 'undesirables,' these two very different manifestations of a commitment to efficiency nevertheless resonate with each other, just as Freud's analyses of psychopathologies foreground aspects of what passes for 'normal' human behaviour. Both Nazism and the consumer society celebrated in the suburb are, of course, marked by a multiplicity of struggles; far from explaining everything, efficiency is one thread among many in the sociohistorical fabric, a thread that is here being isolated for particular attention. In short, this chapter focuses specifically on Auschwitz in order to emphasize the troubling relationship between efficiency and disciplinary power, a relationship that will then be reconsidered in the following chapter, devoted to an analysis of Taylorism as it engulfed corporate culture and the suburban 'good life' it made possible.

The largest concentration camp, Auschwitz-Birkenau has become the primary symbol of Nazi crimes, especially since it was the focus of the 'two main ideological ideas of the Nazi regime: it was the biggest stage for the mass murder of European Jewry, and at the same time a crystallization point of the policy of settlement and "Germanization"' (Steinbacher 3). My emphasis on efficiency is by no means designed to explain the complex phenomenon of Auschwitz in either its historical specificity or its exemplification of state racism: I take Auschwitz as a test case for the conflicted logic of efficiency operating under extreme conditions. The autobiography[1] of the camp commandant, SS-Obersturmbauführer Rudolf Franz Ferdinand Höss, and the most widely read survivor memoir, Primo Levi's *Survival in Auschwitz* (1947),[2] provide us with evidence of the role efficiency played in a 'factory' dedicated to the abuse and murder of innocent victims. Readings of these primary texts are reinforced by information provided in *KL Auschwitz: Seen by the SS* (2007), an anthology of commentaries based on historical documents edited by researchers at the Auschwitz Museum, Jadwiga Bezwinska and Danuta Czech, whose ambition was to present factual accounts of the camp. From these contributions it is possible to uncover strands from the much debated traumatic history of Nazism that speak directly to the most sinister aspects of modernity's commitment to efficiency, both as 'adventure' and as 'routine.'

The emphasis on efficiency reveals first of all that Henry Ford's obsession with the construction of an efficient enterprise to produce as many cars as possible has its perverse counterpart in Höss's ambition to make Auschwitz the most efficient killing machine during his reign there from May 1940 until November 1943. In his autobiography he details the problem-solving skills he employed when faced with the gigantic task of building a concentration camp under difficult and constantly shifting circumstances. Where Ford could rationalize his obsession with efficiency by convincing himself that he offered the masses a much appreciated consumer good, Höss justified his participation in mass murder by describing himself as 'a fanatical National-Socialist' who was 'firmly convinced that our ideals would gradually be accepted and would prevail throughout the world' (Höss 55). These vaguely articulated ideals conceal even from himself the zeal for the active pursuit of efficient outcomes that really motivates him in his daily activities, a zeal that compels us to revise the popular assumption that those supporting the fascist regime were only passively 'doing their duty.' For Höss, the victims at Auschwitz were numbers proving both the efficiency of the camp as a killing machine and its value as the laboratory of socially engineered human beings. With the dispassionate gaze of an anthropologist observing an experiment in social control, he records, in often morally disapproving tones, his reflections on the behaviour of prisoners under the extreme conditions he had himself instituted. The reduction of workers at the Ford Motor Company to mere 'cogs in the machine' finds its extreme extension in the radical dehumanization of prisoners within a system designed to impress on them that they are powerless, dispensable, isolated, and infinitely interchangeable docile bodies under constant threat of death. But it is in Levi's survivor memoir that the extent to which Auschwitz did indeed serve as a deadly experiment in the taming of human bodies is most trenchantly analysed. Efficiency not only served the purpose of expediting the elimination of human 'waste,' but also secured the complicity of both henchmen and prisoners in a dehumanizing social-engineering project. Although the complicity of an investment in efficiency with the murderous violence in Auschwitz should not be conflated with the exploitation of workers at Ford Motor, both the openly repressive and the ideologically internalized mechanisms at work in each case differ more in degree than in kind.

In the end, the analysis of efficiency in this chapter illustrates and confirms Foucault's contention in *'Society Must Be Defended'* (1997; 2003) that we have 'in Nazi society something that is really quite extraordinary: this is a society which has generalized biopower in an absolute sense, but which has also generalized the sovereign right to kill' (*'Society'* 260). The conflicted logic of efficiency analysed in this chapter is thus implicated in the paradox of a regime simultaneously reverting to 'the classic, archaic mechanism that gave the right of life and death over its citizens' and applying 'the new mechanism of biopower' arising with 'discipline and regulation' (260).

Auschwitz: Model City / Death Camp

After the British demographer and political economist Thomas Robert Malthus (1766–1834) published 'An Essay on the Principle of Population' (1798), in which he predicted that the food supply would not be able to keep up with expanding population growth, the management of populations became an urgent issue throughout Europe. In Hitler's extrapolation from Malthusian anxieties, surplus populations were human 'waste' to be reduced and eliminated if society was to be a healthy and efficiently functioning political body. In Hitler's rhetoric, the Jews were referred to as 'weeds' spoiling a potentially beautiful 'garden.' Instead of imperilling the whole German race, it seemed logical to remove only undesirable political opponents and 'alien' groups from within Germany's borders to leave more room for 'pure' Germans. This longing for the accommodation of a privileged population engendered the cry for more territory or *Lebensraum* (living space) in the East to settle the surplus of the ethnically pure to be retained. By sending so-called politicals, asocials, and Jews to Auschwitz, Hitler was able at one fell swoop to cleanse and expand the German realm.

As Sybille Steinbacher reminds us in *Auschwitz: A History* (2004), what is often forgotten is that the Nazis selected the town of Auschwitz not only as the site of a concentration camp but also as a 'model settlement town' (68) for ethnic Germans. Although the ambitious Nazi ideas for the town were only very partially put into practice, 'many plans were worked out to the smallest detail,' conjuring up a 'modern town with public buildings, extensive transport connections and many green spaces' (66). Representatives from various agencies 'sent representatives to Auschwitz, where they spent days and weeks discussing designs, visiting locations and rebuilding the town' (66). These plans for the modernization of Auschwitz were motivated by the desire to entice Germans from the Old Reich to relocate to the East. This relocation served both the policy of 'Germanizing' the East and the need of IG Farben for a 'qualified German workforce' (66). 'With a grand sense of historical purpose, and inspired by the racial task before him,' the architect Hans Stosberg enthusiastically set out to realize 'his vision of modern town planning' (68). In Himmler's vision, Auschwitz was literally to become an ideal garden designed by 'landscape architects and botanists,' a 'research zone' for 'recycling refuse and sewage, for biological waste processing, the growth of plant cultures, and technical innovations in the use of slurry and composting' (75). The settlers arriving in Auschwitz were 'distinguished by a pioneering spirit, a belief in the future, their efforts to bring "German culture" to the East, as well as a high level of business efficiency' (74). In other words, they embodied the spirit of modernity as adventure. Peopled by such enterprising Germans, Auschwitz was thus to become 'the model of the National Socialist "ethnic community"' (69), of which the concentration camp on its doorstep was the sinister symbol of the 'waste' to be eliminated on the road to the 'perfect' society of the future.

The proposed transformation of the town of Auschwitz into an ideal German settlement required that its large Jewish population be expelled and the Polish inhabitants resettled. With the historical emphasis on the enormous number of Jews exterminated at Auschwitz, it is often forgotten that from 1940 to mid-1942 the camp incarcerated mostly Poles.[3] Situated near Kracow in Poland, Auschwitz was chosen as a site for the camp because of 'its suitable location in terms of transportation and because the area can be easily isolated and concealed' (Piper, 'Direct' 17). Bordering on Germany, Poland was also a prime target of Hitler's aim to conquer territory, especially territory 'suitable for agricultural goals' (12), for the resettlement of ethnic Germans. As Heinrich Himmler specified, the Nazis were interested in acquiring *territory* rather than foreign people: 'Our goal is not the complete Germanization of the East in the old sense of the term – i.e., by teaching the population living there German language and law; rather, we are striving for the goal that only people of purely German race will live there' (12). Hitler envisaged not only a Germany free of Jews (*judenfrei*), but also a Poland free of Slavs. Although Poles were sent to Auschwitz for primarily political reasons, the elimination of between 140,000 and 150,000 Poles also served the Nazis' interest in Poland as a 'German "living space," intended for future German settlement' (12). In their conquest of Eastern Europe, the Nazis entertained the liquidation of 'entire nationalities' through solutions ranging from the '"re-Germanization" of a certain portion of the population with a German background, through various measures to decrease the birth rate and raise the death rate, to acts of immediate, physical extermination' (12–13). After the Jews, the Slavs were to be next on the list of inferior races to be eliminated. In the short term, the Poles, who were in the way of the plans for Auschwitz as a model town and a concentration camp, were unceremoniously displaced without any compensation. The Nazis expelled 1600 Poles to make room for the camp's 'area of interest,' creating a 'depopulated region of around 40 square kilometers' (Czech 27) for agricultural and industrial use. In its broadest dimensions, Auschwitz is symbolic of a Nazi experiment in social engineering intent on sweeping away and reorganizing whole national populations.

The most basic function of the death camps was indeed the control and regulation of biologically targeted populations. Symptomatic of this function were the infamous medical experiments carried out on prisoners at Auschwitz. Officially, the criminal experiments 'were meant to serve the needs of the army (some had the goal of improving the state of health of the soldiers), help achieve postwar plans (e.g., in the area of population policy) or provide scientific underpinnings for racial theories (e.g., the superiority of the Nordic race)' (Strzelecka 88). In reality, most were carried out to discover 'a method for the biological extinction of nationalities' (94), which would 'allow the leadership of the Third Reich to destroy lower races using 'scientific methods,' by denying them the ability to reproduce' (91). More specifically, these experiments 'supported Nazi demographic policies by experimenting with potential methods for mass sterilization'; the Nazis were in fact planning the mass sterilization of 'both the Jews and the Slavs' (89). Not surprisingly, individual doctors also took advantage of the criminal licence reigning at the Auschwitz hospital to pursue independent research to further their personal careers. The diary of Johann Paul Kremer,[4] a doctor temporarily posted to Auschwitz, records his scientific interest in 'the changes developing in the human organism as a result of starvation' (Kremer 167). Finding an unlimited supply of research data in the emaciated prisoners regularly murdered at the hospital, he eagerly 'preserved fresh material from the human liver, spleen and pancreas' (167). If he saw a prisoner in an advanced state of starvation, he 'asked the orderly to reserve the given patient for me and let me know when he would be killed with an injection' (167). The selected patient was then 'put upon the dissecting table while he was still alive' (167). Aside from seriously violating his Socratic oath, Kremer displays a callous disregard for the sick, which he dresses up as scientific objectivity. The idea of population control was so generally accepted that the biological experiments at Auschwitz were often performed at the request of reputable pharmaceutical firms and medical institutes.

The Efficient Commandant

If the regulation of populations was a central component of Hitler's political program, Auschwitz emerged as the laboratory for the social control of human beings. It was not that the Nazis sat down and drew up an exact blueprint for the construction and maintenance of a concentration camp to hold surplus or 'expendable' populations. As Höss explains, when he was first sent to make use of existing military barracks for a concentration camp, he was charged with the rather modest construction of a 'transit camp for ten thousand prisoners' (28); however, after 'a visit of the *Reichsführer* [Himmler] in March 1941,' he was given orders to increase the capacity of Auschwitz to hold 30,000 prisoners and was later to build a new camp for another 100,000 prisoners at nearby Birkenau. Although the concentration camps were from the beginning planned to deal with surplus or expendable populations, their design and purpose kept evolving with changing circumstances. Since the early 1930s, political opponents had been the primary inmates of relatively small camps scattered across German-dominated areas. As the Nazis persecuted more and more groups of 'undesirables,' the number of bodies to be secured and starved swelled at a rapid rate. It didn't take much to understand that the supervisory task was made easier with each prisoner who happened to die. In an effort to alleviate overcrowding, poor living conditions and hard labour thus suggested themselves as indirect means for the elimination of inmates.

To speed up this rather slow process of extermination, the Nazis introduced summary executions for putative infractions of so-called regulations and started to kill hospital patients with phenol injections to the heart. Once the Jews began to be persecuted 'under the rubric of "the final solution of the Jewish question"' (Iwaszko 56), mass transports of Jews so swelled the prison population at Auschwitz that existing methods of extermination 'proved insufficient,' leading the Nazis to kill prisoners 'through asphyxiation in the gas chamber' (Piper, 'Political' 163). This solution to a technical problem suggested itself as an adaptation of methods 'applied in Germany at so-called "euthanasia" facilities, used for murdering the mentally ill' (163). The gas chamber became the method of choice on the basis of previously used practices, the availability of suitable buildings, and, finally, the ingenuity of 'engineers' who had tested Zyklon B on Soviet POWs before using it on a massive scale for exterminations, primarily, of Jews. Having 'verified the effectiveness of Zyklon B,' Rudolf Höss and Adolf Eichmann had together decided to employ it 'for killing the Jews' (Piper, 'Living' 166). But the 'success' of this solution presented new problems for the effective disposal of the gassed bodies, especially since the Nazis were concerned to conceal their crimes from the outside world. Initially, the corpses were buried in mass graves; however, to Höss's consternation and dismay, Eichmann's office ordered that 'the mass graves be excavated and the corpses burned' (168). Höss carried out this order first by burning the corpses in the open air and then by rapidly constructing several crematoria. Once the five crematoria were as functional as they would ever be, they were able to burn a total of 4756 corpses a day. As a death factory, Auschwitz was intent on constantly increasing its murderous capacity by improving both the technical equipment and the flow of work; it was an efficient assembly line dedicated to a monstrously irrational outcome.

The parts of Rudolf Höss's autobiography dealing with his years as commandant of Auschwitz provide us with insights into the significance he attributed to his obsession with efficiency in his decisions and behaviour patterns. In his 'Foreword' to the reissued autobiography of Höss (*Death Dealer: The Memoirs of the SS Commandant at Auschwitz*), Primo Levi considers the value, for today's readers, of a book 'filled with evil' and 'narrated with a disturbing bureaucratic obtuseness' (Levi, *Death* 3). As author of *Survival in Auschwitz,* Levi speaks with the authority of a survivor of the death camp of which Höss was in charge. For him, the autobiography is 'one of the most instructive books ever published' (3) for two reasons: it contradicts the claims of revisionists and it illustrates that 'ideologies can be good or bad; it is good to know them, confront them, and attempt to evaluate them' (9). Since Höss was an uneducated man, 'he cannot be suspected of deliberately perpetrating a colossal falsification of history' (3); his book is consequently 'substantially truthful' (3), at least in its reconstruction of what one may cautiously call 'the facts.' In his 'Foreword' to *Auschwitz: Seen by the SS*, Jerzy Rawicz similarly acknowledges that Höss was 'an exceptional figure among the war criminals, as he did not deny responsibility for the millions of Auschwitz victims' (Rawicz 14). Indeed, 'the sincerity' of Höss's statements about the camp cannot be doubted, and his autobiography stands out as 'a reliable testimony of the most unimaginable crimes' (Rawicz 13, 12). However, both Levi and Rawicz express considerable reservations about Höss's often self-exculpatory reflections on his own participation in the events he recounts and the sentiments he retrospectively attributes to himself. When writing about himself, he was clearly less 'credible' (Rawicz 13). Awaiting certain death in a prison cell in Poland, the commandant of Auschwitz is not writing under the illusion that his confession will alter the outcome of his trial;[5] by his own account, he sets down what he remembers to set the historical record straight. What strikes Levi as particularly 'instructive' is that Höss was in every respect an ordinary man who would, under other circumstances, have lived out his life as 'some drab functionary' rather than as 'one of the greatest criminals in history' (*Death* 3). It is, then, 'the autobiography of a man who was not a monster and who never became one, even at the height of his career in Auschwitz, when at his orders thousands of innocent people were murdered daily. What I mean is that we can believe him when he claims that he never enjoyed inflicting pain or killing: he was no sadist, he had nothing of the satanist' (3–4). For Levi, evil can readily replace good; in other words, he recognizes the systemic underpinnings of the horrors perpetrated at Auschwitz. Stressing Höss's strict Catholic upbringing under the disciplinary gaze of his 'fanatic' father, Levi reinforces a narrative insisting that evil was committed out of a sense of duty, unquestioned submission to higher authority, and nationalism. Höss is here assessed as 'a pliant individual' (7) who slavishly and hence *passively* followed orders.

No matter how 'credible' Höss's self-understanding may be, in the autobiography he constructs himself more *actively* as a man obsessed with the efficient execution of challenging tasks the meaning and morality of which he did not consider to be his concern. He had been sent to Auschwitz to construct the largest German concentration camp; the seriously dilapidated Polish army barracks had to be fixed up and converted into prisons, and new buildings had to be planned and constructed. From the perspective of the complaining Höss, 'out of nothing, and with nothing, something vaster than ever before had to be built in the shortest possible time' (Höss 37–8). Speaking of himself as an 'engine, tirelessly pushing on the work of construction and constantly dragging everyone else along with [him]' (36), he sees himself as an embattled technician struggling valiantly against almost insurmountable obstacles. Living in a 'perpetual rush,' he is confronted by always new plans originating from 'the numerous schemes and plans that [Himmler] had laid down for Auschwitz,' plans that required always 'further urgent action' (67). He receives orders, he solves problems, and he writes reports. Painting himself as a dutiful bureaucrat dedicated to the efficient planning and execution of technical tasks, he has no time to ask himself what purpose is being served by his feverish activity.

In his own eyes, Höss was a man whose 'thoughts and aspirations were directed toward this one end' (34) of constructing the camp; he is asking his readers to understand that this all-consuming activity left him no time to worry about the prisoners. If they were being abused, it was neither on his orders nor even in his declared self-interest. On the contrary, it was clear to him 'from the beginning that Auschwitz could be made into a useful camp only through the hard and untiring efforts of everyone, from the commandant down to the lowest prisoner' (28). Not unlike Frederick W. Taylor, he realized that he would have to set a good example if he wanted to get 'the maximum effort out of my officers and men' and that he would only 'get good and useful work out of the prisoners' by 'giving them better treatment' than was the practice in other camps. In an effort to 'obtain the willing cooperation of the prisoners,' he was prepared to house and feed them properly so that he could then 'demand the maximum effort from them' (29). To achieve an increase in overall productivity, he thus proposes a Pareto-efficient scheme predicated on the assumption that everybody would benefit from cooperation. He and his officers would be rewarded by superiors for having achieved an efficient outcome and the prisoners would benefit from better living conditions. The alleviation of hardship for the prisoner can be seen as the equivalent of the Five-Dollar Day incentive offered by Henry Ford. But before Höss has to ask himself if those condemned to death would see the wisdom of his bargain, he discovers that his plan is from the start being torpedoed by inefficient subordinates and incompetent superiors. Given the obstacles he faced, he unfortunately had to leave the prisoners 'entirely in the hands of individuals such as Fritsch, Meier, Seidler and Palitzsch, distasteful persons in every respect' (34). If the prisoners were abused, it was because he himself was often unaware that his officers 'deliberately misinterpreted' what were to them 'unwelcome orders,' orders to which they even gave 'an entirely opposite construction' (32). It follows that his own 'good faith and best intentions were doomed to be dashed to pieces against the human inadequacy and sheer stupidity of most of the officers and men under [him]' (29). In other words, Höss does not seek to justify himself by claiming human sympathy with the plight of the prisoners, but by attributing their victimization to the inability of others to recognize his far-sighted understanding of the utilitarian value they constitute for his own obsession with efficiency.

Höss seeks to convince himself and his readers that he had to sacrifice personal feelings and human emotions to the efficient execution of the difficult tasks he had been assigned. In a parody of our own modern obsession with work, he tells us that he became so 'absorbed' and 'obsessed' with daily routines and problems to be solved that he 'lived only for [his] work' (33). Surrounded by general slovenliness and sheer incompetence, he could not delegate tasks and was forced to do everything himself: 'I had to "organize" the trucks and lorries I needed, and the fuel for them' (32). At the same time, he complains that he had to cope with 'the inefficiency of most of the officials with whom [he] had to deal' (32). In his negotiations with 'various economic offices' and 'local and district authorities' (32), he was consistently frustrated. Although there were 'mountains' of 'urgently needed' barbed wire in a nearby depot, for instance, he could not touch it without first 'getting authority to have it decontrolled from the Senior Engineering Staff in Berlin' (33), a process that would have taken far too long. In the end he had to help himself by pilfering the barbed wire. To make matters worse, his immediate superior was incapable of understanding the difficulties he was facing: 'A very great deal of trouble could have been avoided if Glück's attitude towards me had been different' (66).

Unaware of the ironic response his account would elicit, Höss presents himself as the one who had suffered most at Auschwitz. Disappointed by old comrades who had 'deceived and doublecrossed' him and exposed to the general 'untrustworthiness' surrounding him, he changed from a nice, trusting, and cheerful man into a 'distrustful and highly suspicious' person who saw 'only the worst in everyone' (35). Taking out his frustrations on everybody around him, he became an 'unsociable person' (36) whose intolerable behaviour made even his wife suffer. Circumstances forced him to betray his better nature and to conceal his genuine feelings. Having 'eyes only for my work, my task,' he confesses that 'all human emotions were forced into the background' (35). He could not allow his 'feelings to get the better of [him]' (68) or permit doubts to enter his mind. Although he claims to have asked himself in his 'innermost being' if it was 'necessary to do all this,' to destroy 'hundreds of thousands of women and children,' he always hastens to reassure the reader that he 'dared not admit to such doubts' (78). Since 'everyone watched' him, he had 'to exercise intense self-control in order to prevent [his] innermost doubts and feelings of oppression from becoming apparent' (78). It was not that he was without human feelings and compassion. Indeed, his 'pity was so great' that at one point he 'longed to vanish from the scene' (78) of an officer forcefully carrying a child into the gas chamber. Levi calls this scene the 'least credible lie' (Levi, *Death* 8) in the whole autobiography. Höss is indeed even more 'loathsome' (8) when he asks us to sympathize with his position, to imagine his suffering: 'I had to see everything. I had to watch hour after hour, by day and by night, the removal and burning of the bodies, the extraction of the teeth, the cutting of the hair, the whole grisly, interminable business' (Höss 78). In the end, he came to regard any feelings of pity as 'a betrayal of the *Führer*' (79). Since 'armaments came first,' it was imperative that 'every obstacle to this must be overcome': 'I dared not allow myself to think otherwise. I had to become harder, colder and even more merciless in my attitude towards the needs of the prisoners' (68). But he insists that he did not take pleasure in the extermination of the prisoners, a claim Levi considers to be truthful; it turns out that Höss 'was no longer happy in Auschwitz once the mass exterminations had begun' (80). After having told us of his fondness for the Gypsies, for instance, he turns their suffering and death into his own suffering: 'Nothing surely is harder than to grit one's teeth and go through with such a thing, coldly, pitilessly and without sympathy' (35). Driven by an 'exaggerated sense of duty' (81), he was not a monster but a technocrat focused on the quick and efficient resolution of logistical problems. Throughout the autobiography, he adamantly maintains that his efforts served the German war effort; he can never quite bring himself to acknowledge that the main purpose of Auschwitz was not armaments but mass murder.

Höss was not alone in his commitment to efficiency. The recently opened archives at Bad Arolsen in the former East Germany reveal an obsession with bureaucratic meticulousness that requires twenty-five kilometres of shelves to house the records of 17.5 million of the persecuted.[6] Indeed, the organization and administration of Auschwitz was a model of smoothly functioning efficiency. Divided into three camps (Auschwitz, Auschwitz-Birkenau, and Auschwitz-Monowitz), Auschwitz was administratively organized into three basic groups, each with its own complex set of divisions and subdivisions. On the surface, the camp functioned like any administrative apparatus, to ensure that orders 'were swiftly carried out,' that 'administrative tasks' were coordinated, that communications flowed properly, and that the camp facilities were 'running efficiently' (Lasik 47). What is disturbing about this essentially 'normal' bureaucracy is its apparent disconnection from the criminal purposes it is so clearly intended to serve.

What is even more disquieting is the incitement to murderous violence authorized by a whole juridico-ethical apparatus ostensibly put in place to prevent abuses of power and authority. The SS officers were supposed to be subject to legal proceedings 'for crimes and transgressions' (Lasik 46) committed at the camp. Yet, as has been well documented, they were in a position to 'torture and kill prisoners with impunity' (48). The Political Division was especially notorious for conducting 'an independent reign of terror' (48) under the guise of carrying out normal oversight responsibilities. Ostensibly 'responsible for questions related to the lawful functioning of the camp' (47), this division consisted of units in charge of various registration and documentation processes; it also conducted investigations and interrogations of prisoners and developed a spy network among prisoners. But it was Division III (Prisoners' Camp), the division most directly in daily contact with the prisoners, that was in the best position to abuse and murder prisoners. Immediately responsible for 'directing prisoners affairs,' it was in charge of providing 'order and discipline, proposing punishments for prisoners to the camp commandant, receiving reports about the number of prisoners at the camp and coordinating efforts related to prisoners' provisions and quarters with other units of the camp administration' (48). In constant close contact with the prisoner population, this division prepared 'reports on the number of prisoners during roll-call,' oversaw 'prisoners in their quarters,' and carried out 'executions when called upon' (49). One of its subdivisions was responsible for 'arranging prisoner employment' and 'overseeing prisoners at their work sites' (49–50). Making decisions on the distribution of food and clothing, on living quarters, and on forced labour, the 'bloodthirsty' Prisoners' Camp was in a pre-eminent position to exercise a cruelly 'repressive function' and was particularly instrumental in the 'process of indirect murder at the camp' (50). The sham legal underpinnings and the bureaucratic procedures of this efficient administrative apparatus allowed individual SS officers to terrorize prisoners without having to accept personal responsibility for their actions.

Efficiency was not a sufficient cause for the murderous and often cruel behaviour of SS officers; however, it played a crucial supporting role. According to testimony from surviving former SS men,[7] ordinary Germans seemed convinced by the Nazi rhetoric painting Jews as an enemy to be eliminated if the German nation was to survive. The ideological propaganda machine presented state-sponsored murder as an act of self-defence. It was consequently not only permissible but essential to carry out the genocidal orders of the Führer. But this propaganda would not have been as successful as it was without the appearance of legal procedures and routine bureaucratic processes at Auschwitz; the perpetrators of this crime against humanity must have felt reassured by the appearance of normalcy in the execution of even the most monstrous tasks. Auschwitz gives new meaning to Michel Foucault's argument that laws and prohibitions tend to generate the very transgressions they seek to proscribe; the camp not only normalized transgression as the law but also sanctioned and rewarded its blatant perversion. It was not that every SS officer was a psychopathological sadist; however, the organizational structure at Auschwitz functioned not only to rationalize violence but also to incite it. As Höss puts it, in Auschwitz 'everything was possible' (Höss 45); there were no limits to the indulgence of sadistic impulses. What was not possible was the refusal to participate in the efficient functioning of Auschwitz as a death factory. Moral scruples were no match for the complex machinery that assigned murderous tasks to even the lowest-level bureaucrats. If one considers that the camp maintained a ratio of 'one SS man for every 30–40 prisoners' (Lasik 53), the extent to which supposedly 'normal' bureaucrats and technicians participated in this project of mass extermination is staggering. From its inception in 1940 to its dissolution in 1945, Auschwitz saw roughly seven thousand SS men pass through the camp. These men came from different social, religious, educational, and even national backgrounds.[8] What allowed this 'socially dissimilar group' to cohere was the 'assistance of various social and technical measures,' measures that turned Auschwitz 'into an efficient apparatus for extermination, whose crimes left an enduring mark upon 20th century humanity' (53). Höss might have seen himself as an exponent of 'modernity as adventure' who would most certainly not have wanted to inhabit the murderous edifice of 'modernity as routine' he had constructed.

Inefficient Labour Force / Efficient Method of Extermination

This murderous administrative machine reduced its victims to dehumanized 'cogs' in ways the workers at the Ford Motor Company could never have imagined. The prisoners were not only a surplus population to be exterminated but were also exploited as slave labour and scientifically observed as experimental curiosities in a social-engineering laboratory. If Auschwitz was an exceptionally efficient extermination camp, it was clearly less efficient in the exploitation of prisoners' labour, both inside Auschwitz and outside it, in agriculture, mining, and factories. Since forced labour was part of the indirect extermination of prisoners, Auschwitz was ambiguously efficient in its murderous outcome but not in its productive contributions to the armaments effort. The prisoners themselves were ambiguously treated as both 'waste' material devoid of any claims to human consideration and as docile bodies with human needs to be 'managed.' In either case, they were the 'laboratory rats' for experiments in domination, humiliation, and intimidation. The methods employed were those in use at Ford Motor, where they served far less sinister purposes. The inmates were classified, assigned specific tasks, constantly supervised, spied upon, terrorized at the arbitrary whim of supervisors, and finally either left to die or summarily executed. In a perverse mockery of the Ford workers, they were docile, interchangeable, and expendable bodies subjected to an indifferent system that was paradoxically both totalizing and fragmented. As Piper puts it, 'never in the history of human civilization had anyone planned or carried out atrocities on such a massive scale, never had anyone applied industrial methods to the mass extermination of people' ('Direct' 11). One could add that never had anyone employed industrial methods to such economically non-productive ends. Although the 'final solution' was carried out with considerable problem-solving skills and measurable 'efficient' outcomes, the labour productivity was so suboptimal that the camps were 'efficient' not in terms of industrial output but in terms of a social experiment in the exercise of pure power.

In a reversal of today's priorities, the economic potential of slave labour was clearly secondary to the political demonstration of intimidation and terror. In an anticipation of the amorality of corporate capitalism in the twenty-first century, however, economic calculations seemed to have trumped any moral quandaries. Once again, the complicity of German as well as international industry with the Nazi regime has been well documented; even the Ford Motor Company made an economic pact with the devil in order to increase its overseas production quotas. A link with industry was in fact present at the very inception of Auschwitz. The location of the camp was chosen to accommodate the requirements for a combination of soil and water on the part of IG Farbenindustrie (Farben), a large German chemical concern that planned to construct 'production facilities for synthetic rubber and gasoline at Auschwitz' (Czech 26). It is also known that the bids for four crematoria with gas chambers to be constructed at Birkenau were awarded to the reputable companies of Hoch- und Tiefbau A.G. of Katowice and J.A. Topf und Söhne of Erfurt. With the expansion of Birkenau, Auschwitz increasingly constructed sub-camps that 'either served the SS economy directly (e.g., the farms for livestock and crops in the camp's "area of interest") or, as was normally the case, supplied labor for the factories, mines, and metalworks of such great German concerns as IG Farbenindustrie, Berghütte, Oberschlesische Hydrierwerke Ag, Energieversorgung Oberschlesien AG, Hermann Göring-Werke, Siemens-Schuckert, Rheinmetall-Borsig, and even the German state railway' (Czech 36–7). Such collusion between industry and concentration camps would lead one to suspect that the labour camps would be efficiently organized.

To the undoubted dismay and frustration of the capitalist establishment, slave labour was at best profitable only because it was cheap. As a labour camp, Auschwitz and its satellites were highly inefficient. In contrast with the Ford Motor Company, the 'workers' at Auschwitz were not provided with suitable equipment, appropriate instruction, or rational plans to execute what were often senseless tasks. Violating what Marx and Engels considered the absolute minimum requirement in food and shelter, Auschwitz starved the prisoners, deprived them of even minimally adequate clothing, housed them in cold and overcrowded rooms without sufficient sanitary facilities, and abused them both physically and mentally. Where the dehumanization of workers at the Ford Motor Company was the collateral damage of Henry Ford's obsession with efficient outcomes, at Auschwitz the exploitation of labour was a secondary derivative of the Nazis obsession with the efficient extermination of Jews and other undesirables. Unlike the Ford workers who were being abused in the production of a car for which they were also the intended consumers, the concentration camp inmates were instrumental in the construction of a killing machine for which they were also the intended victims.

In the early years of Auschwitz, there was little enthusiasm for the employment of prisoners in industry; this changed after 1942 when Germany was confronted with 'the need to increase its armaments production and thus mobilize all its reserves, including the concentration-camp prisoners as a potential source of labor' (Piper, 'Prisoner' 107–8). Although firms ranging from 'the great joint-stock companies down to small subcontractors' were keen on 'renting' prisoners, the use of slave labour was regulated by Germany's military requirements; the chemical industry (especially Farben) could claim priority in that its products directly served the armaments industry. In the autumn of 1944, Himmler even 'called an immediate halt to the extermination of the Jews' (Höss 90) because more and more prisoners were required for work in the armaments industry. Höss was ordered to 'resort to Jews who had become unfit for work'; he was to give them 'special care and feeding' so that they could be made employable 'within six weeks' (Höss 91, 92). Calling this order 'sheer mockery' (92), Höss inadvertently draws attention to the economically counterproductive aims of the treatment of prisoners. The economic exploitation of the prisoners was quite obviously not the primary reason for the existence of the camps.

But 'renting out' prisoners as slave labour proved a profitable although carefully controlled sideline. The Nazis were at the very least prepared to allow Farben to construct the Buna Werke adjacent to Auschwitz III and to 'build affiliated camps on or near the grounds of already-existing industrial facilities' (Piper, 'Prisoner' 108). Productivity was primarily maintained through the brutal exploitation of prisoners who could be exchanged for 'new and healthy prisoners' (114) when worn out. Generally in poor health, the prisoners were, of course, far less productive than 'free' or paid labour. What made their employment profitable was their low cost in terms of food, clothing, and housing. In other words, the camps were such a massive reserve army of labour that there was no incentive to provide workers with even the most basic means to sustain themselves. While forced labour was clearly profitable for the capitalist enterprises, it fell far short of its exploitative potential. No matter how unproductive individual prisoners proved to be, the German state, of course, continued to collect the 'rent' paid by employers for the use of prisoners. In spite of being a potentially lucrative and desperately needed financial resource, forced labour was so inefficiently exploited that the reason for the dehumanization of the prisoners must lie elsewhere.

A Social Experiment: The Prisoner as Dehumanized Cog

The treatment of prisoners at Auschwitz served neither the economy nor even the express purpose of genocidal population control. Although the Nazis developed increasingly efficient methods for mass murder, this tendency is insufficient to explain the function of abuse and torture to which prisoners were exposed throughout the years Auschwitz was in operation. It is most often assumed that people are so evil by nature that they will engage in the most inhumane practices simply because there are no limits to their indulgence of 'primitive' instincts. However, the tolerance and incitement to violence in the camps could perhaps also be explained as a reversion to a pre-modern form of sovereignty. In *Discipline and Punish,* Michel Foucault analyses the role torture played in the historical shift in the administration of penal punishment. While those subject to punishment in Foucault's analysis may not have been guilty of a crime, they were at the very least considered a real threat to sovereign power. In the case of concentration-camp inmates, this held for the criminals and politicals who made up the early populations of Auschwitz but for neither the Jews nor the gypsies. Parading as 'progress,' this 'regress' to a pre-modern form of violence constituted a monstrous intensification of an earlier political rationale. Auschwitz symbolizes a return to the assertion of absolute sovereign power as a 'hold on the body,' which Foucault identifies with 'public executions' (*Discipline* 10) in sixteenth- and seventeenth-century Europe. Although the Nazis were intent on concealing their crimes, they did so mostly out of fear of repercussions should they lose the war. The concentration camps were widely known to be death camps; the threat of being deported to one inspired fear in everybody under Nazi rule. It was thus enough to know that executions took place for them to take on a public character. Foucault's point is that torture in pre-modern Europe was not meant as a legal punishment but as a political spectacle. In direct correlation with Auschwitz, torture is to be seen 'as the effect of a system of production in which labour power, and therefore the human body, has neither utility nor commercial value that are conferred on them in an economy of an industrial type' (54). The tortured body was a political instrument designed to affirm the sovereign power's absolute right to establish who is to be punished and for what crime. According to 'the ordinance of 1670,' the accused had no access to the evidence against them, while the magistrate was empowered to 'accept anonymous denunciations, to conceal from the accused the nature of the action, to question him, with a view to catching him out, to use insinuations' (35). Those deemed enemies of the sovereign were subjected to sham legal proceedings, which 'brought to a solemn end a war, the outcome of which was decided in advance, between the criminal and the sovereign' (50). The torture and execution of defenceless prisoners at Auschwitz was thus a necessary aspect of the Nazis exhibition of power.

At a time when the growing influence of corporate capitalism threatened to displace the decision-making power of political institutions, the Nazis sought to reaffirm politics over economics through ceremonies and spectacles ranging from the Nuremberg rallies to the display of unlimited power over docile bodies in the camps. If the body in the camps had to be tortured before it was killed, it was, in Foucault's terms, 'to manifest the disproportion of power of the sovereign over those whom he had reduced to impotence. The dissymmetry, the irreversible imbalance of forces were an essential element in the public execution,' as it was in the 'ceremonies by which power [was] manifested' (*Discipline* 50, 47) at, say, Auschwitz. The 'policy of terror' underpinning absolute sovereign power meant that everybody was to be made aware, 'through the body of the criminal, of the unrestrained presence of the sovereign' (49). The function of the public execution, as of torture at Auschwitz, was not to 'reestablish justice' but to 'reactivate power' (49). The emphasis in Auschwitz on the efficient administrative organization, the search for efficient technical solutions, and the installation of an efficient apparatus of rules and regulations were thus merely ceremonial adjuncts to the reaffirmation of an arbitrary and unlimited sovereign power over the body that had, in Foucault's history, long been superseded by different penal mechanisms that punished the 'crime' rather than the 'criminal.' Annulling the modern assumption that the 'law must now treat in a 'humane' way an individual who is 'outside' nature,' the Nazis reverted to the old justice of treating 'the 'outlaw' inhumanely' (92). The torture of prisoners at Auschwitz should not be dismissed as the deplorable atavistic behaviour of individual SS officers; it was a systemic form of punishment that had in earlier times been 'carried out in such a way as to give a spectacle not of measure, but of imbalance and excess; in this liturgy of punishment, there must be an emphatic affirmation of power and of its intrinsic superiority' (49). In *'Society Must Be Defended,'* Foucault reiterates this logic by linking it directly to Nazi society. With its emphasis on disciplining and regulating individuals and whole populations, 'Nazism was in fact the paroxysmal development of the new power mechanisms that had been established since the eighteenth century.' At the same time, it was also 'a society which unleashed murderous power, or in other words, the old sovereign right to take life' (*'Society'* 259). What makes the role of efficiency in Auschwitz so paradoxical is precisely that it simultaneously reinforces both the old murderous sovereign power and the modern disciplinary apparatus. When Primo Levi asks us to 'consider that the Lager was pre-eminently a gigantic biological and social experiment' (Levi, *Survival*), he draws our attention to Auschwitz not only as a killing factory but also as a social-engineering project intent on constructing the docile bodies needed to demonstrate to the German people and to the world at large the murderous power of the Nazi Third Reich.

In *Survival in Auschwitz* (1947; 1996), Primo Levi recounts in harrowing detail the effects of the Nazi 'social experiment' on himself and his fellow prisoners. His descriptions of the successful process of dehumanization carried out by the Nazis are confirmed in *Fateless* (1975), the semi-autobiographical novel of Nobel Prize–winner Imre Kertész, and are also in many ways mirrored in the anthropological gaze that Rudolf Höss trains on the prisoners described in his autobiography. Constituting the main focus of Holocaust studies, the victimization of concentration-camp inmates is undoubtedly the most extensively documented aspect of the Nazi period. As has often been pointed out, the prisoners were reduced to subhuman creatures so as to discourage resistance on their part and to make it easier for the SS men to abuse and kill them. From the perspective of efficiency, what is striking in this social-engineering experiment of the Nazis is the *inefficient* use of the same methods that Ford Motor employed to ensure the *efficient* utility of workers. The camp instituted systems of classification and procedures that are in themselves examples of rational organization. Prisoners were identified not only by numbers but also by the colour of the triangles and other insignia on their uniforms. In addition to the yellow star for Jewish prisoners, the red triangle 'designated a political prisoner,' the green triangle indicated 'professional criminals,' the black triangle referred to 'antisocial prisoners,' the violet one was for 'Jehovah's Witnesses,' and the pink one for homosexuals. Some prisoners were further classified by letters like 'E' for 'reformatory prisoners' and 'SU' for Soviet prisoners-of-war (Iwaszko 64–5). Although every concentration camp assigned numbers to its inmates, Auschwitz was alone in the practice of tattooing identification numbers on the prisoners' forearms. This practice only became established after 1942 when the camp became inundated with the arrival of large Jewish transports. To this day, the tattooed identification number is the primary symbol of the dehumanization of prisoners at Auschwitz. This kind of dehumanization through standardization was reinforced by a complex bureaucracy of record keeping. Upon first entering Auschwitz, the prisoners had to undergo a registration process requiring them to fill out 'a blank form (*Häflings-Personalbogen*),' which 'served as the basic source of information regarding the prisoner' (Iwaszko 60) throughout his or her days at the camp. As the twenty-five kilometres of shelves in the newly opened archive in Bad Arolsen confirm, the Nazis were meticulous bureaucrats, entering information as insignificant as a single louse found in a prisoner's hair. The standardization of workers at Ford Motor finds its caricatured extension at Auschwitz, where prisoners will no longer have a name but will be known by the numbers tattooed on their forearms.

But these essentially industrial methods of standardization are perverted in the camp to serve the non-productive end of the dehumanization and destruction of non-threatening 'enemies' of the state. There is then an ironic disconnection between industrial and political efficiency that is particularly telling in the persecution of Jews. Even before German Jews were deported either to ghettoes in the East or to concentration camps, they were incrementally reduced to 'bare life' as they were 'legally' stripped of the very rights of citizenship that the law was instituted to ensure. In a particularly galling irony, the Nazis appealed to 'protective custody' orders to rationalize the illegal seizure of Jewish property and the incarceration of Jews in concentration camps. Under measures ostensibly instituted to 'protect' Jews from the violence the SS and SA unleashed against them during Kristallnacht ('night of broken glass') in November 1938, the Nazis sent 30,000 Jewish men to concentration camps. As Höss explains, '"for their own protection, and to save them from the wrath of the people," all Jews who played any part in trade, industry or the business life of the country were arrested and brought into the concentration camps as "protective custody Jews"' (Höss 55–6). The 'wrath of the people' was, of course, a myth perpetrated by the SS and SA who beat to death many Jews and systematically ransacked their homes and businesses. Before 1942, when Jews began to be persecuted 'under the rubric of "the final solution of the Jewish question"' (Iwaszko 56), the phony 'legal' processes of 'protective custody' and 'preventive custody' were used to justify the torture and murder of those so 'protected.' In spite of what he has witnessed, Höss persists in referring to the chief female supervisor at Auschwitz as 'the commander of the protective custody camp' (Höss 61). The ironic reversal of legal processes to rationalize the very crimes they were meant to prevent is one of the most 'effective' methods the Nazis used to illustrate their unlimited power to make 'everything possible' at Auschwitz.

Where the oppression of workers at Ford Motor could be explained as an unfortunate by-product of the company's obsession with efficient outcomes, the abuse of prisoners at Auschwitz was a spectacular display of unlimited power by the privileged over those deprived of citizenship. Upon arrival at the concentration camp, the prisoners were ordered to relinquish 'all underwear and civilian clothes,' were forced to have all hair shaved, and had to endure showers that frequently sprayed 'hot or ice-cold water' on them (Iwaszko 59). The introduction to life at Auschwitz was disorienting and humiliating. What happens to Levi and the fifteen-year-old protagonist of *Fateless* makes no rational sense to them. Before the prisoners enter the shower, for instance, Levi notices that a man with a broom sweeps away all the prisoners' civilian shoes: 'He is crazy, he is mixing them all together, ninety-six pairs, they will be all unmatched' (*Survival* 23). Forced to wait naked outside in the cold, he is told that everybody will be given 'other shoes, other clothes' (25), namely, the 'poorly-fitting, often dirty and louse-infested' camp uniforms and wooden clogs that made it 'difficult to move, especially on the icy roads in winter' (Iwaszko 60). Before entering the shower, the boy in Kertész's story dutifully ties together his shoes only to discover that he was later asked 'to choose from a mountain of strange, wooden-soled, canvas-inlaid shoes, not tied by shoe strings' (Kertész 72). He, too, was given 'pants suitable for old men' (72) and a worn-out shirt. Both Levi and Kertész's Köves are dressed in the standard striped uniforms made of material far too thin for the cold winters in Poland; the clogs caused sores that easily became infected, making prisoners sick and unfit for work, and leading to their death. Food was insufficient and lacking in basic nutrients. The barracks were cold and overcrowded; prisoners were forced to sleep so tightly packed together that they could not turn around. Sanitation was woefully inadequate. Weakened by such appalling living conditions, the prisoners were forced to perform hard labour in all kinds of weather, with inadequate equipment, and under the supervision of sadistic overseers. They were employed in the building and maintenance of the camp itself, in agriculture to produce food for the SS staff and themselves, and in nearby mines and factories. In their ill-fitting striped uniforms and with their increasingly emaciated faces under shaved heads, the prisoners found it more and more difficult to recognize themselves in the mirror. They were so changed in appearance that even relatives often could not find each other, and the Italians Levi meant to meet 'every Sunday evening' stopped seeing each other as it was too distressing to observe how they had all become 'ever more deformed and more squalid' (*Survival* 37). The starvation diet and the exhaustion caused by forced labour reduced prisoners to uniformed skeletons devoid of all individuality.

Every aspect of life at Auschwitz was meticulously organized for the sole purpose of disciplining or taming the inmates. Throughout every day spent there, prisoners were exposed to the arbitrary chicanery of *kapos* (criminal prisoners) and other supervisors. Any SS officer could abuse and torture prisoners at his whim, and the cruelty of the *kapos* towards other prisoners was not only condoned but implicitly encouraged. Upon his arrival at Auschwitz, Levi witnesses the prisoners returning from work who 'walk in columns of five with a strange, unnatural hard gait, like stiff puppets made of jointless bones; but they walk scrupulously in time to the band' (*Survival* 30). There is a macabre irony in this sight of cadaverous men marching in immaculate formation. To reinforce the camp's coercive regime, the new arrivals and returning prisoners were arranged 'according to a precise order; when the last squad has returned, they count and recount us for over an hour' (30). The disciplinary routines and open abuses to which the prisoners were subjected constitute in Levi's eyes 'a great machine to reduce us to beasts' (41): 'I push wagons, I work with a shovel, I turn rotten in the rain, I shiver in the wind; already my own body is no longer mine: my belly is swollen, my limbs emaciated, my face is thick in the morning, hollow in the evening; some of us have yellow skin, others grey. When we do not meet for a few days we hardly recognize each other' (37). Although prisoners labouring at different industrial facilities were often 'engaged in very strenuous construction work' (Piper, 'Prisoner' 111), work inside factories was coveted for being less deadly: 'Work outdoors, constantly supervised by the SS men and kapos, presented more opportunities for hitting and mistreating the prisoners than work carried out with machines, whose rhythm and tempo was regulated to a greater extent by technical processes' (113). Being a 'cog' chained to machines at Ford Motor would here be considered preferable to being a 'cog' exposed to the 'rule-of-thumb' whims of human beings. At Auschwitz, prisoners deteriorated rapidly. A mere fortnight after his arrival, Levi 'already had the prescribed hunger,' learned not to let himself 'be robbed,' and discovered on the back of his feet 'those numb sores that will not heal' (*Survival* 37). Work, far from making the inmates 'free' (the camp motto, copied from Dachau, was *Arbeit macht frei*), was thus a 'proven and effective method of extermination'; if there was 'no sensible task to be done, the prisoners received senseless tasks' (Piper, 'Mass' 144). Dehumanized and disoriented, Levi acknowledges having been transformed into the criminal and subhuman creature fantasized by the Nazis. Instead of assisting others, he is prepared to steal a spoon he finds 'lying around' or a 'piece of string' (*Survival* 37). The inefficiencies of the system work efficiently to control and destroy what imbues Levi with human dignity.

The regulations and procedures necessary for the smooth functioning of the Ford Motor Company are at Auschwitz employed to abuse the prisoners and to rationalize their torture. According to Levi, the 'rules of the camp' were 'incredibly complicated' and the 'prohibitions [were] innumerable: to approach nearer to the barbed wire than two yards; to sleep with one's jacket, or without one's pants, or with one's cap on one's head; to use certain washrooms or latrines which are "*nur für Kapos*" or "*nur für Reichsdeutsche*"; not to go for the shower on the prescribed day, or to go there on a day not prescribed; to leave the hut with one's jacket unbuttoned, or with the collar raised; to carry paper or straw under one's clothes against the cold; to wash except stripped to the waist' (*Survival* 34). In addition to these arbitrary rules, the 'rites to be carried out were infinite and senseless' (34). The forced labour, too, was 'a Gordian knot of laws, taboos and problems' (35). Moreover, the prisoners 'did not know what regulations, if any, applied to their behavior,' exposing them to the enforcement of unknown or constantly reinterpreted rules through 'a wooden staff' (Piper, 'Mass' 146) or a shot in the back of the head. The explanation for this senselessness is 'repugnant but simple: in this place everything is forbidden, not for hidden reasons, but because the camp has been created for that purpose' (Levi, *Survival* 29). The effects of this simulation of an efficient system are for Levi measurable in the successful transformation of ordinary human beings into docile and dehumanized bodies. It did not take long for Levi to feel that he and his comrades 'had reached the bottom': 'It is not possible to sink lower than this; no human condition is more miserable than this, nor could it conceivably be so. Nothing belongs to us any more; they have taken away our clothes, our shoes, even our hair; if we speak, they will not listen to us, and if they listen, they will not understand. They will even take away our name: and if we want to keep it, we will have to find ourselves the strength to do so, to manage somehow so that behind the name something of us, of us as we were, still remains' (26–7). Dehumanization is the intended outcome of a system that merely *mimics* the organizational structure of an efficient industrial model.

At the camp, it is no longer possible to distinguish between what is real and what is not. It is a simulacral order designed to disorient the inmates so as to make them infinitely pliable, exchangeable, and disposable. The appearance of a standard bureaucratic order plays on the normalizing power of people's internalized commitment to efficiency and order. From the moment they are apprehended, the prisoners comply 'all by themselves' with practices and rituals that interpellate them as 'subjected' subjects. Confronted by senseless conditions, they fall back on what is familiar to normalize an utterly confusing experience. Resistance to the violent regime is made difficult not only because the SS guards threaten to beat and shoot prisoners at the slightest provocation, but also because the simulacral order they face removes any solid anchoring point from which to mount an opposition.

The question has often been raised as to why victims who were aware of the fate awaiting them did not oppose their oppressors. Although there were both spontaneous and organized attempts to resist at Auschwitz, such rebellions were indeed relatively rare and never successful. Aside from the effective methods of surveillance by the SS, the prisoners were habituated to observe rules and regulations. Typical of this tendency is the episode of an inmate asking a guard if there were 'certain regulations that apply to prisoners at the camp' so that he could 'avoid any unpleasantries' (Piper, 'Mass' 146). For his troubles he was treated 'like an idiot' and 'hit in the face' (146). The assumption that good behaviour leads to rewards often persists even in the face of incontrovertible evidence to the contrary. Speaking of a fellow inmate, Levi comments: 'He still thinks ... oh no, poor Kraus, his is not reasoning, it is only the stupid honesty of a small employee, he brought it along with him, and he seems to think that his present situation is like outside, where it is honest and logical to work, as well as being of advantage, because according to what everyone says, the more one works the more one earns and eats' (*Survival* 132). What such episodes illustrate is not only the wilful exercise of power on the part of the SS staff but also the willingness of the victims to play by the rules, no matter how senseless or counterproductive such behaviour may prove to be.

In *Fateless*, Kertész resorts to irony for his dramatization of his young protagonist's internalization of his own victimization. In his youthful innocence, Köves wants to trust the Nazis who are in charge of the camp. Exposed to the selection process carried out by a medical doctor whose approval he craves ('I felt that he liked me' [Kertész 63–4]), he comments admiringly: 'Everything was working well, everything ran smoothly, everyone was in his place and attended to his function, precisely, cheerfully, like a well-oiled machine' (64). Not realizing that he, too, would soon have his head shaved and wear the camp uniform, he assumes on first arrival that the inmates he sees are 'real convicts, clad in the striped suits of criminals, with shaven heads and round caps' (58). His initial reaction to the Jewish prisoners is truly harrowing in its irony. In a desperate effort to disavow a common fate with the inmates, he describes them as if he were a superior outsider: 'Their faces, too, were not particularly trustworthy: they had widespread ears, protruding noses, and sunken, cunning, tiny little eyes. Indeed, they looked like Jews in every respect. I found them suspicious and altogether strange' (58). In a further irony, the boy, who will soon have the same sunken features as the prisoners he disdains, looks to the highly efficient Nazis for a sense of normalcy and security: 'I noticed, however, that out here German soldiers, with their green caps, green collars, and eloquent hand movements, were pointing out directions, keeping an eye on everything: I was somewhat relieved on seeing them because they seemed well-cared-for, clean, and neat, and in this chaos they alone seemed to exude solidity and calm' (59). Depicting life in the camp through the eyes of a young boy, Kertész not only defamiliarizes the dangerously perverse by making it appear comfortingly normal to the speaker, but also accentuates the crucial role that efficiency plays in Köves's internalization of the Nazi gaze.

A Social Experiment: The Commandant as Anthropologist

In his autobiography Höss makes observations about the behaviour of prisoners that confirm Levi's suspicion that Auschwitz was an experiment in social engineering. When the commandant of Auschwitz is not rationalizing his own participation in mass murder, he sets himself up as an objective anthropologist who is reporting on the customs of a strange tribe of human beings. Responding to Höss's 'horrifying descriptions of mass murder in the gas chambers' in his autobiography, Kazimierz Smolen comments that they 'read like a factual account written by a completely disinterested observer. On the witness stand, Höss answered directly and categorically, without any hesitation' (Smolen 301). In his account, the commandant seems infinitely fascinated by the reactions of prisoners to the extreme social conditions under which they are forced to operate. Writing while he himself is awaiting execution in a Polish prison, he asserts that in 'no place is the real "Adam" [primitive man] so apparent as in prison' (Höss 39). Acting on the instinct for self-preservation, prisoners exhibit a 'crass egotism' (38) that gives Höss much cause for thought. What distressed him first of all is that the prisoners seem to be lacking in all solidarity. Put in charge of their fellow inmates, the *kapos* blatantly abused their power: 'One would have thought a common fate and the miseries shared by all would have led to a steadfast and unshakable feeling of comradeship and cooperation, but this was far from being the case' (38). This seems to Höss particularly illogical since he quite rightly realizes that the 'masses of prisoners could not be easily supervised' (38) without the cooperation of the *kapos,* a point he later reinforces by stressing more generally that 'it would not have been possible to control or direct these thousands of prisoners without making use of their mutual antagonisms' (45–6). While he acknowledges that 'it is the guards and the supervisors who create the opportunity' for behaviour he deplores and that the 'system of terror' prevailing at the camp discouraged anyone from intervening on behalf of another, he is amazed to find that 'even people, who in ordinary life outside the camp were at all times considerate and good-natured, became capable, in the hard conditions of imprisonment, of bullying their fellow-prisoners mercilessly, if by so doing they could make their own lives a little easier' (38–9). In his opinion, the 'treatment they received from the guards, however brutal, arbitrary and cruel, never affected them psychologically to the same extent as did this attitude on the part of their fellow inmates' (39). Survivors have reported acts of spontaneous and organized cooperation among prisoners; however, Levi also confirms Höss's observation that self-preservation did indeed drive him and his fellow-inmates to take advantage of each other. What is particularly offensive in Höss's anthropological gaze is the position of moral superiority from which he judges victims like Levi.

This superior anthropological gaze sweeps across the camp's main ethnic groups to focus on differences in reactions to a shared plight. Commenting on frictions among the three main political groups of Polish prisoners, he disapprovingly describes how the more brutal and corrupt among them would gain positions of control and influence. Once a prisoner held an 'important position,' he would 'see to it that his friends were put wherever he wished them to be' (Höss 45). Overall, Höss concluded, the prisoners demonstrated a fierce and merciless determination to achieve supremacy. The Russian POWs came in for particular criticism: 'Overcome by the crudest instinct of self-preservation, they came to care nothing for one another, and in their selfishness now thought only of themselves,' and would in fact 'beat each other to death for food' (Höss 47, 48). Without any self-consciousness about his own participation in the dehumanization he documents, Höss concludes: 'They were no longer human beings. They had become animals, who sought only food' (48). With obvious moral opprobrium, Höss tells us: 'I never got over the feeling that those who had survived had done so only at the expense of their comrades, because they were more ferocious and unscrupulous, and generally "tougher"' (48). In contrast, he was more positively inclined towards the Gypsies and the Jehovah's Witnesses. The Gypsies are singled out as an anthropological curiosity. He approves of the Nazi decision to preserve members of two specific tribes of Gypsies because 'they were the direct descendants of the original Indo-Germanic race, and had preserved their ways and customs more or less pure and intact.' Himmler 'now wished to have them all collected together for research purposes. They were to be precisely registered and preserved as an historic monument' (49). What Höss liked about them was that their 'moral attitude is also completely different from that of other people.' He finds it endearing that they 'do not regard stealing in any way wicked. They cannot understand why a man should be punished for it' (53). Once again, Höss seems entirely blind to the irony of this moral stance when it comes from the pen of the commandant of Auschwitz, who was responsible for the mass murder of millions of innocent people. Where he approves of the Gypsies for their spirit of independence, he favours female Jehovah's Witnesses for their meek compliance. In his opinion, knowing that they suffered in captivity 'for Jehovah's sake,' they were quite 'contented with their lot,' even agreeing that 'it was right that the Jews should now suffer and die, since their forefathers had betrayed Jehovah' (60). This ideological complicity with their own victimization meant that they could be trusted to assist his wife with the household and the care of his own children.

Höss reserves his most critical anthropological observations for the Jewish prisoners. Although he emphasizes that he has 'never personally hated the Jews,' he nevertheless acknowledges that he 'looked upon them as the enemies of our people' (56). He consequently detects in the Jewish prisoners behaviour he considers 'typically Jewish': they bribed fellow prisoners, they were 'mainly persecuted by members of their own race, their foremen or room seniors' (54), and they shirked work. A Jewish block senior, for instance, is singled out as 'the "devil" incarnate'; he 'showed a repulsive zeal towards the members of the SS, but was ready to inflict any kind of iniquity on his fellow-prisoners and members of his own race' (54). Discussing an anti-Semitic weekly circulating in the camp, Höss claimed to be opposed to *Der Stürmer* because of its 'disgusting sensationalism' (55). But then he blamed the Jews for this sensational anti-Semitism by adding that it was later learned that a Jew had 'edited the paper' and also wrote 'the worst of the inflammatory articles it contained' (55). Moreover, contrary to all evidence, he avers that the Jews in the camp were in possession of a lot of money, allowing them to enjoy advantages through bribery. If he decided to have 'their money taken away from them,' it was only to protect them from being 'plundered' (56) by the criminals. Revealing his anti-Semitism, he contends that the Jews 'did their best to do each other in the eye whenever they could' by wrangling a position or even inventing 'new posts for themselves so as to avoid having to work' (56). He further accuses them of making false accusations against other Jews if they could thereby 'obtain a nice, easy job' (56). Once they were in a privileged position, he stresses, they surpassed even the criminals in the way they harried and persecuted 'their own people quite mercilessly' (56). In spite of all this, he reassures the reader, he made no distinctions between Jews and other prisoners, treating them 'all in the same way' (56). Since the Jews knew that, 'without exception,' they were 'condemned to death,' they fatalistically accepted their fate, submitting 'patiently and apathetically' to all the 'misery, distress, and terror' (57) of camp life. If they deteriorated faster than other prisoners, it was entirely because of the 'psychological state' induced by their treatment at the hands of other Jews rather than because of the physical hardships and abuses they suffered at the hands of the Nazis.

The disconnection between Höss's professed 'inner feelings' and his ruthless butchery of millions of people he knew to be innocent of any criminal wrongdoing appears most strikingly in the impression he conveyed to those observing his testimony at his trial in Poland before the

Supreme National Tribunal, 11–29 March 1947. As Kazimierz Smolen comments, 'Höss only demonstrated a certain liveliness when he testified about the mass exterminations' (Smolen 301). Speaking of mass murder as if it had been a manufacturing process, the commandant gave eyewitnesses Janusz Gumkowski and Tadeausz Kulakowski the impression that he was

> some factory director explaining certain difficulties in how his factory works and striving to justify them in terms of objective difficulties: transports that were too large, the low productivity of the crematoria, tie-ups caused by damage to the facilities. It was as if he wanted to explain to the Tribunal that he had done everything in his power to carry out the 'special action' [mass extermination] in the most effective way possible, and that if he had failed to do so, if the results did not reach the level that Himmler and Eichmann thought possible to achieve – then it was not his, Höss's, fault ... What seemed particularly strange was the calmness with which he spoke about the extermination campaigns. Listening to him, one got the impression that he had sought to eliminate completely from his consciousness the issue of human life – which was playing the dominant role here – in carrying out his assignment. At times, one could sense in his voice a tone of gentle pride when he spoke about the obstacles that he had been able to overcome, and at other times, once again, a tone of guilt and shame that he had not been able to carry out his duties and had failed the trust placed in him by his superiors. (in Smolen, 301–2)

It was not that Höss simply and slavishly obeyed orders; he had internalized the ideology of efficiency which demanded that he exert himself to secure the most optimal outcome possible under the circumstances confronting him and to do so without consideration for the *purpose* served by this commitment to efficiency.

Bureaucrats rather than Monsters

Documents illegally smuggled out of Auschwitz 'gave the names and descriptions of the several dozen SS men who were the most vicious criminals at the camp. Commandant Rudolf Höss headed the list' (Jarosz 227). Can we simply dismiss Rudolf Höss as an evil, unfeeling man? Or was he, like the Adolf Eichmann who emerged from Hannah Arendt's analysis of his trial in *Eichmann in Jerusalem: A Report on the Banality of Evil* (1963), a more or less ordinary functionary who carried out bureaucratic tasks the consequences of which he deemed to be beyond his personal moral responsibility? With her emphasis on the 'banality of evil,' Arendt's study of Eichmann was initially highly controversial in that it dispelled the comforting notion that the Nazi perpetrators of systematic mass murder must simply have been 'monsters' or sociopaths. But a by now famous experiment carried out in May 1962 at Yale University by the psychologist Stanley Milgram anticipated Arendt's findings by concluding that 'cruelty is not committed by cruel individuals, but by ordinary men and women trying to acquit themselves well of their ordinary duties' (Bauman 153). Arendt and Milgram seek to throw light on the circumstances that compel people to obey orders that are often against their better judgment. Although their focus is elsewhere, their investigations reinforce the ideological significance of efficiency in the ability of perpetrators of evil to diassociate their actions from their sense of self. Where Milgram is primarily interested in the relationship between the subject administering pain to an apparently innocent victim and the authority figure who gives the orders and assumes responsibility for their consequences, Arendt is more concerned with the systemic conditions that made possible the emergence of the concentration camps as 'killing factories.'[9] In his excellent study *Modernity and the Holocaust* (1989) Zygmunt Bauman argues that it was 'the rational world of modern civilization that made the Holocaust thinkable' (13) by drawing to some extent on Arendt's analysis of Eichmann and by devoting a whole chapter to Milgram's experiments. A crucial aspect of modern civilization that was 'not the Holocaust's *sufficient* condition' but was 'most certainly its *necessary* condition' (13) was the ideology of efficiency underpinning both the Ford Motor Company and the death camps.

The 'banality of evil' documented in Arendt's study shows that Eichmann was not a 'monster' but a man who 'could organize' (Arendt 45) anything. Like Höss, he was an ordinary technocrat who committed acts against humanity in the course of an ordinary day at his desk. As Arendt points out, Eichmann had in fact 'never belonged to the higher Party circles; he had never been told more than he needed to know in order to do a specific, limited job' (84). For him the evacuation, deportation, and finally extermination of Jews had become 'routine business' (84). Removing Jews from German soil was for him a logistical rather than a moral problem. In spite of Höss's contention that Eichmann 'was completely obsessed with the idea of destroying every single Jew that he could lay his hands on' (Höss 79), Eichmann seems to have been driven by an investment in efficiency rather than by personal hatred for Jews. In the early stages, he even tried to negotiate with the Jewish community to minimize the harm that was to come to them. At one point, he convinced both his Nazi bosses and the Jewish community that emigration, rather than incarceration, was a desirable option; Germany would be free of Jews and the Jews would find a new homeland somewhere else. When Heydrich informed him that Hitler had ordered the physical extermination of the Jews, Eichman recalled thinking: 'I had never thought of such a thing, such a solution through violence. I now lost everything, all joy in my work, all initiative, all interest; I was, so to speak, blown out' (Arendt 84). However, no matter what his personal feelings may have been, he enthusiastically approached forced emigration in the first place and later extermination as technical problems to be solved.

In the case of forced emigration, Eichmann set out to remove obstacles to the bureaucratic process by instituting an efficient system to speed up the necessary issuing of passports: 'He imagined 'an assembly line, at whose beginnings the first document is put, and then the other papers, and at its end, the passport would have to come out as the end product" (Arendt 45). Once the 'assembly line was doing its work smoothly and quickly' (45), Eichmann expected the Jewish functionaries with whom he had negotiated to be pleased. Their appalled response to Eichmann's 'solution' highlights the disjunction between technical and moral apprehension: 'This is like an automatic factory, like a flour mill connected with some bakery. At one end you put in a Jew who still has some property, a factory, or a shop, or a bank account, and he goes through the building from counter to counter, from office to office, and comes out at the other end without any money, without any rights, with only a passport on which it says: "You must leave the country within a fortnight. Otherwise you will go to a concentration camp"' (in Arendt, 46). As the 'solutions' to the Jewish question moved from 'forced emigration' to 'forced evacuation,' Eichmann established himself as an expert in the organization of collecting and transporting Jews to the camps; he instigated the deportation, registration, selection, and extermination of prisoners documented in Levi's memoir and Höss's autobiography. Arendt stresses that the murderers participating in 'the extermination machinery' were 'not sadists or killers by nature; on the contrary, a systematic effort was made to weed out all those who derived physical pleasure from what they did' (114, 105). Interestingly enough, Eichmann 'did not see much' (89), even on repeated visits to

Auschwitz. The camp commandant 'spared him the gruesome sights. He never actually attended a mass execution by shooting, he never actually watched the gassing process, or the selection of those for work' (89–90). From Höss's testimony, it appears that Eichmann was in fact part of a group of 'high-ranking Party leaders and SS officers' who came to Auschwitz 'so that they might see for themselves the process of extermination of the Jews.' Eichmann and the others were not only present but were 'deeply impressed by what they saw,' expressing some concern for Höss and his men who had to 'go on watching these operations' (Höss 79). Arendt's point still holds in that it is unlikely that Höss would have made these important Nazis watch one of the more gruesome executions.

In the selection of those Nazis who were forced to 'see much,' the Nazis were faced with the problem of 'how to overcome not so much their conscience as the animal pity by which all normal men are affected in the presence of physical suffering' (Arendt 106). In Arendt's interpretation, the Nazis resorted to the highly effective strategy of 'turning these instincts around, as it were, in directing them toward the self' (106). Instead of blaming themselves for the suffering they had caused, they said to themselves: 'What horrible things I had to watch in the pursuance of my duties, how heavily the task weighed upon my shoulders!' (106). As we have seen, Höss employs this strategy to great effect, asking us to pity him for the horrible sights he had to witness. The banality of evil documented in *Eichmann in Jerusalem* and Höss's memoir presents us with stark lessons about the motivations and rationalizations made available to those priding themselves on being efficient organizers and managers. From its inception at the Ford Motor Company to optimize the efficiency of machine parts, the assembly line offered itself as the most suitable technical means for carrying out the Final Solution ordered by Hitler.

As Bauman points out in *Modernity and the Holocaust,* inhumanity is not confined to 'occasional breakdowns' (154) in rational behaviour, but proves to be fully compatible with our desire for rational order. Cruelty can no longer be simply relegated to the psychopathology of individuals, but must be understood as emerging from the specific social conditions within which individuals enact their roles. As 'social relationships' become 'rationalized and technically perfected, so is the capacity and the efficiency of the social production of inhumanity' (154). The introduction of the assembly line at Ford Motor exemplifies an exponential alteration in social relations on the factory floor. The rationalization of the material process of production brought with it a rationalization of the social relations that resulted in the dehumanization of both workers and managers. It could be said that the efficiency gains celebrated in the production of Model Ts were accompanied by efficiency gains in 'the social production of inhumanity.' The pursuit of efficiency licensed cruelty to others in that it allowed the perpetrators to distance themselves from their victims. For Bauman, the concentration camp guard was able to listen to Mozart while Jews were being gassed because he, like Eichmann and Höss, understood himself as a technician carrying out an assigned task. The death camps are simply the most extreme manifestation of inhuman behaviours that arise with the kind of rationalizing processes adopted by the Ford Motor Company to increase productivity.

Milgram's videotaped experiments provide some useful insights into the psychological processes that account for the disjunction between technical and moral behaviours when people find themselves subjected to the gaze of an authority figure who instructs them to act against their conscience. While Milgram seemed mostly interested in establishing how easily people bow to authority, he also shows the crucial role played by the kind of social distance or mediation one could trace back to the division of labour. In a setup that continues to raise ethical questions, Milgram invited unsuspecting people into his laboratory to act as a 'teacher' who corrects the mistakes a 'learner' makes by administering increasingly severe electric shocks. The 'teacher' is told that the aim of the experiment was to test the relationship between memory and learning; in actual fact, it was the 'teacher' who was the subject of the experiment, while the 'learner' was part of the setup. Milgram himself was astonished to find that many ordinary people were prepared to inflict pain to the point of apparently causing the 'learner' to lose consciousness or even die of a heart attack. Not surprisingly, perhaps, when asked to anticipate the outcome of the experiments, most people 'were confident that 100 per cent of the subjects would refuse to co-operate as the cruelty of actions they were commanded to perform grew, and would at some fairly low point break off' (Bauman 154). The percentage of those willing to administer shocks to the point where they feared that the 'learner' might die differed according to the distance between the subject and the authority figure on the one hand and to the victim on the other. Although compliance was highest when the 'expert' in the white lab coat was physically present, a lower percentage carried out orders even when the orders were conveyed by telephone or in the form of a taped message. Reassured by the institutional legitimacy of the experiment, the subjects were more likely to inflict pain on victims they could hear but not see. Physical proximity to the victim caused the subject to withdraw sooner from the experiment. When the subjects were 'told to force the victims' hands on to the plate through which the electric shock was allegedly administered, only 30 per cent continued to fulfill the command till the end of the experiment' (155). The percentage rose to over 60 per cent once the subject administered the shock by a lever to a subject placed out of sight. Physical and emotional distance from the victim made it easier for the 'tormentor' to inflict pain. Yet even the subject, featured prominently in the videotape, who was clearly disturbed by his actions, repeatedly administered a shock to a man he not only pitied but also considered to be in grave physical danger.

No evil intentions or psychopathological tendencies are needed for people to contribute to the destruction of other human beings or the natural environment through the performance of routinized actions. For Bauman, then, 'the meaning of Milgram's discovery is that, immanently and irretrievably, the process of rationalization facilitates behaviour that is inhuman and cruel in its consequences, if not in its intentions. *The more rational is the organization of action, the easier it is to cause suffering* – and remain at peace with oneself' (155). More importantly, though, it transpired that willingness to cooperate increased when cruel actions were introduced sequentially and incrementally. Where the first steps seemed innocuous enough, the more steps were added the harder it was for the subjects to back away from actions they were clearly uncomfortable performing. Although neither Höss nor Eichmann was particularly anti-Semitic, they were both so committed to solving logistical problems that they carried out orders even when they initially baulked at 'a solution through violence.' Once Henry Ford began to invest in maximizing efficiency gains, he, too, was able to close his eyes to the abuses perpetrated by Harry Bennett. 'In modern society,' concluded Milgram, 'others often stand between us and the final destructive act to which we contribute' (155). Extrapolating from the findings of Arendt and Milgram, we could say that Henry Ford occupied the position not of the expert in the white lab coat but of the 'teacher' inflicting pain against his real wishes. Although he never intended to harm his workers, he was reviled by them as a ruthless and heartless oppressor. However, instead of pitying the workers, he felt betrayed by them, illustrating Arendt's contention that men like Eichmann and Höss pitied themselves for the heavy burden they were asked to carry. Committed to the pursuit of efficiency gains, Ford, Höss, and Eichmann were no longer 'free' to follow their own human impulses. They were not the men in control of the 'experiment,' but had submitted to the gaze of the ideology of efficiency, which had successfully interpellated them. Far from seeing themselves as 'monsters,' they considered themselves to be ordinary men who were adept at the efficient organization of machines and human beings.

An analysis of 'Auschwitz' from the perspective of an investment in efficiency ultimately leads us beyond the standard interpretation that perpetrators of evil rationalize their actions by appealing to a sense of duty and claiming to have followed orders implicit in the Milgram experiments. To do one's duty or to follow orders denotes a passive acquiescence in immoral practices that is contradicted by the active desire to be as efficient as possible in the planning and execution of tasks that are seen as challenges to be met rather than duties to be endured. Aside from serving conflicting functions, an obsession with efficiency for its own sake thus allows for a dangerous disassociation of technical means and humanly meaningful ends.

4 Efficiency and Disciplinary Power: The Iron Cage and the Suburb

The desire to optimize efficient outcomes shifts the locus of control over self and others from external sources to the internalization of ideological imperatives. The openly coercive tactics at the Ford Motor Company and in the Nazi death camps were limited in their applicability and effectiveness; in contrast, Taylor's concept of scientific management 'adapts the way a virus does, fitting in almost everywhere' (Kanigel 499). Although Henry Ford's assembly line was a revolutionary innovation, it was restricted to the sphere of modern industrial production. Ford's strategies have been regarded as 'the special case' (498) of Taylor's more universal articulation of management principles. By replacing the 'direct coercion' we associate with Ford's approach with the 'socialization of individual workmen into company goals' (Merkle 25), Taylor's 'mental revolution' introduced supervision and surveillance to assist workers in the internalization of labour discipline. From the 'visible' hand of Ford, efficiency transmogrified itself into the 'invisible' hand of Taylor. The impact, not always recognized, of this 'virus' on both individuals and the social order was enormous. Once scientific management spread from the factory to public and private offices, the most significant consequences were undoubtedly the often disavowed reification of the individual human consciousness and the mostly overlooked depoliticization of the public sphere.

Max Weber's analysis of the bureaucratization of political life illustrates the depoliticizing consequences of the rational or efficient organization of any administrative apparatus. Alert to the rationalizing processes in both economic and political arenas, Weber appears to be impressed by bureaucracy as the embodiment of rational efficiency at the same time as he deplores it for its depersonalizing tendencies.

Bureaucracy, for him, constitutes a totalizing system or 'iron cage' from which there are few viable lines of flight. Concerned with the material means of political power, he investigates the *structures* and *institutions* of economic production and political patterns from a systemic theoretical perspective. Long before Michel Foucault produced his influential studies on disciplinary power, Weber understood that the commitment to rational efficiency produced the ideologically tamed bodies so central to Taylorism. Scientific management can then be seen as a 'special case' of Weber's broader analysis of bureaucracy as a challenge to political power. In this chapter, Taylor's 'mental revolution' is first of all discussed in the context of Weber's crucial concept of bureaucratization. For a more concrete illustration of the depersonalized and depoliticized individual posited by Weber and generated by Taylor, we will finally turn to William H. Whyte's sociological treatise *The Organization Man* (1957) and Sinclair Lewis's novel *Babbitt* (1922), both works once highly acclaimed and now almost forgotten. In these texts the ideology of efficiency is shown to have been thoroughly *internalized* by figures who are prepared to disavow their entrapment in Weber's 'iron cage' in exchange for the security metaphorically guaranteed by the 'gated community' of the suburb. In contrast to the previous chapter on the complicity of efficiency with the openly murderous violence symbolized by the extreme case of Auschwitz, this chapter examines the logic by which efficiency 'works by itself' as an often invisible coercive force under the most ordinary circumstances of work, home, and leisure.

Bureaucratization and Politics

Taylor (1856–1915) and Weber (1864–1920) were almost contemporaries; however, where Taylor celebrated the efficient rationalization of institutions, Weber, despite its appeal, was ultimately hostile to its impact on the individual consciousness and the social order. It is not surprising that there is no evidence of Taylor having been aware of Weber's work, especially since it had not been translated in Taylor's lifetime. But Weber does refer to the 'American system of "scientific management,"' which is for him a disciplinary mechanism intent on reducing the individual human being to a mere 'function' within a system. He clearly objects to 'the psycho-physical apparatus of man [being] completely adjusted to the demands of the outer world, the tools, the machines' (Weber, 'Sociology' 261). Weber seeks the sources of the rationalizing tendencies of institutions not in the factory but in the concept of bureaucracy, the development of which can be traced to Roman, Greek, medieval, and early oriental social orders. However, he argues that bureaucratization intensified during the modern period and links it to the money economy, that is, to the capitalist production process revolutionized by Ford and Taylor. 'The development of the *money economy*' is for him 'a presupposition of bureaucracy' in that it provides 'a stable system of *taxation*' that is 'the precondition for the permanent existence of bureaucratic administration' (Weber, 'Bureaucracy' 204, 208). Focused more on the bureaucratization of public offices than of private capitalist enterprises, Weber's theoretical reflections illuminate what has often been called 'the second industrial revolution' of scientific management as it spread from the factory to private and public offices. Concerned primarily with the professional bureaucrat, Weber implicitly deals with a powerful middle class that, in the United States, developed with the introduction of the 'efficiency expert,' a figure that made possible a new professionalism that 'expanded the market for that formerly excessive commodity, the college graduate' and thus 'promoted the interests of the middle class' (Merkle 75). With the deskilling of labour, the ranks of managers who told workers how to perform their tasks kept swelling. This constant increase in the control of middle-class management over labour was part of the class struggle waged against the proletariat. In her useful survey of the repercussions of Taylorism in *Management and Ideology: The Legacy of the International Scientific Management Movement* (1980), Judith A. Merkle shows that scientific management was able to sweep across the United States precisely because it empowered the middle class. But, in an ironic twist, the middle class itself became the victim of scientific management: 'The lower levels of office management lent themselves particularly to Taylorization' (Merkle 79). Scientific management thus 'eats its own white-collar children, as office operation is first quantified, then systematized, and then reduced, as far as possible, to a minimum of mental and physical movements' (79). The 'efficiency craze of the 'teens and 'twenties' (62) proved Taylor's point that scientific management was not merely an industrial but also a distinctly mental revolution.

Replacing subjective whim with rational rules, Taylor claimed that scientific management instituted an objective system; however, he failed to grasp that its value-neutral logic was open to appropriation by conflicting economic and political interests. He did not foresee, for instance, that his emphasis on expertise would empower the middle classes. Analysing the swelling ranks of bureaucrats in public administration, Weber stresses that once bureaucracy has established itself, it becomes an efficient machine that is so stable and enduring that it proves 'practically unshatterable' ('Bureaucracy' 228). The power of decision-making politicians is being eroded in favour of bureaucrats who passively reproduce the interests of the administrative apparatus. Indifferent to political agendas, this apparatus tolerates masters of any political persuasion. Although the system as such is indeed value neutral, it can 'easily be made to work for anybody who knows how to gain control over it' (229). It withstands revolutions and conquests: 'A rationally ordered system of officials continues to function smoothly after the enemy has occupied the area; he merely needs to change the top officials' (229). A case in point is the architect of a unified Germany, the powerful chancellor Otto von Bismarck. After his resignation, Bismarck discovered to his dismay that the system he had put in motion continued to function as 'if he had not been the master mind and creator of these creatures, but rather as if some single figure had been exchanged for some other figure in the bureaucratic machine' (230). It could be added that even Hitler relied on the existing state apparatus to consolidate his power base; contrary to his rhetoric, he had no intention of dispensing with either industrial capitalism or established bureaucracies. Through a process called *Gleichschaltung,* he purged the public administration through decrees that installed Nazis in positions his 'enemies' had been forced to vacate or through a duplication of offices that established the SS, for instance, as an organization paralleling the older police forces. Hitler reconfigured and bent the existing bureaucracy to serve new functions; he did not radically reform it. Once the Nazi system had taken root, Hitler himself was in reality no longer necessary for its perpetuation.

The history of scientific management confirms that an efficiently rationalized system can take root in very different cultural and political traditions. In France, for instance, Taylorism found a fertile ground in that this country's philosophical tradition had always put a premium on reason and had experimented with rational administrative reforms from the Code Napoléon of Louis Bonaparte to the socialist utopianism of Henri de Saint-Simon. In Germany, scientific management was above all popular as 'an industrial doctrine' but also for its 'emphasis on authoritarian' or 'social-organic' (Merkle 173) aspects. Not surprisingly, Taylorism was also entirely compatible with the 'so-called "totalitarian" economic system of the Third Reich, involving a combination of public control and planning with the maintenance of private ownership in the form of huge cartels' (199). More surprisingly, perhaps, Taylorism was not only compatible with fascism but also with communism, fascism's arch enemy. In its 'practical fusion of socialism, centralism, and bureaucratism' (103), scientific management proved most compatible with the centralized planning ambitions at the heart of Leninist Marxism. Since the already backward Russian economy was in dire need of rationalization, Lenin assumed that he could appropriate aspects of American Taylorism to increase productivity in 'a nonexploitative fashion suitable to the workers' state' (106).[1] In its virus-like insinuation into social spaces, scientific management was thus able to combine what seem to be politically highly incompatible features, which suggests that it gave rise to a politics of administration enabled by a highly adaptable ideology of efficiency.

Aside from opening itself to political appropriation, the efficient rationalization of private and public bureaucracies was in itself a political move destined to undermine especially democratic forms of politics. At the beginning of the twentieth century, Americans became sufficiently disenchanted with democracy as an inefficient process favouring special-interest groups to debate the merits of applying scientific management to political institutions with marginalized groups like the technocratic movement (Technocracy, Inc.) and the Taylor Society. If national planning were left to technically trained professionals, then society could be organized according to objective rational principles. The 'spoil system' infecting the tradition of political office would also be replaced by a merit system supposedly equally open to all talented ranks of society. In reality, of course, this merit system clearly favoured the mobile middle classes; the 'talented' were not likely to emerge from the ranks of the socially and economically disenfranchised, who were ill prepared to impress future employers with their educational credentials or able to pass the objective examinations required for entering the civil service. Although these marginalized advocates of the scientifically managed society did not prevail, the argument that the state ought to be rationalized so as to achieve an optimal level of social distribution had considerable weight with Franklin Delano Roosevelt in his drafting of the New Deal. As Merkle concludes, the New Deal may have been Keynesian in its 'predominant economic influence,' but in administrative organization it was marked by 'a new concept of "state Taylorism,"' manifesting itself in 'the process of policy-making' (260). At this time in American history, the basic idea of a society administered by experts rather than socially privileged owners of property had in many quarters been received as a desirable progressive move.

Since the bureaucratized apparatus assigned positions according to competency rather than personal connections, it appeared to reinforce the process of democratization. With the introduction of the professional expert, the old patronage system in both Germany and the United States began to crumble. But Weber convincingly argues that democracy made possible the very bureaucratic organization that also disempowers it. History shows that bureaucratic organization 'has usually come into power on the basis of a leveling of economic and social differences' ('Bureaucracy' 24). On the face of it, the merit system meant that anyone could learn the rules and thereby acquire the expertise necessary for a given vocation. It seemed, then, that bureaucratic efficiency benefited both capitalism and democracy. Where 'non-bureaucratic forms of domination' are contradictorily marked by 'strict traditionalism' as well as 'free arbitrariness and lordly grace,' bureaucratic forms offer the 'optimum possibility for carrying through the principle of specializing administrative functions according to purely objective considerations' (217, 215). Such objectivity has to be understood as the 'discharge of business according to *calculable rules* and "without regard for persons"' (215). But this reduction of the 'person' to a 'function' empties democracy's guarantee of individual freedom of all content. Since the emphasis on 'calculable rules' has to expel 'emotional elements' (216) as waste, the freedom promised by democracy to realize one's human potential has been seriously compromised.

Weber further warns that bureaucratization has often been 'carried out in direct alliance with capitalist interests' ('Bureaucracy' 230) and often in opposition to democracy. Bureaucracy works against mass democracy in that it instals a permanent administrative structure directed by experts who have neither been elected by the people nor are being held accountable to them. In its demand for 'the universal accessibility of office,' democracy is in conflict with the tendency of bureaucracy to develop a stable and 'closed status group of officials' (226). In other words, bureaucracies reintroduce a hierarchy of privileged elites. Weber thus insists that 'one has to beware of believing that "democratic" principles of justice are identical with "rational" adjudication (in the sense of formal rationality)' (217). Although democracy has in many ways 'produced' bureaucratization, it 'inevitably comes into conflict with the bureaucratic tendencies,' especially in the '*leveling of the governed* in opposition to the ruling and bureaucratically articulated group, which in its turn may occupy a quite autocratic position, both in fact and in form' (226). Beyond this question of justice, the exclusionary 'authority of officialdom' restricts the 'sphere of influence of "public opinion"' (226) that is the cornerstone of the democratic principle of political participation and debate. The complicity of democratization with bureaucratization means that the masses remain politically and economically disenfranchised in spite of the principle of formal equality before the law. For all these reasons, Weber concludes that 'democracy as such is opposed to the "rule" of bureaucracy, in spite and perhaps because of its unavoidable yet unintended promotion of bureaucratization' (231).

The development of efficient bureaucracies was not only in conflict with politics but threatened to displace it. The priority accorded to levels of administrative bureaucracy indicates that an economic-industrial form of organization was replacing earlier political models of decision-making. The shift to a Taylorized administrative state was in effect paramount to 'the depoliticization of politics' (Merkle 68). In the United States, political and corporate organizations were marked by a seemingly unlimited growth of their bureaucracies. Political decisions were increasingly motivated and constrained by the interests of entrenched administrative apparatuses. In the public sphere, efficiency was no longer touted simply as a 'business-oriented, cost-cutting approach' but as a 'rival theory of politics' (265, 281) intent on perpetuating itself as a permanent and pervasive social force. In the United States, the separation of powers was at least capable of curbing the excesses of an increasingly powerful administrative bureaucracy by making it accountable to the executive arm of government. In contrast, administrative organization and political operations in fascist regimes were 'purposely welded together' (262), making it difficult to establish 'the degree to which such organizational techniques were directly responsible for the excesses of the totalitarian regimes which they served, whether they led inevitably to massive bureaucratic growth, impersonality, and finally to the casting off of all social morality as an irrational barrier to the planning process, or whether they simply gave new life and strength to destructive political regimes' (262). Once social morality has been castigated as a brake on the pursuit of efficient outcomes, the behaviour of Rudolf Höss and other Nazi functionaries can be explained as a 'rational' response to 'irrational' tendencies. But even in democratic societies, the public response to efficiently engineered administrative apparatuses shifted 'from one of opposition to "dictatorship" to one of approval for "efficiency"' (274). In other words, the 'fascist' aspects of the tendency to make efficiency an unquestioned virtue are easily, if not wilfully, disavowed. In 'Politics as a Vocation,' Weber famously deplores the ascendancy that 'administrative officials' and 'professional politicians' gained over 'political officials' and 'occasional politicians' (83, 90). Before an 'expert officialdom' challenged 'the prince's autocratic rule' (Weber 88), occasional politicians were either men of property or rentiers with independent financial means who were in the enviable position of being able to make political decisions without concern for their career prospects. Distinguishing between 'living "for"' and 'living "off"' politics (84), Weber fears that the conscience of politicians is incurably being compromised by pragmatic self-interest. The professional politician is essentially a civil servant who depends for his livelihood on the perpetuation of the bureaucracy he serves. The bureaucratization of politics vitiates the 'ethical locus' (117) of politics as a 'calling' or a vocation. Since Weber disapproves of the old 'patronage system,' he is left with 'the choice between leadership democracy with a "machine" and leaderless democracy, namely, the rule of professional politicians without a calling' who form an elite obeying '"the rule of the clique"' (113). The 'three pre-eminent qualities' he considers to be 'decisive for the politician' are 'passion, a feeling of responsibility, and a sense of proportion' (115), emotional qualities that are beyond the efficiency calculus of bureaucratic administrations. Faced with the irreversible march of the rationally efficient bureaucratization of the political apparatus, Weber concludes that the conditions were no longer favourable to 'the management of politics as a vocation' (114). It seems, then, that the emergent social investment in efficiency at stake in both Taylor and Weber promoted a shift from a politically centralized system to ideologically flexible and dispersed disciplinary mechanisms designed to manage individuals and administer organizations.

The Efficient Corporation

The depoliticization of politics can be seen to serve the interests of corporate capitalism; the ideological adaptability made possible by Taylorization and bureaucratization mirrored and empowered the indispensable mobility of capital. Although the modern corporation is a monster initially created by politicians, it has now taken on a life of its own and swallowed the political master. It is not coincidental that the modern corporation emerged in the middle of the nineteenth century in order to organize the production and distribution of the new consumer goods that efficient machines made available for marketing. While various forms of corporate association can be traced back to the beginnings of trade, the specific formation of the modern corporation coincides with the machine age and with the modern sense of the term 'efficiency.' The nineteenth century saw the invention of the railways, which were such 'mammoth undertakings' that they required 'huge amounts of capital investment,' thereby combining for the first time 'the capital, and thus the economic power, of unlimited numbers of people' (Bakan 10, 8). Where previous enterprises were financed by socially privileged individuals or families, the railways opened the door to investors from all ranks of life in both Britain and the United States: 'By the middle of the century, with railway stocks flooding markets in both countries, middle-class people began, for the first time, to invest in corporate shares' (11). However, rogues and speculators plagued joint-stock companies as far back as the early eighteenth century, when the infamous bubbles created by the South Sea Company in England and the Mississippi Company in France finally burst in 1720. The craze for railroad stocks eventually produced another bubble that was punctured on 'Black Thursday,' 18 September 1873, resulting in so many railroads going bankrupt that investor confidence was seriously shaken. It was at this point that politicians stepped in to give shape to the modern corporation through legislation limiting an investor's liability to the amount invested. Where the railroads provided the material conditions for the creation of modern stock exchanges, the political decision to introduce the concept of 'limited liability' provided the impetus for the rapid development of the modern corporate system.

The law of 'limited liability' initially sought to protect individuals from being '*personally* liable, without limit, for the company's debts' (Bakan 11). Ironically, though, it ended up once again privileging the system over the individual. After heated debate, 'limited liability was entrenched in corporate law, in England in 1856 and in the United States over the latter half of the nineteenth century' (13). The concept of limited liability made it possible for corporations to raise large amounts of capital, thereby encouraging them to grow not only in number but also in size. In the wake of further legal changes, mergers and acquisitions concentrated economic power in the hands of a relatively small number of large corporations: 'In less than a decade the U.S. economy had been transformed from one in which individually owned enterprises competed freely among themselves into one dominated by a relatively few huge corporations, each owned by many shareholders' (14). Stock owners now enjoyed greater protection under the law but, as individuals, they were too 'dispersed to act collectively' and thus lost their capacity to 'influence managerial decisions' (14–15). At the same time, the corporations themselves were increasingly being controlled by 'large armies of full-time managers,' figures arising with the establishment of the railways, who 'didn't own the organizations they worked for but nevertheless devoted their entire careers to them.' They had a 'high sense of their calling' and 'pioneered many of the tools of the modern corporation,' most significantly 'the accounting and information systems needed to control the movement of trains and traffic, to account for the funds they handled, and to determine profit and loss for the various operating units' (Micklethwait and Wooldridge 60–1). This 'separation of ownership and management and the financing of corporations by sales to the public of shares' is also what characterizes the 'late stage of capitalism' (Gerth and Mills 68) for Weber. It follows that the owners of corporations were no more in absolute control of their enterprises than were the professional politicians of the administrative apparatus over which they presided. At the same time, 'shareholders had, for all practical purposes, disappeared from the corporations they owned' (Bakan 15). While they were no longer being held personally responsible for the businesses they financed, they were also no longer actively engaged in the management of these corporations.

Most ironically, perhaps, with the introduction of 'limited liability,' the law decreed that the corporation itself had to 'assume the legal rights and duties' (Bakan 15) that would normally apply only to persons. Although the corporation has neither body nor soul, it has the rights and freedoms of an independent, autonomous being. But, unlike an individual, the corporation is not held to moral standards. Since the corporation is meant to be the most efficient profit-generating machine possible, its sole moral and social responsibility is to 'make as much money as possible for [its] shareholders' (34). In Bakan's provocative terms, the corporation is thus a legal person exhibiting highly pathological traits: 'The corporation's legally defined mandate is to pursue, relentlessly and without exception, its own self-interest, regardless of the often harmful consequences it might cause to others' (1–2). Henry Ford, for instance, was taught the lesson that any actions taken to further an agenda beyond the profit motive were in effect illegal. Believing that 'his Ford Motor Company could be more than just a profit machine' (36), Ford increased the wages of his workers and slashed the price of the Model T to make it more affordable for the masses. '"I do not believe that we should make such awful profits on our cars," he is reported to have said. "A reasonable profit is right, but not too much"' (36). But when Ford 'canceled dividends and thereby deprived John and Horace Dodge of the funds they needed to start their own company' (36), his former partners took him to court. The Dodge Brothers argued that 'profits belong to shareholders' so that 'Ford had no right to give their money away to customers, however good his intentions' (36). The courts agreed with the Dodge brothers; *Dodge v. Ford* entrenched 'what has come to be known as "the best interests of the corporation" principle' (36). Any manifestation of corporate social responsibility is thus 'illegal' (37); moreover, corporations are 'compelled to cause harm when the benefits of doing so outweigh the costs' (60). They are encouraged to unload the cost of the harm they do on third parties, on 'workers, consumers, communities, the environment' (60). Instead of taking responsibility for the damage they cause, corporations 'externalize' harmful side-effects by making government pay for cleaning up polluted rivers or caring for sick workers. Once a collectivity is accorded the legal rights formerly reserved for individuals, the behaviour of this 'legal person' is for Bakan psychopathological. He calls the corporation irresponsible, manipulative, grandiose, lacking in empathy, exhibiting asocial tendencies, incapable of feeling remorse, and tending to relate to others superficially (57). In classical Prisoner's Dilemma terms, the self-interest pursued by the company as a 'legal person' is thus for society collectively self-defeating. Although the creature of political decisions, the corporation was in effect 'the first autonomous institution in hundreds of years, the first to create a power center that was within society yet independent of the central government of the nation state' (Drucker in Micklethwait and Wooldridge, 54).

Today's corporate culture seems so ubiquitous that sites of resistance are difficult to locate. Although the psychopathology of global capitalism has been graphically illustrated by antiglobalization activists like Noam Chomsky and Naomi Klein and philosophers like Michael Hardt and Antonio Negri or Benjamin Barber, these critical voices are hard pressed to offer any viable alternatives. The corporation confronts us as an apparently natural force we can do little to resist. However, as Micklethwait and Wooldridge point out, 'No matter how much modern businessmen may presume to the contrary, the company was a political creation. The company was the product of a political battle, not just the automatic result of technological innovation' (53–4). Bakan concurs, stating that, 'beguiled by the "natural entity" conception of corporations, the notion that they are *independent* persons, we tend to forget that they are entirely *dependent* upon the state for their creation and empowerment' (154). Indeed, in the nineteenth century, governments were not afraid to revoke the charters of misbehaving companies. As late as 1911, for instance, the U.S. Supreme Court ordered Standard Oil to be broken up into smaller companies. However, government deregulation of companies has proved costly; the neo-conservatism prevailing in the second half of the twentieth century empowered corporations to pressure governments through lobbying and the competitive threat of globalization to retreat from its watchdog position. By the end of the twentieth century, explains Bakan, 'the corporation had become the world's dominant institution' (139). However, government has not so much retreated as adapted to the economic conditions created by the corporate culture: 'Overall, however, the state's power has not been reduced. It has been redistributed, more tightly connected to the needs and interests of corporations and less so to the public interest' (154). We have, he concludes, 'created a difficult problem for ourselves. We have over the last three hundred years constructed a remarkably efficient wealth-creating machine, but it is now out of control' (159). Dominating both private and public institutions, society's investment in efficiency as an unquestioned value emerges as a significant factor in the empowering of corporations and the depoliticization of politics deplored by Weber.

The Iron Cage and Depersonalization

The mental revolution Taylor advocated to increase the efficiency of economic production had consequences for the relationship between the individual and the social collectivity far beyond the workers and owners he admonished in *Principles of Scientific Management*. The emphasis on rational organization meant that both governmental and corporate administrations were progressively being reshaped 'along the depersonalized and increasingly structured and specialized lines of the machine-model of bureaucracy' (Merkle 279–80). These bureaucratized social structures necessitated the emergence of a new type of personality, the depoliticized and depersonalized modern individual. Although Weber is famous for his argument that the 'Protestant ethic' prepared the ground for the 'spirit of capitalism,' his analysis of the bureaucratizing process supplies us with a description of the transmogrification of the characteristics required of the individual from early-industrial to late-consumer capitalism. His metaphor for bureaucratic society under capitalism is 'the iron cage,' the 'modern economic order' that is 'now bound to the technical and economic conditions of machine production which today determine the lives of all the individuals who are born into this mechanism' (Weber, *Protestant* 181). He concludes *The Protestant Ethic and the Spirit of Capitalism* (1904–5) with the complaint that economic considerations have displaced all other human concerns: 'In Baxter's view the care for external goods should only lie on the shoulders of the "saint like a light cloak, which can be thrown aside at any moment." But fate decreed that the cloak should become an iron cage' (181). Ironically, this iron cage came about as the unintended consequence of a spiritual 'calling' adamantly opposed to its materialist ideology. Although Weber depicts the rationalized social order as an inescapable totality, he stresses the inability of this totality to eliminate residues of waste that enabled it but are also always poised to disrupt it. Weber's conception of the rationalizing process is that of the thermodynamic engine which aspires to a totalization that must logically always remain incomplete.

In *The Protestant Ethic and the Spirit of Capitalism* Weber contends that ascetic Puritanism supplied the personality best suited to the logic of capitalism. Although capitalism was anathema to religious spirituality, Puritanism, in its emphasis on thrift and 'the systematic rational ordering of the moral life as a whole' (126), created the conditions propitious for capitalist accumulation. The 'ascetic compulsion to save' and the 'restraints which were imposed upon the consumption of wealth' served to increase acquisitive activity and hence made possible 'the productive investment of capital' (172). Out of religious asceticism and inner spirituality had emerged a 'specifically bourgeois economic ethic' (176). Weber's contention that Protestantism functioned as a preformed system for the surfacing of the capitalist mode of production has been usefully summarized by Harvey Goldman:

> For *modern* capitalism to have developed as it did, Weber argued, a *new kind* of person must have existed, a person with special qualities and capacities for work, with a natural inclination for the new kind of rationalized labor that capitalism as a system brought with it. But, he argued, these new men possessed that inclination and capacity *before* capitalism

> was established as a system capable of imposing such labor through the pressure of its material demands, and they derived this strength from non-economic – in this case, religious – sources. (Goldman 19)

Crucial to Calvinism was the idea of vocation as the 'abnegation of the self' (Goldman 42); the natural self must be tamed through systematic self-control and ascetic discipline.[2] The 'sanctification of works [was] raised into a *system,*' into the 'ethical praxis of everyday man' (43). The 'subjection of feeling' in Puritanism is thus at the 'core of what we can call the "rationalized" personality' (43). At the same time, the ideal of ascetic self-denial did not allow Puritans to enjoy the fruits of their labour. According to Weber, since possessions could not be enjoyed, economic activity led to accumulation and investment for their own sake. Moreover, since worldly success was for Calvinists a sign of being elected, it was 'a product of Providence' which meant that one did not have to worry inordinately about the 'unequal distribution of wealth' (47). Not only did 'the spirit of ascetic religion unintentionally [give] rise to economic rationalism,' but the '*ascetic compulsion to save*' (46) prepared the soil for the capitalist system. It follows that a 'capitalist *form* can survive with a traditionalist *spirit*' (27) that is in fact antithetical to it. For Weber, then, Protestantism introduced the *systematic* rationalization of self and world: 'Beyond capital accumulation and the religious defense of exploitation of labor, Protestant asceticism furthered "the tendency toward a bourgeois, economically *rational* conduct of life," because it was "its most essential and above all, its single consistent carrier [*Träger*]." Thus "it stood at the cradle of the modern economic man"' (48). What Weber demonstrates above all is that the intentions of the Puritan ethos are incommensurate with the bourgeois-capitalist ends they ultimately served. In a sense, Weber's rational analysis of capitalist rationalization conveys precisely the unpredictable (and hence irrational) nature of systems.

Weber is undoubtedly best known for his depiction of the depersonalizing trends he identifies in his depictions of the 'iron cage' in the essays on bureaucracy. In the economic arena, the bureaucratic machine strives to increase its efficiency by focusing on categories that make 'paid bureaucratic work' not only 'more precise' but also 'cheaper than even formally unremunerated honorific service' ('Bureaucracy' 214). The fragmentation of work into specialized tasks in the factory finds its extension in administrative apparatuses. The same principles – the fixed distribution of duties, the stable and strictly delimited authority 'to discharge these duties,' and the proper qualifications for 'the regular and continuous fulfillment of these duties' – mark both the 'bureaucratic management' of economic domination and the 'bureaucratic authority' of 'public and lawful government' (196). Categories central to Henry Ford's obsession with efficiency recur in Weber's characterization of bureaucratic apparatuses: 'Precision, speed, unambiguity, knowledge of the files, continuity, discretion, unity, strict subordination, reduction of friction and of material and personal costs' (214). In language that is reminiscent of the assembly line, Weber describes the bureaucrat as a 'single cog in an ever-moving mechanism which prescribes to him an essentially fixed route of march' (228). Actions are not motivated by the subjective intentions of individuals but by the functions these individuals perform within a dynamic, anonymous, infinitely adaptive, and ultimately oppressive system.

This privileging of the system means that those who believe themselves to be in positions of power and control are in reality also victims of the iron cage. In his description of the bureaucrat 'who controls the bureaucratic apparatus,' Weber clarifies that he who exercises power 'cannot squirm out of the apparatus in which he is harnessed' ('Bureaucracy' 228). If anything, the bureaucrat in power seems more chained to the 'machine' he is supposed to control than the labouring 'cog' operating the assembly line at Ford Motor. His integration into the mechanism is more complete than the factory worker's because he has internalized his dependence on the system that rewards him with career incentives. By entering an office in the public or private sector, the bureaucrat has accepted 'a specific obligation of faithful management in return for a secure existence' (199). It seems that he has freely entered into a mutually beneficial social contract; however, he has thereby sacrificed the individual autonomy that Weber considers to be essential for the realization of his human potential. Although freed from what Taylor called 'rule-of-thumb' arbitrariness, a bureaucracy acting 'without regard for persons' (215) reduces the individual to an infinitely interchangeable and depersonalized fragment in an indifferent system. Bureaucracy's 'specific nature, which is welcomed by capitalism, develops the more perfectly the more the bureaucracy is "dehumanized," the more completely it succeeds in eliminating from official business love, hatred, and all purely personal, irrational, and emotional elements which escape calculation' (215–16). The 'iron cage' is for Weber increasingly populated by the routinized creature created by the rationalized society. But this creature is no longer characterized by the personality traits Weber traces to the Puritan ethic; with the introduction of the Model T, Ford shifted the emphasis from the producing to the consuming creature of late capitalism. This new figure is encouraged to behave hedonistically rather than ascetically. The 'iron cage' turns out to be the gilded or 'gated' suburb dramatized in Sinclair Lewis's novel *Babbitt* (1922) and analysed in Whyte's sociological study *The Organization Man* (1957).

The Efficient Middle Classes

Weber's routinized creatures are clearly not the violently oppressed cogs on the factory floor of the Ford Motor Company. They are the ideologically constructed individuals who 'work all by themselves' to serve the needs of corporate capitalism. Their commitment to the efficiency calculus spreads beyond their working lives to include areas of private life previously thought to be exempt from its impact. As Taylorization expanded to rationalize offices in both public and private enterprises, it created masses of individuals who 'spend eight of their waking hours every day internalizing bureaucratic values, creating the possibility that they may see them as appropriate ones to apply in activities outside of their official duties' (Merkle 282). In *The Organization Man* Whyte focuses on the consciousness of individuals who embrace their total integration into a thoroughly homogenized consumer society. This 'brave new world' (Whyte 324) is no longer dominated by the emphasis on thrift in the Protestant ethic, but thrives on a new social ethic dedicated to the wasteful consumption of goods required for the efficient functioning of the market economy. The modern individual analysed by Whyte and dramatized by Lewis is the depoliticized and depersonalized creature inhabiting Weber's iron cage. Babbitt embodies the features and behaviour patterns Whyte attributes to the organization man who embraces the suburb as his natural environment. In both texts the emphasis is on the threat the apparently harmonious group poses to the earlier conception of the unique individual empowered to choose freely. While the notion of 'free' individuals entirely conscious of who they are or what they want has been deconstructed by postmodern discourses theorizing the constraints operating on human self-consciousness and social agency, the nostalgic yearnings in the writings of Weber, Whyte, and Lewis nevertheless foreground a shift in the relationship between the individual and the collectivity under late capitalism. As these critics of modern society are no doubt aware, the possibility of recovering the 'unique individual' implicit in their targeting of the conformist new personality is no longer a historical option. Rather, the unique or free individual constitutes an 'ideal type' against which these thinkers measure the deterioration of a meaningful human existence in the standardized modern society.

The ideology of efficiency is most obviously referenced in Lewis's depiction of Babbitt as a thoroughly standardized personality. Through satiric exaggeration, Lewis's novel captures more vividly than any social document the ideological power of Taylor's mental revolution as it compels people to internalize conditions serving capitalist interests. Instead of objecting to what the workers at Ford Motor experienced as dehumanizing, Babbitt embraces uniformity as a reassuring safety net. By adapting to social norms, he feels socially secure. His sense of self is formed by consumer goods that mark him as an upwardly mobile middle-class 'modern.' Lewis takes great pleasure in the satirical depiction of Babbitt's delight in technical inventions designed to increase the efficiency of gadgets from the 'electric car-lighter' as 'a priceless time-saver' (Lewis 51) to 'the best of nationally advertised and quantitatively produced alarm-clocks' with their 'cathedral chime, intermittent alarm, and a phosphorescent dial' (Lewis 3). Taking pride in the car he drives, the house he calls home, and the office in which he works, Babbitt is devoted to 'the God of Progress' and of 'Modern Appliances' (8, 5).[3] The bedroom in his modern house in the upscale 'residential settlement' of Floral Heights sports the 'standard designs of the decorator,' the 'standard electric bedside lamp,' and the 'standard bedside book' (Lewis 14, 13). A 'masterpiece among bedrooms, right out of cheerful Modern Houses for Medium Incomes,' the bedroom 'settled instantly into impersonality' (14, 13). Although the rest of the house was 'as competent and glossy as this bedroom' it is a monument to suburban self-congratulation and hence 'not a home' (14). Along the same standardized lines, Babbitt spends his working hours in the Reeves Building, a modern office complex 'as fireproof as a rock and as efficient as a typewriter,' located in the 'business-center of Zenith,' the middle-sized city that is said to be indistinguishable from any other city in 'Oregon or Georgia, Ohio or Maine, Oklahoma or Manitoba' (30, 49). In his eagerness to fit into the 'mechanical civilization' symbolized 'by the Ford car' (Friedman 66), Babbitt adopts the standard dress code for a man of his social class and standing. His 'uniform' of the 'Solid Citizen' consists of a 'standard' 'gray suit,' which was 'completely undistinguished,' and 'standard boots,' which were 'extraordinarily uninteresting' (Lewis 8–9). The emphasis on standardization in Henry Ford's drive for efficiency gains has so infiltrated the social fabric that Babbitt heartily approves of 'dominating movements like Efficiency, and Rotarianism, and Prohibition, and Democracy' as 'our deepest and truest wealth' (82).

But where the rationalization of the production process at the Ford Motor Company resulted in the increased output of actual cars, Babbitt's commitment to efficiency proves to be empty of content. Modelled on Ford's industrial efficiency, standardization is for Lewis's character allied with speed. Dressed to the standards of 'the perfect office-going executive,' he was one of the men 'in motors' who were 'hustling to pass one another in the hustling traffic' in a city that embraced hustling 'for hustling's sake' (26, 149). However, after having 'hustled back to his office,' he typically sat down 'with nothing much to do except see that the staff looked as though they were hustling' (149). The figure most representative of those devoted to 'earnest efficient endeavor' is Babbitt's new secretary, the 'impersonal,' 'pale,' and 'industrious' Swede who is described as 'a perfectly oiled and enameled machine' (177, 355). Although she 'took dictation swiftly' and 'her typing was perfect,' even Babbitt 'became jumpy when he tried to work with her' (355). Performing her duties with admirable dedication, she is no more interested in the purpose of her activities than her employer is in what he sells. In constant motion, Babbitt privileges empty process over meaningful product. Although Ford is known for innovations not only in production but also in the marketing of cars, he encouraged the consumption of a tangible material good. In contrast, Babbitt is a 'Go-getter' dedicated to 'the cosmic purpose of Selling – not of selling anything in particular, for or to anybody in particular, but pure Selling' (138). He was 'the plump, smooth, efficient, up-to-the-minute and otherwise perfect modern' who replaced his father-in-law and business partner, 'the old-fashioned, lean Yankee, rugged, traditional stage type of American business man' (66) typified by Ford the tycoon. In conformity with the consumer society, his display of efficiency is wasteful rather than productive.

Babbitt typifies the new type of personality that emerges for Weber with the rationalization of bureaucratic apparatuses and for Whyte with the highly specialized and professionalized 'organization man' occupying the offices of private and public enterprises. Contrary to popular opinion, the society of late capitalism relies not on the ruthless individualist but on the mediocre 'well-rounded man' (Whyte 147) who willingly adapts to the needs of the system. He is a Babbitt who voluntarily subordinates his own desires to the demands of the corporate society for standardized, adaptable, and interchangeable cogs. It is not that the organization man is entirely unaware of his conformity. For him to persist in his adaptive behaviour, he has to disavow the absence of any meaningful purpose to his activities. Babbitt's dedication to pure Selling finds its equivalent in the allegiance of organization men to 'The Organization itself' (Whyte 308) rather than to any particular organization; they are fixated on the 'development of their professional techniques' (308) rather than on promoting company goals. They market loyalty not to an employer or to an idea but to a highly adaptable technical competence that gives them the mobility to change jobs. Reduced to a mere function in the system, these men realize that personal qualities are extraneous if not actually harmful to their career progress. Young, upwardly mobile middle-class conformists are for Whyte guilty of an 'idolatry of the system' that compels them to embrace the 'self-imposed tyranny' (189, 166) of group solidarity leading to a 'surrender' of individualism. The logic emerging from Whyte's analysis suggests that the need for interchangeable workers in factories and offices resulted in feelings of rootlessness that paradoxically created the very conditions for the 'total integration of the individual' (189) within the capitalist system. The spirit of 'modernity as adventure' characterizing the enterprising tycoon and robber baron has given way to the normalizing tendencies of 'modernity as routine' typified by the middle-class conformist.

For the conformist, the collectivity is no longer an alien machine but the source of the individual's security and happiness. It is not enough for the organization man that 'he belong; he wants to belong *together*' (Whyte 52). He accepts unquestioningly that '*the group is superior to the individual,*' embracing team spirit as the precondition for 'a *harmonious* atmosphere' that will 'bring out the best in everyone' (53, 54). Eschewing overt coercion, he accepts that 'the secret of happiness' consists in acquiring the 'skills of adjusting oneself' to the group and the 'skills of managing other people' (102). Where the Protestant ethic emphasized work and competition, the new social ethic is intent 'on managing *others'* work and on co-operation' (124). Instead of oppressing others, the organization man assumes that it is virtuous to manipulate others for their own good; the putative goal of managing others is thus to make them happy. 'Fitting in' presents itself as the road to social security and personal happiness. Seduced by the promise of this destination, he allows himself to be tamed by means of obvious disciplinary techniques intended to 'train' and 'observe' not only his technical performances but also his social attitudes. This emphasis on training and surveillance concentrates on the 'whole man' (191) rather than, as in the case of Ford or Taylor, on a particular set of aptitudes or skills. It is no longer a question of Ford's or Taylor's desire to get 'things done' (191), but of creating a 'docile body' to be seamlessly integrated into a smoothly functioning totality. Testing not only for appropriate skills but also for suitable personality traits, corporations and public bureaucracies bar entry to potential 'deviants' whose corrosive influence might disrupt the team spirit. If a 'deviant' individualist should circumvent this initial censorship, surveillance techniques will weed him out at later stages. Although formal aptitude and personality testing is today no longer the norm, interviewing techniques and reference checks continue to screen for an applicant's 'degree of radicalism versus conservatism, his practical judgment, his social judgment, the amount of perseverance he has, his stability, his contentment index, his hostility to society, his personal sexual behavior' (191). The extent of such surveillance compelled 'one trainee' to remark that 'not only can't you get lost, you can't even hide' (137). To his horror, Whyte discovered that the 'average trainee' talked most 'enthusiastically' about surveillance' (137). There was little resistance to disciplinary regimes asking 'a man' to 'testify against himself' (222).

At the same time as Lewis's Babbitt submits to 'booster' norms, he declares himself to be an advocate of 'free' market competition. He consequently rationalizes his manipulation of others by unwittingly using and abusing the logic of Taylor's Pareto-efficient equilibrium model. Justifying his shady dealings with the Zenith Street Traction Company, he argues that whatever the company 'desired to do would benefit property-owners by increasing rental values, and help the poor by lowering rents' (Lewis 24). On the surface, it appears as if everybody would be better off than before. However, the profits for the company are, of course, much higher than the amount by which the rents are lowered. Moreover, the company quite clearly engages in a property swindle to fill its own pockets. Exemplifying the free-rider problem inherent in equilibrium models, Lewis criticizes Babbitt for acting in a self-interested way that proves collectively self-defeating. In an effort to rationalize his business practices, Babbitt claims that his cheating was not some base 'trickery' but a defence against being 'suckered' by others; he thus argues that 'most folks are so darn crooked themselves that they expect a fellow to do a little lying, so if I was fool enough to never whoop the ante I'd get the credit for lying anyway!' (44).

In another scene, Babbitt uses the appeal to Pareto efficiency in a land-speculation deal in which he defrauds Archibald Purdy. When Purdy objects to being forced to pay twice the value of a property, Babbitt defends himself and his partner by arguing: 'Supposing Lyte and I were stinking enough to want to ruin any fellow human, don't you suppose we know it's to our own selfish interest to have everybody in Zenith prosperous?' (Lewis 47). In this scene, he disingenuously appeals to the mutually beneficial formula of Pareto efficiency in order to violate its premises by gaining an unfair advantage. At another moment he seriously embraces this formula but with a significant modification: 'He almost liked common workmen. He wanted them to be well paid, and able to afford high rents – though, naturally, they must not interfere with the reasonable profits of stockholders' (170). In his wily deceptions of others and his self-justification, he makes elastic use of Taylor's Pareto-efficient model and the free-rider problem endemic to its dependence on cooperation.

The 'boosters' are no doubt at their most hypocritical in their rationalizations of class privilege; in their attitude to labour they have no use for Taylor's equilibrium model. Babbitt strenuously objects to class cooperation as a threat to capitalism by complaining that 'efficiency-experts' and 'scientists' want to 'replace the natural condition of free competition by crazy systems,' such as 'cooked-up wage-scales and minimum salaries' (Lewis 81, 297). This complaint typifies the hostility of management to Taylorism as an empowerment of the working classes. Instead of seeking an optimally efficient distribution of benefits and responsibilities, members of the Good Citizens' League[4] reinforce their privileged position by contending that 'the true-blue and one hundred per cent American way of settling labor-troubles was for workmen to trust and love their employers' (369). The efficiency of the system demands that workers cooperate with employers and employers compete with each other. The unfair economic distribution is further obscured by the political rhetoric of democratic equality. To preserve their economic privileges, the good citizens of Zenith all 'agreed that the working classes must be kept in their place; and all of them perceived that American Democracy did not imply any equality of wealth, but did demand a wholesome sameness of thought, dress, painting, morals, and vocabulary' (369). Dramatizing Weber's assertion that efficiency does not necessarily promote democracy, Lewis reinforces Weber's warning that 'economic power and the call for political leadership of the nation did not always coincide' (Gerth and Mills 35). The good citizens assume that they are in political control of the standardized society and enjoy its economic benefits. In exchange for this privileged economic position, they are willing to sacrifice democratic debate to the bureaucratic administration of citizens.

Whyte's sociological investigations make clear that the pact with the devil that efficiency offers us is security and predictability at the expense of individual freedom and privacy. The efficient organization promotes team effort as the embodiment of Taylor's Pareto-efficient privileging of cooperation over competition. But in contrast to Taylor's assumption that the cooperative behaviour of class antagonists would advance the long-term interests of the collectivity, the team players in Whyte's study remain narrowly self-interested and compete with each other. Since the integration of organization men into a particular company or institution is always temporary and provisional, they do not have the necessary commitment to an organization's goals to sacrifice their self-interest. Moreover, the surface cohesion of a team conceals the competitive moves each player plans or executes against other members of the team. The organization man competes first of all with others for inclusion in the team and then manoeuvres to maintain or improve his position. Engaged in an oxymoronic 'co-operative competition' (Whyte 176), each member of the team is pitted against everybody else. The security promised by the pact with the devil proves on closer inspection to be highly unstable because 'for what, and against whom' (176) the organization man is competing is never clear. The much vaunted team spirit succumbs to the race to the bottom endemic to cooperation: 'The best defense against being surpassed, executives well know, is to surpass somebody else, but since every other executive knows this also and knows that the others know it too, no one can ever feel really secure' (177). This insecurity is exacerbated by the 'absence of tangible goals,' combined with 'plenty of activity to get there' (176). Once we accept with Taylor that the system is first, we are implicated in a process compelling us to internalize as well as disavow the very disciplinary regimes that isolate us from others, fuel our insecurities, and condemn us to the status of docile bodies.

In his commitment to efficiency and career progress, Whyte's organization man is the precursor of today's psychopathological workaholic. Having internalized the gaze of the system, he proudly asserts that 'he cannot distinguish between work and the rest of his life' (Whyte 164).

If his working life is devoid of purpose, then he condemns himself to a meaningless personal existence that he then seeks to gloss over with plenty of further activity. As Whyte asks pointedly, 'Why, when the purpose of our vast productive apparatus is the release of man from toil, do the people in charge of it so willfully deny themselves the fruits of it?' (166). Babbitt exemplifies this paradox by investing in the 'sane and efficient life' without seriously asking himself why he 'was so busy that he got nothing done whatever' (Lewis 176, 158) and without seriously analysing why he experienced feelings of discontent. Transferring his purposeless hustling at work to his home life, he spends no meaningful time with his family, barely registering the presence of wife and children.[5] Instead of enjoying his weekends, he contaminates his leisure time with the same dedication to mindless activity he displays at work: 'Every Saturday afternoon he hustled out to his country club and hustled through nine holes of golf as a rest after the week's hustle' (Lewis 149). In short, Babbitt has 'tragically' lost his 'capacity for pleasure' and 'cannot really be entertained' (Grebstein 42). What matters to Babbitt is not the freedom to think for himself or to discover and follow his own desires, but the security that the routines of home, office, and leisure time seem to guarantee. Far from being appalled by the homogenization of society and the reification of his own consciousness, he applauds the 'extraordinary, growing, and sane standardization of stores, offices, streets, hotels, clothes, and newspapers throughout the United States' (Lewis 178). As critics complain, Lewis's protagonist is a surface phenomenon, a commodified consciousness, a fact the author himself accentuates: 'These standard advertised wares – toothpastes, socks, tires, cameras, instantaneous hot-water heaters – were his symbols and proofs of excellence: at first the signs, then the substitutes, for joy and passion and wisdom' (92). In postmodern terms, he inhabits a mediated or hyperreal world.

In his comfortable suburban life in 'Floral Heights,' Babbitt submits to the tyranny of the group Whyte outlines in his analysis of the planned community of Park Forest as a 'social laboratory' (Whyte 367). Physically planned to facilitate neighbourliness, Park Forest arranges friendships according to 'incidental connections' (366) of locality rather than personal compatibilities and loyalties. Upwardly mobile, the inhabitant of this suburb is depicted as a standardized, interchangeable, and highly adaptable transient: 'He is the one who is quick to move out, but as soon as he does another replaces him, and then another' (311). Created in the image of the organization man, suburbia is the 'ultimate expression of the interchangeability sought by organizations' (330). Consisting of uniform houses, Park Forest testifies to a new tolerance for homogeneity that had earlier seemed quite unacceptable to assembly-line workers at the Ford Motor Company. Trained by the organizations that reward them, the suburbanites are 'adaptable to the constant shifts in environment' (437) beyond their immediate control. The organization man is quite prepared to relinquish control even over his finances; he is as passive as 'Pavlov's dog' and expresses an 'urge for the organized life' that strikes Whyte as an 'entrapment' (358, 363). The emphasis on the 'happy group' experience inculcates a sense of belonging that excludes 'the deviant' and imprisons 'the normal.' The personality under construction in the socially engineered suburb was suitably outgoing, more active than contemplative, tolerant of enforced intimacy, and prepared to sacrifice privacy (388–90). As one suburbanite put it, 'You are never alone, even when you think you are' (389). Having at first 'advertised Park Forest as housing,' they now began simply 'advertising happiness' (314). While the happy group considers itself to be tolerant and egalitarian, in reality it is 'very intolerant of those who aren't tolerant' (395). Acting simultaneously as tyrant and friend, the happy group subjects the 'deviant' to the violence of exclusion and 'breeds neuroses' (441) in those submitting to its insistence on normalcy.

The middle classes living in the suburbs and working in the offices of corporations and public institutions allowed themselves to be 'scientifically managed' and hence normalized without much struggle. The disciplinary strategies in the office and the suburb were not deeply concealed; they operated on the surface of people's everyday lives. What was particularly shocking to Whyte was the discovery that the conformity he describes was 'not *unwitting* conformity' (366). These new men and women were not entirely duped by the system to which they voluntarily submitted. Instead of rebelling against the normalizing tyranny of the group, they willingly accepted 'becoming the interchangeables of our society' (437). In exchange for at least the impression of enjoying material security and 'happiness,' they no longer needed the illusion of earlier generations that individuals are, or ought to be, 'in control of their destinies' (437). They seem comfortable with the notion that they are 'determined as much by the system as by themselves,' that they are 'more acted upon than acting' (437). Securely installed in efficiently run systems that kept them too busy for contemplation, they refused to ask themselves why they should conform or cooperate. Disavowing what they knew, they refused to acknowledge that 'the old authoritarian' only wanted your 'sweat' while the 'new man wants your soul' (440). They offered up their souls in the mistaken belief that their 'interests and those of society can be wholly compatible' (440). Indicting their yearning for 'peace of mind' as 'surrender' to the organization, Whyte objects above all to their sanguine acceptance of the 'overriding importance of equilibrium, integration, and adjustment' (448, 444), that is, of categories we recognize as central to the investment in efficiency in Fordism and Taylorism.

Anticipating Louis Althusser's identification of the church and the school as primary sites of ideological interpellation, Whyte argues that these institutions reinforce the disciplinary process of social adaptation and Lewis similarly targets religion and education for their normalizing practices. According to Whyte, the 'teacher strives not to discipline the child directly but to influence all the children's attitudes so that as a group they recognize schools work so hard to offset tendencies to introversion and other suburban abnormalities' (425). The 'attitudes' to be inculcated by the schools are for Babbitt geared towards efficient and utilitarian outcomes. His son convinces Babbitt that correspondence schools are more efficient educational institutions than universities, which lose 'a whole lot of valuable time' on 'studying poetry and French and subjects that never brought in anybody a cent' (Lewis 82). Although Babbitt has some appreciation for university education as 'cultural capital' (Bourdieu), he rails against 'irresponsible teachers and professors' who refuse to 'help us by selling efficiency and whooping it up for rational prosperity' (180, 181). Reading an advertisement for correspondence courses, he was impressed by a testimonial to their efficiency: 'There were eight simple lessons in plain language anybody could understand, and I studied them just a few hours a night, then started practicing on the wife' (76). In the end, he even speculates that 'these correspondence-courses might prove to be one of the most important American inventions' (82). As the history of scientific management has shown, educational institutions were particularly eager to adopt Taylorized assessment instruments and supervision.

The churches are for Whyte equally productive in fostering conformity through practices and rituals that create the desired ideological responses. With their interminable meetings and social activities, the churches reinforced Forest Park as 'a hotbed of Participation' (Whyte 317) designed to ensure that people were kept busy, strengthened the appropriate social bonds, and refrained from critical thinking. In

Babbitt religious figures tend to be ridiculed for their provincialism, narrow-mindedness, hypocrisy, and salesmanship; however, their parodic depiction also criticizes the infiltration of Taylorism into a domain supposedly far removed from its industrial roots. The distinguished evangelist Reverend Mr Monday is described as 'the world's greatest salesman of salvation,' who assumes that 'by efficient organization the overhead of spiritual regeneration may be kept down to an unprecedented rock-bottom basis' (Lewis 95). But it is the pastor of Chatham Road Presbyterian, the Reverend John Jennison Drew, who is the main target of Lewis's satire. Being 'eloquent, efficient, and versatile,' he advocated a 'practical religion' that was 'beneficial to one's business' (196, 199). With a sidelong glance at Taylor's principles of scientific management, Lewis explains that Drew saw no reason why the Sunday School could not be organized 'scientifically' (202). Babbitt is more than prepared to 'take part in a real virile hustling religion,' a sort of 'Christianity Incorporated' (203), which, at his suggestion, was organized according to military ranks. At the Chatham Road Church, then, 'everything zips,' with the 'subsidiary organizations [being] keyed to the top-notch of efficiency' (212). As Whyte makes clear, the Protestant churches were particularly suited to keep people so thoroughly busy and organized that they found little time for self-contemplation.

Although Sinclair Lewis was the first American to win the Nobel Prize for literature (1950), and *Babbitt* was initially 'one of the greatest international successes in all publishing history' (Schorer 114), praise for the novel as social satire was from the start tempered by the complaint that Lewis had created a 'type' rather than a fully developed character. The novel is marked by a struggle to overcome the conflict between 'two types of literary exposition,' between the 'presentation of a sensitive, humane Babbitt' and the 'parody Babbitt, whom Lewis was unable to humanize' (Hutchisson 49). A related complaint is that Babbitt is a 'static character' who fails to develop in the course of the narrative. He seems to make readers uncomfortable because he succumbs to the environment that entraps him. Even critics otherwise sympathetic to Lewis's satire complain that the novel offers no alternatives and hence no hope or escape from the system.[6] Babbitt's rebellion against Zenith, the novel's dramatic centre, is short lived and leaves him more thoroughly integrated into the system than before. Instead of heroically struggling against his dehumanization, he is accused of easily caving in to social pressures. His moral failure is typically attributed to psychological factors: 'His yearning to break with everything decent and normal by Zenith's standards is frustrated by his personal inadequacy' (Schriber 103). However, blaming personal weakness for the distorting impact of industrialism avoids confronting the systemic conditioning of the 'modern' individual at stake in *Babbitt.* If the protagonist's experiences leave him essentially unchanged, this lack of character development may be a violation of aesthetic expectations, but it reflects a determining socio-economic environment that is intentionally depicted as virtually foreclosing individual growth.

Structurally a comedy, *Babbitt* begins with a character who is at one with his society, finds himself temporarily at odds with it, and ends up solidly reintegrated into it. His temporary mid-life crises takes the self-indulgent form of a nostalgic escape with Paul Riesling into the woods of Maine, a brief desultory affair with an empty-headed adventuress, and a half-hearted attempt to stand his ground against the solid citizens of Zenith. During this period of rebelliousness, Babbitt allowed previously repressed feelings of dissatisfaction with what a critic calls 'the smallness, the pettiness, the triviality, and the lack of joy and freedom in the existence of a typical member of a moneyed society' (Grebstein 38) to come to the surface. However, unable to locate any viable alternative, his defiance of the social norms is either shallowly escapist or undermined by his fear of consequences. In the end, he realizes that he has so internalized everything his social group stands for that 'he could never run away from Zenith and family and office, because in his own brain he bore the office and the family and every street and disquiet and illusion of Zenith' (Lewis 286). With a name conjuring up 'Babbitt metal,' an 'antifriction alloy used for bearings' (Hilfer 88), he and his fellow boosters are metaphorically depicted as the alloy that keeps the wheels of industrial capitalism oiled. They are not the cogs forcefully chained to the assembly line, but the 'scientifically managed' or 'disciplined' bodies tamed to perform their ideological functions 'all by themselves.' It may well be the case that Lewis's sympathy for his character and 'the proud gusto and pleasure behind his caricatures' make the satire 'so funny – and so comfortable' that the novel's critical edge is undermined, leading one reader to suspect Lewis of being 'fundamentally uncritical of American life' (Kazin 100). In spite of such limitations, *Babbitt* remains a fictionalized 'social anthropologist's field report' (Schorer 111) that provides insight into a truly banal 'banality of evil' uncovered not only by Hannah Arendt but also by Weber and Whyte.

Babbitt typifies Whyte's suburban organization man who is prepared to sacrifice privacy and intimacy in order to present to the world the self-confident and optimistic demeanour the group expects in order to reassure itself that it thrives on happy togetherness. He is both victim and perpetrator of the group tyranny that keeps everybody busy in the often aimless pursuit of efficient outcomes for their own sake. Whether he is at home, in the office, at the Athletic Club, at church, or on the golf course, he hustles to be more efficient, popular, and successful than his fellow boosters. He is portrayed as a superficial character who readily adapts his ideas and behaviours to meet the expectations of others he seeks to imitate. If he is amiable, it is because he seeks approval from the group. Engaged in 'the frenzied activities of the innumerable clubs and organizations which the system provides to keep its citizens from too much thinking' (Grebstein 39), Babbitt is justifiably scorned by critics as a despicable conformist who has decided that in 'a noisy confraternity lies the best resort of orthodoxy' (Hoffman 46). He is the extrovert who rarely allows himself to admit to discontent with the efficient mechanical life. The characteristics of group tyranny appear most glaringly when he decides to defy the booster solidarity by entertaining his friend, the 'deviant' Paul Riesling, privately at a table for two at the Athletic Club, the heart of the booster's claim of being 'a world-force for optimism' (Lewis 246). Although 'privacy was very bad form' (576), he risks social opprobrium for the freedom to confess to Paul that the morning's business triumph left him 'kind of down in the mouth' (57). On their train journey to Maine, Babbitt is uncomfortably aware of Paul's 'deviant' behaviour. Sharing a compartment with a 'Clan of Good Fellows' (138), Paul 'failed to join the others' while Babbitt happily participates in the group's raucous boosterism. Withdrawing from the group to read 'a serial story in a newspaper,' Paul is contrasted with Babbitt, who is unable to 'read with absorption' or to 'essay the test of quietness' (134, 89, 224). Instead of assessing a passing steel mill for its business potential, Paul further embarrasses his friend by commenting on its aesthetic appeal. As the other men 'stared at him' (138), Babbitt hastened to reassure the boosters that Paul was in fact a solid businessman. In his unostentatious affirmation of his right to privacy, contemplation, and unhappiness, Pauls stands for values Babbitt is willing to sacrifice in exchange for the social security he locates in group solidarity. The contrast between Babbitt and Paul dramatizes Whyte's point that group tyranny breeds neurosis in the conformist and subjects the deviant to exclusion.

Since for Babbitt the Athletic Club is 'Zenith in perfection' (Lewis 51), it is not surprising that its ideological hold is not easily broken.

Thoroughly normalized, he dreams of escape routes that, when offered, he cannot pursue. Even the mundane decision to accompany Paul to Maine fills him with 'primitive terror' because he has no idea 'what he could do with anything so unknown and so embarrassing as freedom' (126). If Babbitt is indeed 'a creature of fear,' as one critic puts it, he has been rendered so by an 'economic structure' that threatens 'the dissenter with exile and hunger' (Lewisohn 20). The supposedly joyous group is, in reality, a desperate community of victims seeking refuge from unpredictability and contingency. The perfect society with its efficient organization has to be protected at all cost from any possible disruptions.

Disruptions are most effectively averted or rectified by means of disciplinary mechanisms of which surveillance is the most visible form. Throughout his rebellion, Babbitt is terrified of the threat his actions pose to his security. He is therefore highly susceptible to strategies ranging from persuasion to threats that the good citizens of Zenith employ to bring the 'deviant' back into the fold. Representative of the group as tyrant, Vergil Gunch tries to persuade Babbitt that his defence of the strikers supported by Seneca Doane is ill conceived; playing on Babbitt's fear, he reminds him that they are engaged in an epic 'struggle between decency and the security of our homes on the one hand, and red ruin and those lazy dogs plotting for free beer on the other' (Lewis 328). A far more successful tactic is Gunch's exploitation of the middle-class ideology Babbitt has for a long time now internalized. All he has to do is train his normalizing gaze on the misbehaving Babbitt, who only allows himself to admit to himself that his way of life is 'incredibly mechanical' whenever there is 'no Vergil Gunch before whom to set his face in resolute optimism' (224). Already 'vaguely frightened' at his temerity to disagree with Gunch during his mid-life crisis, Babbitt is distinctly unnerved when 'Gunch said nothing, and watched; and Babbitt knew that he was being watched' (300). When he later 'saw that from the sidewalk Vergil Gunch was watching him' (301), the already terrified rebel further crumbled under the other's disciplinary gaze. His affair with Tanis Judique essentially disintegrates under the watchful eyes of Gunch, who occupies a neighbouring table at a restaurant: 'All through the meal Gunch watched them, while Babbitt watched himself being watched' (326). The good citizens have no need of coercion; they use amiability to enforce conformity.

The disciplinary power of Gunch's surveillance is reinforced by not so veiled threats to Babbitt's financial welfare and secure position in the community. His father-in-law pointedly reminds him, 'You can't expect the decent citizens to go on aiding you if you intend to side with precisely the people who are trying to undermine us' (Lewis 352). While such threats frighten Babbitt, he continues to defy the community by refusing to be bullied into submission. For some time, he stubbornly closes his ears to his father-in-law, who accuses him of 'wrecking the firm' (354) on which his family's security depends. However, as he finds himself increasingly snubbed by the good citizens of Zenith, he begins to crumble under the strain: 'Before long he admitted that he would like to flee back to the security of conformity, provided there was a decent and creditable way to return' (357). His wife's appendectomy presents him with a 'creditable way' to reintegrate himself into the group. Once the boosters are certain that Babbitt has been tamed, they invite him to join the Good Citizens' League, which he had earlier defied: 'Then did Babbitt, almost tearful with joy at being coaxed instead of bullied, at being permitted to stop fighting, at being able to desert without injuring his opinion of himself, cease utterly to be a domestic revolutionist' (368). To his credit, Babbitt realizes with some dismay that 'he had been trapped into the very net from which he had with such fury escaped and, supremest jest of all, been made to rejoice in the trapping' (375). The criticism that Babbitt does not develop as a character, that he is the same at the end as he was at the beginning, is precisely the novel's point. Taylorization has infiltrated the social fabric to such an extent that there is no escape from its normalizing gaze. If there is a perceptible tendency among critics to denigrate Lewis's depiction of Babbitt as the comic-satirical exaggeration of a stereotypical American inhabiting a limited historical moment in time, it may well have arisen from our refusal to acknowledge that we have not so much outlived as incorporated Babbitt.

The commitment to efficiency in *Babbitt* illustrates above all how successfully not only the body but also the soul of those trapped in the normalizing social conditions has been tamed. Whyte's analysis of these normalizing regimes anticipates Michel Foucault's more recent contention that the innermost self or 'soul' in earlier discourses does not pre-exist social inscriptions but is produced by them. In *Discipline and Punish,* Foucault makes this point in terms that seem particularly pertinent to the material situation satirized in *Babbitt:*

> It would be wrong to say that the soul is an illusion, or an ideological effect. On the contrary, it exists; it has a reality; it is produced permanently

> around, on, within the body by the functioning of a power that is exercised on those punished – and, in a more general way, on those one supervises, trains, and corrects; over madmen, children at home and at school, the colonized; over those who are stuck at a machine and supervised for the rest of their lives. This is the historical reality of this soul, which, unlike the soul represented by Christian theology, is not born in sin and subject to punishment, but is born rather out of methods of punishment, supervision, and constraint. (Foucault, *Discipline* 176–7)

Lewis painstakingly documents how the modern soul emerged out of the socio-economic and technological conditions that are partly symbolized by Ford's assembly line and by Taylor's scientific management. Not long after Ford Motor seemed to reduce men to mere 'cogs in the machine,' Lewis understood that the consolidation of industrial capitalism induced men and women to internalize the normalizing practices poised to rob them of the illusion that they were unique individuals. Submitting to group tyranny, Babbitt is the standardized American who is paradoxically not the stable individual of modernity but the decentred and mediated self of postmodernity. Far from being an individual oppressed by a hierarchical power structure, he voluntarily accepts his role as an interchangeable fragment in a totalized structure, a flexible 'part' willing to adapt to the constantly shifting configurations of an increasingly globalized capitalism.

Most tragically, though, Babbitt's pact with the devil condemns him to be a tamed social cog without for all that providing him with the secure sense of self he craves. In the end, he cannot articulate who he is or what 'he wanted to do' (Lewis 233), because he has disintegrated into multiple and incompatible selves: 'He was, just then, neither the sulky child of the sleeping porch, the domestic tyrant of the breakfast table, the crafty money-changer of the Lyte-Purdy conference, nor the blaring Good Fellow, the Josher and Regular Guy, of the Athletic Club. He was an older brother to Paul Riesling, swift to defend him, admiring him with a proud and credulous love passing the love of women' (55). He exemplifies the decentred subject of postmodern theory at stake in debates on whether the socially constructed self should be celebrated as an emancipatory escape from the Descartian *cogito* or deplored as the loss of a psychologically stable and socially anchored sense of self. In Derrida's terms, Babbitt is not portrayed as a Nietzschean affirmer of 'freeplay' but as the mournful Rousseauist nostalgically yearning for a lost authenticity. Babbitt's swift return to the haven of boosterism demonstrates the difficulty experienced by the 'modern' to find lines of flight from Weber's 'iron cage.' For both Weber and Lewis, the social subject is both too fragmented and too integrated into a totality to find viable alternatives. In spite of such cultural despair, Weber proposes art, eroticism, and charisma as potential sites of resistance to the 'disenchanted' modern world.

Lines of Escape from the Iron Cage

In many ways, Weber seems to express the spirit of his age when he proffers aestheticism and eroticism as possible escape routes from the deadening effects of the efficiently rationalized social order. In the literary canon, at least, love and personal relations as well as art and culture recur as mostly nostalgic retreats from modernity. If rationalization was widely considered to be an 'evil,' then the irrational presented itself as the most logical opposite. Weber's attitude to eroticism and aestheticism was suitably attentive to complexities and hence highly ambivalent. Alert to the dangers inherent in the irrational, he was nevertheless embracing it not so much for its own sake but for its ability to break through the superficialities of the normalized life into the deeper recesses of supposedly more authentic self-knowledge. Sex and eroticism were for him 'non-routinized' (Weber, 'Religious' 344) sources of genuine sensations and experiences. As Lawrence A. Scaff clarifies in *Fleeing the Iron Cage* (1989), Weber's search for lines of escape from the culture of efficiency highlights problems that as we will see, also bedevil the alternatives touted by novelists resistant to modernity. While expressing reservations about eroticism as mere hedonistic pleasure-seeking, Weber 'reconceived' it 'as the ultimate source of "life" – that is, of life's most *irrational* force' (Scaff 109). As the 'sublimated expression' of 'sexual love,' eroticism owes its appeal to its capacity to 'press beyond the everyday to the limits of the extraordinary' (108). The more thoroughly rationalization suffused the social fabric, the more limited and extreme were the forms resistance would have to take. It is in comparison with 'the mechanisms of rationalization' that 'eroticism appeared to be like a gate into the most irrational and thereby real kernel of life' (Weber, 'Religious' 345). The aim of this release from an increasingly rationalized civilization is not self-indulgent hedonism but a heightened self-awareness of the life-denying tendencies of the iron cage.

Along the same lines, the aesthetic is for Weber a prime site of resistance to the efficiency calculus. Discussing art as a substitute for religion, Weber writes: 'Art becomes a cosmos of more and more consciously grasped independent values which exist in their own right. Art takes over the function of an inner-worldly salvation, no matter how this may be interpreted. It provides a *salvation* from the routines of everyday life, and especially from the increasing pressures of theoretical and practical rationalism' ('Religious' 342). Arguing for art as an autonomous sphere, Weber seems to side with movements of 'art for art's sake' that flourished around the turn of the century. Art is not privileged for any socially useful function or ethical imperative but for its radical opposition to instrumental rationalism: 'The aesthetic comes to be endowed with such significance *because* it is grasped by the mind in dialectical opposition to the grinding forces of routinization and rationalization' (Scaff 104). Indeed, as the 'most "inward" form of human expression,' music, for instance, 'can conceivably set itself against moral and scientific culture even as a "diabolical" power, as the finest and highest expression of amoral and irrational interiority' (104). It is for its ability to break with the banality of routinized existence that art owes its centrality in Weber's thought. The aesthetic is for him aligned with Nietzsche's Dionysian axis of artistic creation; works of art are measured according to their ability to capture an authentic life force. As with all of Weber's lines of flight, the alternative risks falling back into the logic it seeks to subvert. Wagner's opera dramas, for instance, came close to capturing the desired life force, except that they were at times either too self-indulgently subjectivist or too artificial. Moreover, fearing the rationalization of the aesthetic sphere, Weber was suspicious of the privileging of form over content in movements of art for art's sake. In their self-consciousness, experiments in form risked reinforcing the intellectualizing tendencies poised to support rather than subvert the instrumentally rational social order.

Although both aestheticism and eroticism are capable of changing the way subjects think about themselves and their place in the social world, they offer predominantly personal retreats from rationalization, leaving the economic and socio-political sphere largely untouched. Given his preoccupation with the rationalization of the administrative state apparatus, it is not surprising that Weber concerned himself with 'the effects of bureaucratic usurpation of political decisions' (Scaff 156), that is, with the threat the scientifically managed personality, forged by industrial capitalism, posed to the 'free' individual and to the autonomy of the political sphere. With the bureaucratization of society, political decisions were increasingly being made by impersonal functionaries protecting their specialized sphere of influence. Political vision was being replaced by economic self-interest. The modern state thrives on a rule of law that 'has come about through the "expropriation" of the means of finance, warfare, and administration' (167) that used to be the domain of politics. Whether democratic or socialist in orientation, the modern state is now for Weber 'a compulsory association which organizes domination' ('Politics' 82) rather than a genuine community protecting individual freedom. The rational *efficiency* of the modern state is thus seen by Weber as incommensurate not only with patriarchal domination but also with the principle of *equality* at the core of both democracy and socialism. Under the impact of rational bureaucratization, 'economic power and the call for political leadership of the nation did not always coincide' (Gerth and Mills 35). In his search for a principle capable of challenging both tradition (patriarchy) and the 'iron cage' (bureaucracy), Weber resorted to the figure of the 'charismatic leader' to restore politics to its rightful decision-making place.

The problematical concept of 'charisma' appeals to Weber by virtue of an affiliation with the irrational, also found in his privileging of aestheticism and eroticism. Unlike the functionary who acts without 'regard to person,' the charismatic leader owes his power precisely to his personal authority: the 'charismatic figure' is a 'natural leader' endowed with 'specific gifts of the body and spirit' (Weber, 'Meaning' 245). Breaking with traditional norms and economic interests, he assumes power not through election from below or appointment from above but through the strength of a personality capable of inspiring loyalty in a group of devoted followers. Weber endorses charisma on the same basis as he approves of eroticism and aestheticism: manifesting itself in both 'divine' and 'diabolical' forms, charisma embodies the irrational as a site of resistance to the instrumental rationality of the iron cage. Using 'charisma' in a 'completely "value-neutral" sense.' Weber includes in his list of historical examples figures ranging from the heroic 'Homeric Achilles' and 'the Irish culture hero, Cuchulain' (245) to evil conquistadores and robber barons as well as suspect mystics like Joseph Smith, the founder of Mormonism. In other words, for a leader to qualify as 'charismatic,' he need not embrace 'edifying' (246) ideas or practices. Power based on 'personal charisma' depends not on democratic debate but on an ability to convince 'those to whom he addresses his mission to recognize him as their charismatically qualified leader' (246, 247). In contrast to the depersonalized bureaucratic functionary, the charismatic leader owes his position to 'his *personal* qualification and to his *proved* worth' (247). The most pertinent requirement of 'charismatic domination' is that it should be 'the very opposite of bureaucratic domination' (247). It is consequently imperative that charisma 'always rejects as undignified any pecuniary gain that is methodical and rational. In general, charisma rejects all rational economic conduct' (247). Charismatic leaders may 'seek booty and, above all, gold,' but cannot be party to processes of 'economic exploitation by the making of a deal' (247). Disdaining 'all ordered economy' (248) and all social-contract negotiations, the charismatic figure kicks away the crutches of rational calculation and efficient routines and puts his trust in his own strength and convictions.

Just as eroticism has to be distinguished from hedonism and aestheticism from empty formalism, so genuine charisma is not the self-indulgent hedonism of subjectivist culture but the acceptance of a responsibility not unlike the Puritan 'calling' of self-denial, ascetic labour, isolation, self-control, and service to an ideal. Self-appointed, the charismatic leader rejects external authorities and norms and 'knows only inner determination and inner restraint' ('Meaning' 246). Standing 'outside of routine occupations, as well as outside the routine obligations of family life,' such a leader 'makes a sovereign break with all traditional or rational norms' (248, 250). While autonomous and separated from 'this world,' he is nevertheless 'responsible precisely to those whom he rules' (248, 249) and to the mission he embodies. Although his decisions are arbitrary and absolute, they are constrained by the knowledge that he has to satisfy the expectations of those he rules. If he fails those who look to him for inspiration and leadership, he 'faces dispossession and death, which often enough is consummated as a propitiatory sacrifice' (249). In this emphasis on freedom from tradition, on the genius of the unique individual, and on an ethic of responsibility, charisma seems ultimately marked by a nostalgic yearning for the values of old forms of liberalism. This point is convincingly driven home in the similarity of 'dichotomies' that Gerth and Mills uncover between charismatic leadership and the 'heritage of liberalism' in their mutual opposition to 'the everyday life of institutions': 'Mass *versus* personality, the "routine" *versus* the "creative" entrepreneur, the conventions of ordinary people *versus* the inner freedom of the pioneering and exceptional man, institutional rules *versus* the spontaneous individual, the drudgery and boredom of ordinary existence *versus* the imaginative flight of the genius' (Gerth and Mills 53). Those entranced by a charismatic figure may be responding less to strictly *personal* charisma than to the ultimately anarchic premises of the older liberal tradition with its hostility to regulatory constraints.

Weber implicitly acknowledges the anachronistic and nostalgic orientation of charisma by according it only limited and ephemeral political influence. By the early twentieth century, the 'iron cage' had indeed become so totalized that eruptions of charismatic leadership diminished in frequency and intensity. In light of Weber's own demonstration of the ubiquity and inevitability of the rationalist way of life, it is difficult to conceive of an oppositional strategy or tactic that would not always already be contaminated by rational calculation or co-opted by it. Unlike Taylor, Weber did not believe in social engineering as a means of effecting change. As he had already discovered in *The Protestant Ethic and the Spirit of Capitalism,* systems prove susceptible to the logic of unintended consequences. The emphasis on asceticism in Puritanism unintentionally provided the personality best suited to the economic rationalism of capitalism that was most clearly antithetical to religious spiritualism. According to this logic, history is punctuated by discontinuities, for no matter how totalized the 'iron cage' might appear, it was always necessarily open to disturbances and reversals. It is in such gaps that charismatic figures could arise to temporarily disrupt and reorient the totalizing tendencies of the 'iron cage.' Their historical interventions are always revolutionizing but also always 'unstable' (Weber, 'Meaning' 248), temporary and predisposed to 'routinization' (250). Just as charismatic leadership emerged serendipitously, it risked falling victim to its own unintended consequences. Paradoxically, then, it is precisely by being successful that charismatic leadership undoes itself. Ultimately, it is the 'fate of charisma, whenever it comes into the permanent institutions of a community, to give way to powers of tradition or of rational socialization' (Weber, 'Sociology' 253). As charisma wanes, the 'iron cage' reaffirms its efficient regime of rationalization, bureaucratization, and scientific management all the more rigorously.

In his thorough analysis of the efficiently bureaucratized modern society, Weber proved a perceptive diagnostician; however, he failed in his endeavour to prescribe an effective cure. In the first place, charismatic leadership proved too unstable to withstand the juggernaut of rationalization for long. In effect, in a dialectical progression, temporary eruptions of charisma were more likely to strengthen rather than weaken the 'iron cage.' In the second place, the distinguishing features of charismatic leadership proved susceptible to incorporation into the efficiency calculus. As Weber realizes, 'everything and especially these "imponderable" and irrational emotional factors, are rationally calculated – in principle, at least, in the same manner as one calculates the yields of coal and iron deposits' ('Sociology' 254). In other words, depending for its survival on 'a rationally intended "success"' (254), charisma is in the final analysis but another instrument in the arsenal of rational planning. Most disturbingly, perhaps, charisma is entirely compatible with its opposite, the rational disciplines: 'discipline as such is certainly not hostile to charisma or to status group honor' (253). This complicity emerged most nakedly in the case of Hitler's simultaneous reliance on personal charisma and the efficient rationalization of the administrative apparatus. When critics ask themselves if Weber 'might have turned Nazi,' they tend to point to the sociologist's 'Machiavellian attitude' to politics and to the celebration of the irrational in his 'philosophy of charisma' (Gerth and Mills 43). For Gerth and Mills, there is no question that Weber's 'humanism, his love for the underdogs, his hatred of sham and lies, and his unceasing campaign against racism and anti-Semitic demagoguery' would have made him a 'sharp' critic of Hitler (43). But Weber might have been an even sharper critic of the disciplinary regimes informing Hitler's conception of the efficiently organized Third Reich. At the same time, he might not have been surprised by Hitler's ability to exploit, at one and the same time, the strength of his personal gifts in the promotion of a broadly appealing nationalistic mission and the pre-existing impersonal bureaucracy operating on the basis of rational discipline. At the same time as Hitler's rhetoric decried the 'iron cage' of modernity, his actual practice conformed to Weber's insight that 'the blind obedience of subjects can be secured only by training them exclusively for submission under the disciplinary code' ('Sociology' 253). In his conflation of irrational charisma and rational efficiency, Hitler demonstrated a complicity that makes it difficult to set one up in opposition to the other. Weber's fear that the 'iron cage' was indeed inescapable would no doubt have been significantly heightened had he lived to observe Hitler's rise to power.

PART TWO

The Culture of Efficiency in Fiction

5 Efficiency and Population Control: Wells, Shaw, Orwell, Forster

In a letter dated 18 May 1931, Aldous Huxley made it clear that *Brave New World* was, at least in part, intended as an attack on H.G. Wells: 'I am writing a novel about the future – on the horror of the Wellsian Utopia and a revolt against it' (Huxley, *Letters* 348).[1] As we will see, the opening scene of Huxley's novel satirizes eugenics as an absurd extension of the logic informing the obsession with efficiency symbolized by the assembly line at the Ford Motor Company. In Huxley's imagination, the assembly line gives rise to eugenic engineering experiments the aim of which was the totalitarian control of society. This train of thought was not as fanciful as it may strike us today. For all too long, it has been customary to dismiss eugenics as the obviously evil and misdirected aberration of Nazi ideologues. This comforting narrative has recently been punctured by Donald J. Childs's *Modernism and Eugenics: Woolf, Eliot, Yeats, and the Culture of Degeneration* (2001), which illustrates that an interest in eugenics had deep cultural roots and was widespread in England around the turn of the nineteenth and twentieth centuries. But Childs's primary concern with the discourse of degeneration touches only incidentally on the link between efficiency and eugenics at issue in *Brave New World.* In this chapter this link will be examined in attitudes voiced by literary figures who explicitly confronted the eugenic implications of social engineering projects that Huxley associates with the assembly line at the Ford Motor Company.

The two most prominent voices promoting eugenics were those of George Bernard Shaw and H.G. Wells. Their dates – Shaw lived from 1856 to 1950 and Wells from 1866 to 1946 – make them contemporaries who experienced both the high hopes the Great Exhibition of 1851 placed in technology and the social dislocations arising with mass industrialization. Closely associated with movements celebrating efficiency, these two prominent literary figures debated the social benefits of machines and speculated on the eugenic possibility of selectively breeding for a superior human being as well as on the desirability of eliminating those deemed unfit for propagating the species. Their preoccupation with efficiency manifests itself in the thinking of literary figures born in the late nineteenth century. Aldous Huxley (1894–1963) and George Orwell (1903–50) are famous for anti-utopian novels acknowledged to be a direct response to the utopian longings typical of Wells and, to a lesser extent, of Shaw. Predating Huxley's *Brave New World* (1932) and Orwell's *Nineteen Eighty-Four* (1949)[2] by quite some years, E.M. Forster expressed his unambiguous opposition to social efficiency in the short story 'The Machine Stops' (1909). But in his novels, Forster (1879–1970) joins other modernists of his generation – Ford Madox Ford (1873–1939), Virginia Woolf (1882–1941), D.H. Lawrence (1885–1930), Joseph Conrad (1857–1924)[3] – who no longer considered the theme of efficiency to be sufficiently topical to deserve the attention it had received a few years earlier. For these modernists, efficiency either imposes itself in specific scenes or suffuses the social fabric in which characters dramatize themes seemingly quite unrelated to the efficiency calculus. This chapter will focus on both explicit proponents of efficiency (Shaw and Wells) and equally explicit detractors of it (Orwell and Forster in 'The Machine Stops') in order to foreground the social tensions that emerged with the consolidation of the machine age around the turn of the century. The other explicit confrontation with efficiency, Huxley's *Brave New World,* is far too complex to fit into this preliminary discussion of efficiency among literary intellectuals, and will be analysed in a separate chapter.

Although Britain put up a more concerted resistance to Taylorism than any other European country, the efficient management of public institutions and the social engineering of society to perfect its political organizations played a prominent role in intellectual debates around the turn of the nineteenth and twentieth centuries. Prominent figures embraced efficiency as a panacea for Britain's social and political woes. Belonging to overlapping social groups, George Bernard Shaw, H.G. Wells, and Bertrand Russell joined both the Fabian Society, founded in 1881 by Frank Podmore and Edward Pease, and the 'Co-Efficients,' a dining club founded in 1902 by Fabians Sidney and Beatrice Webb.[4] Hostile to capitalism, they embraced a socialism that rejected the Marxist-inspired revolution of the working classes in favour of a more gradual transformation of society through education and, more problematically, through eugenics. Promoting heredity and an elite education at Oxford, Cambridge, or Edinburgh, their socialism proved entirely compatible with the preservation of Britain's conservative caste system. Enmeshed in spiritualism,[5] socialism, and a eugenically engineered utopianism, the early Fabians were intellectual elitists with rather poorly defined political ambitions. However, participating in meetings of the Co-Efficients, they showed themselves to be more pragmatically 'anxious to expound (over their eight hotel meals a year) how "each department of national life can be raised to its highest possible efficiency"' (Holroyd 133). Through membership in the Co-Efficients, the Fabians were thus in close contact with the National Efficiency Movement which was primarily concerned with the elimination of waste in public institutions.

In literary circles, the two most prominent advocates for efficiency were Shaw and Wells. Writing before Ford's assembly line accentuated the homogenizing impact of standardization, these two literary figures focused their attention more on social than technical efficiency. It is not that they entirely ignored scientific advances and technological innovations: Wells is said to have possessed an 'irrepressible delight in technical efficiency and ingenuity' (Haynes 119), and his novels often highlight the achievements of machines. On a more whimsical note, scandalized by the irrational English spelling system, which he insisted involved an enormous waste of 'time, labour, and money' in writing, printing, and education, Shaw bequeathed a small fortune to help create a new alphabet along the economical lines he proposed. Although their alphabet was smaller, Russians, he complained, could spell the name 'Shaw' with only two letters, whereas 'I have to spell it with four letters,' a '100 per cent loss of time, labour, ink, and paper' (Shaw, 'Language' 71, 65). Shakespeare, he conjectured, 'might have written two or three more plays in the time it took him to spell his name with eleven letters instead of seven' (83). Shaw campaigned throughout his life for efficiency and against inefficiency. But the main impetus behind the promotion of efficiency in the writings of both Shaw and Wells was the efficient management of human beings.

In spite of their socialist leanings, Shaw and Wells assumed that society should be engineered so as to increase its overall efficiency; their conception of efficiency reflects Ford's total-output model rather than Taylor's equilibrium alternative. Their insistence on optimizing efficient *outcomes* is clearly at odds with the optimal *distribution* of efforts and benefits central to socialist thinking. Applying the logic of the machine metaphor for *technical* efficiency to the logic of social-engineering projects focused on *social* efficiency, they were driven to embrace politically highly suspect solutions to the ills they diagnosed. Although both Shaw and Wells exhibited a typically British hostility to (American) capitalism, they had little to say about the affront that the standardized society posed to the culturally sensitive intellectual. They had no time for Huxley's later complaints about the deplorable impact of the assembly line on human consciousness. For them, the efficient organization of social and political life constitutes an improvement over the inefficiencies troubling a Britain that was for Shaw a ship drifting aimlessly without captain or pilot at the helm and was for Wells plagued by 'semi-accurate, muddled thinking' (Haynes 112). In the case of Wells, the desire 'for order amounting almost to an obsession' has been explained as a reaction to his own upbringing: 'His family background of a disordered and incompetently run home, an unprofitable, ill-managed shop and an apparently ineffectual mother, came to symbolise for him all that disorder and concomitant waste which undermined the social structure at every level' (112). Born in 'a shabby little house on Dublin's south side,' Shaw came from a similarly 'eccentric and disorganized family,' with a father who was 'pulling his family into ruin by his amiable fecklessness, his incompetent management of his grain business, and his habit of consoling himself with the bottle' (MacKenzie and MacKenzie 31). Their enthusiasm for efficiency seems to have grown out of a discontent that is not unlike Taylor's impatience with the wasteful 'rule-of-thumb' approach at Bethlehem Steel. But, unlike Taylor, Shaw and Wells targeted parliamentary democracy as a particularly irritating source of inefficiency. Their commitment to efficiency thus fuels anti-liberal attitudes that also permeated Germany during the early decades of the twentieth century, contributing to the failure of the Weimar Republic that opened the door to Hitler's Nazis.

George Bernard Shaw: Selective Breeding

The theme of efficiency dominated Shaw's thinking throughout his long life and career. He wanted an efficient road and railway system, an efficient police force, an efficient postal service, and an efficient military. Advocating what amounts to a Taylorized administrative apparatus, he believed that public officials should derive their authority from 'their ability and efficiency' (Shaw, 'Preface to *Too True*' 23). But, as Taylor knew, this apparatus requires an 'expert' to ensure its smooth functioning. For Shaw, this expert was the efficient political leader; he approved of men like Nelson and Wellington but also of Stalin and Mussolini, whom he described as 'the most responsible statesmen in Europe.' Since they 'had no hold on their places except their efficiency' (23), he maintained that their power was not absolute but curbed by accountability. 'If,' wrote Shaw, 'a ruler can command men only as long as he is efficient and successful his rule is neither a tyranny nor a calamity: it is a very valuable asset,' far better than the rule of 'people too ignorant to understand efficient government, and taught, as far as they are taught at all, to measure greatness by pageantry and the wholesale slaughter called military glory' ('Preface to *The Millionairess*' 126). While one might be sympathetic to his dislike of 'pageantry' and' military glory,' his elitist disdain for the 'people' rings a strange note for a writer calling himself a socialist. More disturbing, though, is his praise for Mussolini who, preferring 'discipline' to freedom and democracy, had done away with 'elected municipalities, replacing them with efficient commissioners of his own choice, who had to do their job or get out' (117). The society Shaw envisages thus resembles the Ford Motor Company under the strong leadership of Henry Ford, who allowed Harry Bennett to decide who had done a good job and who was to 'get out.'

Shaw's approval of the dictators of the 1930s is perhaps motivated less by his love of efficient tyranny than his opposition to inefficient parliamentary democracy. We need to remember that in all his plays and prefaces, he opposed the industrious and the thrifty to the idle and the wasteful; the prevalence of waste – waste of human life, of time, of effort, of talent, of money – is a fact that roused him to fury. If he argues that 'able despots,' including Stalin, 'have made good by doing things better and much more promptly than parliaments' ('Preface to *Geneva*' 22), it is because he considers 'our pseudo-democratic parliamentary system' with its 'delays,' 'evasions,' 'windy impotence,' and 'anarchic negations' to be highly inefficient (in Gibbs, 398). The alternative would be what Shaw called 'substantial democracy,' that is, the 'impartial government for the good of the governed by qualified rulers' ('Preface to *Geneva*' 22). He quite correctly attributed the success of 'Fascist movements, led by such figures as Mussolini and Hitler' to the 'failure of parliamentary democracy' (in Gibbs, 397). Expressing his views on Hitler in an interview published in 1931, Shaw argued that 'The Third Reich (the Hitlerites' name for their proposed State) owes its existence and its vogue solely to the futility of liberal parliamentarism on the

English model' (in Gibbs, 397). Since their inefficiency caused democracies to be 'swept into the dustbin of Steel Helmets, Fascists, Dictators, military councils, and anything else that represents a disgusted reaction against our obsolescence and uselessness' (in Gibbs, 397), the remedy was to 'repair our political institutions' (397) along more efficient lines.

Obsessed with efficiency, Shaw disavows the totalitarian political control implicit in the desire for total outcomes. Although a celebrated pacifist, Shaw was willing to tolerate violence to promote an efficient political system. Acknowledging certain 'revolting incidents of the Fascist terror' (in Gibbs, 398), he continued to approve of Mussolini and Hitler by convincing himself that they acted in the 'interest of effecting necessary social and political change' (398). While he supported Hitler's efficient methods, he disapproved of his anti-Semitism, maintaining that 'Judophobia' was a form of insanity. Yet his investment in efficiency undoubtedly contributed to blinding him to evidence emerging after the war of the systematic extermination of millions of Jews and so-called 'undesirables.' In the words of his protective housekeeper, he 'could not bring himself to believe in the German concentration camps like Dachau and Belsen' (in Chappelow, 30). Confronted with images of 'heaps of corpses,' he thought that these 'were more the consequences of incompetent guards and food shortages than the inevitable product of Hitler's policies, most of which he had approved' (Nathan 219). Unable to face the fact that efficient methods made large-scale genocide possible, Shaw preferred to think that only a few people had died at the hands of a small number of 'callous toughs' (in Nathan, 235) among the concentration-camp guards.

If Shaw could not admit to himself the extent of Nazi atrocities, it was in part because he himself had endorsed eugenics as an efficient means of perfecting society. In *Bernard Shaw's Remarkable Religion* (2002), Stuart E. Baker calls 'simply stupid' those people who 'believe that because eugenics was evoked to justify Nazi genocide, eugenics must necessarily be evil' (169). Explaining that Shaw's main aim was to combat 'artificial inequality produced by property and rent,' Baker defends the dramatist's ambition to improve the human race through 'better breeding' (169). Since Shaw often expressed extreme views in a tongue-in-cheek fashion, it is impossible to establish for certain just how seriously one ought to take his language of race regeneration. In a famous lecture he delivered on 3 March 1910 to the Eugenics Society, Shaw reportedly 'spoke of revising the normative view of the sacredness of human life,' advocated 'abolishing marriage,' and, most shockingly, favoured implementing 'eugenic measures' (Stone 127). The *Daily Express* attributed to him the view that 'a part of eugenic politics would finally land us in an extensive use of the lethal chamber. A great many people would have to be put out of existence simply because it wastes other people's time to look after them' (127). *The Daily Express* apologized to 'its readers for printing such material,' while other newspapers, 'equally repulsed by Shaw's talk,' tended to report it 'drily, without comment' (127). In a defensive posture that was to become rather typical in the critical literature on Shaw, the *Birmingham Daily Mail* declared, 'This is all very shocking, but it is also Shavian, and as some centuries must elapse before Society has fitted itself for such a wildly "ideal" doctrine as this, no one need trouble himself seriously about it' (128). According to Stone, the press either recognized the talk as 'a skit on the dreams of the eugenicists' (128) or expressed dismay at its more outrageous proposals. In retrospect, what is troublesome is that the lethal chamber and the stud farm became so central to Nazi eugenics policy: 'the "negative" policy of genocide – the Holocaust – and the "positive" policy of the *Lebensborn,* Himmler's nascent project to promote "sound breeding" among SS members' (129).[6] Focusing primarily on the 'lethal chamber' rhetoric, Stone means to point out that 'in England, decades before the Nazis began gassing Jews to death by the million, the fantasy of the lethal chamber was already being mooted' (129). It could then be argued that Shaw, no matter how seriously or perversely he advocated outrageous positions, contributed to an intellectual atmosphere conducive to highly suspect social-engineering projects.

An analysis of Shaw's political and dramatic writings reveals a consistent, if also constantly qualified support, for both the 'positive measures' of eugenics, 'such as the encouragement of "hygienic marriage,"' and 'negative' measures, 'such as the "sterilization" of the unfit' (Stone 115). Claiming not to have 'the smallest rational objection to a human stud farm,' Shaw, nevertheless, expressed reservations about our ability to 'know what to breed' (Shaw 1971, 186; in Baker, 170). He consequently objected to the Eugenic Society's argument that selective breeding should start with the elimination of diseases like 'tuberculosis, epilepsy, dipsomania, and lunacy' (Shaw 1971, 186; in Baker, 170). Since Shaw's conception of the 'Superman' was influenced by Nietzsche's elitist call for an intellectual aristocrat rather than a physically or biologically superior specimen, he is 'quite likely to be a controlled epileptic ... and he will certainly be as mad as a hatter from our point of view' (Shaw 1971, 186; in Baker, 170). The play in which Shaw explores ideas of selective breeding is, of course, *Man and Superman* (1903), an unorthodox work that consists of an 'Epistle Dedicatory,' a four-act play, and the political pamphlet called 'The Revolutionist's Handbook.'

In this play the main character, John Tanner (who metamorphoses into Don Juan), contends that neither nature nor cultural influences can be left to their own devices if we want to ensure the regeneration of the human stock. While selective breeding would be an ideal solution, he is unclear on what would constitute both a fail-safe method and an optimal outcome: 'The proof of the Superman will be in the living; and we shall find out how to produce him by the old method of trial and error, and not by waiting for a completely convincing prescription of his ingredients' (*Man* 218). By 'trial and error,' he does not mean a biological experiment but a social engineering project. In 'Simple Truth,' Shaw proposes that we 'must trust to nature: that is, to the fancies of our males and females' (Shaw 1971, 186; in Baker, 170). *Man and Superman* is an extended exploration of the ways in which society could exploit these 'fancies' to serve the state. In other words, Tanner advocates regulating human sexuality through social reforms. Social engineering was to direct and control 'nature,' or what Shaw, under the influence of Henri Bergson and Samuel Butler, called 'Creative Evolution,' 'the Life Force,' or 'the evolutionary appetite.' As Don Juan laments, the life force 'wastes and scatters itself' by raising up 'obstacles to itself' and thus tending to destroy 'itself in its ignorance and blindness' (*Man* 144). The life force is powerfully 'irresistible' but also frustratingly random; it needs 'a brain' because, in its 'ignorance,' it will otherwise risk resisting itself (144). The 'brain' would presumably be supplied by a Platonic philosopher-king who would combine Shaw's scintillating intellect and Mussolini's charismatic political leadership.

In *Man and Superman* Shaw makes what was in his time a highly radical proposal: sexual reproduction should be uncoupled not only from marriage but also from the bonds of love. Seemingly convinced that sexual attraction is too fickle a base to guarantee an optimal reproductive outcome, he contends in 'The Revolutionist's Handbook' that society has to intervene in the bedroom: 'And so, if the Superman is to come, he must be born of Woman by Man's intentional and well-considered contrivance' (*Man* 221). If our aim is 'to replenish the earth' (158), then marriage may prove to be the least effective method: 'There is no evidence that the best citizens are the offspring of congenial marriages, or that a conflict of temperament is not a highly important part of what breeders call crossing' (222). Targeting monogamous marriage, he argues that women should be encouraged to have children by different fathers, and breeders with a proven track record should be rewarded for efficient outcomes: 'If a woman can, by careful selection of a father, and nourishment of herself, produce a citizen with efficient senses, sound organs, and a good digestion, she should clearly be secured a sufficient reward for that natural service to make her willing to undertake and repeat it' (253). Far from advocating a sexual revolution of 'free love,' Shaw means to circumscribe and constrain who is to populate the utopian society.

Procreation is to become the subject of an openly disciplinary regime; sexuality is to be regulated for the sole purpose of improving the population. The emphasis is not on the biological stud farm but on the 'evolutionary appetite,' which has been defined as 'an insatiable craving for improvement, for bettering the race – in other words, an irresistible urge toward greater and greater efficiency on the largest scale' (Couchman 17). Along these lines, the play even proposes the establishment of 'a State Department of Evolution, with a seat in the Cabinet for its chief, and a revenue to defray the cost of direct State experiments, and provide inducements to private persons to achieve successful results' (*Man* 253). While we cannot be certain just how seriously Shaw entertained some of the more extreme proposals in *Man and Superman,* the logic of the argument leads to the suggestion of a 'joint stock human stud farm (piously disguised as a reformed Foundling Hospital or something of that sort)' that might, 'under proper inspection and regulation, produce better results than our present reliance on promiscuous marriage' (*Man* 254). Although it is highly doubtful that Shaw entertained the stud farm as a practical method, it suggested itself tantalizingly as the perfect utopian solution. The efficient planning of reproduction takes us inexorably from a paternalistic program of encouraging the mating of socially desirable men and women to the total control over reproduction in the eugenic laboratory of the stud farm.

Although Shaw is said to have favoured positive over negative eugenics, the two are mutually implicated and hence not easily disentangled. The flip side of selective breeding is the removal of social barriers and the elimination of those unfit for procreation. As we have already seen, the institution of marriage is a particularly vexing obstacle to a successful breeding program. In the 'Epistle Dedicatory,' Shaw blames marriage for unregulated breeding practices the dire consequences of which need to be prevented through social engineering: 'The resolve of every man to be rich at all costs, and of every woman to be married at all costs, must, without a highly scientific social organization, produce a ruinous development of poverty, celibacy, prostitution, infant mortality, adult degeneracy, and everything that wise men most dread' (*Man* xvii). But this attack on marriage as a destructive social constraint is not inspired by the Freudian intention of liberating modern men and women from sexual repression. On the contrary, the desired 'replacement of the old unintelligent, inevitable, almost unconscious fertility by an intelligently controlled, conscious fertility' requires 'the elimination of the mere voluptuary from the evolutionary process' (231). Invested in sexual pleasure for its own sake, the voluptuary exemplifies precisely the randomness of the life force that Shaw seeks to direct into predictable channels. The married couple and the libertine are equally dangerous to the evolutionary process of improving humankind through selective breeding.

In Shaw's opinion, democracy constitutes yet another barrier to the social-engineering project he envisions. Privileging individual rights over collective interests, parliamentary democracy seeks to safeguard the very random choices that impede the establishment of the perfect society. In 'The Revolutionist's Handbook,' he dismisses democracy in a wittily cynical aphorism: 'Democracy substitutes selection by the incompetent many for appointment by the corrupt few' (258). In its current state, democracy is hindered by 'the level of the human material of which its voters are made' (232) and its future is foreclosed by the commitment to universal suffrage. Unlike democracy, socialism does at least allow for the possibility of social control. What Shaw endorses is not the emancipation of the proletariat but the prospect of creating conditions conducive to 'the socialization of the selective breeding of Man' (252). For this program to be realized, we 'must eliminate the Yahoo, or his vote will wreck the commonwealth' (252). If society is to be improved, he informs us, we 'must either breed political capacity or be ruined by Democracy' (xxv), a system hampered by its inability to distinguish between the Superman and the Yahoo. Socialism was to assist in the process of Creative Evolution, a process allowing the mind to 'triumph over matter and shape it in a desired direction' (MacKenzie and MacKenzie 294). As Norman and Jeanne MacKenzie remind us, the 'vision of the superman, using will to bend destiny, to "eliminate the yahoo" and create a new and superior race, was one that had been implicit in Shaw's life and work' (294). Driven by a totalizing desire to improve society, Shaw advocated efficient methods that privileged rational organization over basic human rights.

Shaw clearly endorses efficiency in the organization and administration of the political order; however, in his treatment of the economic system, he ignores the industrial-capitalist problem of standardization in favour of an equilibrium model premised on a socialist distribution of wealth. Unlike Hitler, Shaw did not just pay lip service to socialism; eugenics was for him the bedrock on which to build a society of equals. For him, the race could only be improved under conditions of equality: 'You cannot equalize anything about human beings except their incomes' (Baker 170). His equilibrium formula assumes that the '"pressure of self-interest" works to restrain antisocial behavior but only when the victim is on an equal footing with the culprit' (170–1). Not unlike Taylor, he proposes 'promotion by merit' (171) to achieve an equal distribution of wealth. However, the assignment of merit would presumably be placed in the hands of the 'exceptionally intelligent, clever, talented, and self-assured women and men' with whom, Baker claims, Shaw was 'fascinated' (173). Once the nation's gene pool had been improved, education would be asked to complete the project of creating the useful citizen. Contrary to what one might expect from an intellectual elitist, Shaw dismissed traditional liberal education as a sham; compared to technical instruction, it struck him as grossly inefficient. The process of Creative Evolution leading to a superior race was for Shaw the solution to a problem that had plagued him throughout his life: how to reconcile individual freedom with the demands of society. In the efficiently organized utopia, the individual will have been eugenically predisposed and socially cultivated through education to respond naturally to the needs of the perfect society.

When Baker resists tarnishing Shaw with the Nazi brush, he overlooks the fact that efficient social-engineering projects tend to shade over into programs of exclusionary social-control experiments. In a 1938 letter to Beatrice Webb, Shaw seems quite sanguine about a humane 'weeding out' of the 'undesirables': 'I think we ought to tackle the Jewish question by admitting the right of States to make eugenic experiments by weeding out any strains that they think undesirable, but insisting that they should do it as humanely as they can afford to, and not shock civilization by such misdemeanours as the expulsion and robbery of Einstein' (Nathan 226–7). It seems, then, that the equal distribution of wealth applies only to those deemed meritorious by a superior political figure. In Shaw's opinion, the new socialism he envisages cannot arise from the raw material currently available to the social engineer: 'As to building Communism with such trash as the Capitalist system produces it is out of the question. For a Communist Utopia we need a population of Utopians; and Utopians do not grow wild on the bushes nor are they to be picked up in the slums: they have to be cultivated very carefully and expensively. Peasants will not do' ('Preface to *On the Rocks*' 161). Before a new order can be created, the old 'garden' has to be free of the 'weeds' that impede its growth.

It was thus only a short logical step from the argument that human beings had to be made more efficient to the position that the inefficient had to be eliminated. Declaring himself 'a hater of waste and disorder' ('Preface to *Major Barbara*' 243), Shaw maintained that every government is 'obliged to practise [extermination] on a scale varying from the execution of a single murderer to the slaughter of millions of quite innocent persons' ('Preface to *On the Rocks*' 143). His most extended discussion of the necessity of extermination occurs in the preface to *On the Rocks* (1933), published the same year Hitler effectively assumed power in Germany. Although Shaw is known to relish being provocative, the callous attitudes he displays in the preface should not be explained away as the semi-jocular comments of the licensed court-jester. In the name of efficiency, Shaw prefers to 'disable' a dangerous criminal by means of execution rather than imprisonment, which would 'waste the lives of useful and harmless people in seeing that he does no mischief' (144). Along similar economic lines, he proposes to shoot a runaway convict so as to 'save the trouble of pursuing and recapturing him' (151). When tenants of land 'become economically superfluous and wasteful,' the owner of the private property is encouraged to 'exterminate them' (148). The context indicates that Shaw may merely have meant that they should be 'evicted.' However, his tendency to use 'extermination' in the sense of 'eradication,' 'abolition,' 'eviction,' 'expropriation,' and 'firing' speaks to a preoccupation with the 'weeding' out of the inefficient who mar the perfect garden. In Russia, he notes, the liquidation of 'the sort of people who do not fit' into 'a certain type of civilization and culture' (146) is known as 'weeding the garden' (159), a term Hitler was also fond of using when speaking of the need to exterminate the Jews. In a statement combining Shaw's desire for a socially engineered society and disapproval of Hitler's anti-Semitism, Shaw has no trouble with the 'extermination of whole races and classes' that 'has been not only advocated but actually attempted' at the same time as he dismisses the 'extirpation of the Jew' as figuring only 'for a few mad moments in the program of the Nazi party in Germany' (145). When he disapproves of Hitler's anti-Semitism, he does so because he cannot see how the elimination of the Jews serves the logic of efficiency. It is at least partly for the same reason that he is strongly opposed to the unnecessary infliction of pain: 'Killing can be cruelly or kindly done; and the deliberate choice of cruel ways, and their organization as popular pleasures, is sinful; but the sin is in the cruelty and the enjoyment of it, not in the killing' (143).[7] Aside from disliking 'cruelty, even cruelty to other people,' he is primarily concerned to put extermination 'on a scientific basis' (153, 143), the same appeal to objectivity we find in Taylor's utopian thinking. Although it is quite possible that Shaw's 'outspoken views on the need for stringent eugenic measures' may have been 'delivered with somewhat mischievous intentions,' he continued well into 1922 to back such extreme positions as the painless killing of 'incorrigible villains' in the 'lethal chamber' (Stone 78, 130).

The most concentrated dramatization of efficiency in Shaw's works occurs in act 4 of *John Bull's Other Island* (1904), in a debate between Tom Broadbent, an English land developer, and Peter Keegan, a defrocked Irish priest, about the future of some land in Irish Roscullen. 'The world belongs to the efficient' ('John' 173), says Broadbent while lecturing the resistant Keegan on the benefits to be gained from putting society on a more efficient footing. A casual reading of the scene may suggest that efficiency is being denigrated. But a closer look reveals that Keegan's objections to Broadbent's development plans are aimed at the economic inequalities arising with capitalism rather than at the efficient political order Shaw apparently supports in prefaces and letters when he praises the leadership of dictators in the 1930s. What emerges in this play is a disjunction between economic and political investments in efficiency that Shaw and other writers often find difficult to recognize, let alone reconcile.

Allowing for the Shavian mix of inversions, reversions, and subversions of the individual viewpoints dramatized, the backbone of the play remains the stark contrast drawn between the Irish and the English temperaments embodied by Keegan and Broadbent respectively. Both men are speculators, but in very different senses. Keegan, the older man, represents the Irish stereotype: he is a man who lives in a world of dreams and imagination, a man criticized for losing contact with the real world. As a defrocked priest, he tends to see the world in religious terms, dividing men and women into two classes, the saved and the damned. He finds the 'real world' deeply troubling and unsatisfying; however, the English John Bull stereotype, the business-like Broadbent, finds the world 'a jolly place' ('John' 152), and feels quite at home in it. Introduced in a stage direction as 'a robust, full-blooded, energetic man in the prime of life,' Broadbent regards himself as a man of reason and common sense, professing to 'see no evils in the world – except, of course, natural evils – that cannot be remedied by freedom, self-government, and English institutions' (72, 152). Chief among his values is efficiency. 'There are,' he stoutly proclaims towards the end of the last act, 'only two qualities in the world: efficiency and inefficiency, and only two sorts of people: the efficient and the inefficient' (170). Not that all Irishmen are inefficient, Broadbent concedes; it's just that they are 'duffers,' whereas 'I know my way about' (170). The play never quite allows us to establish whether Shaw's sympathies are with Keegan, who is imbued with a superior and more imaginative mind, or Broadbent, who is efficient and gets things done.

It is fairly certain that Shaw means us to approve of Broadbent's ambition to improve Ireland through a social-engineering project. As an entrepreneur and developer, as well as a civic engineer, his aim is to take a country he thinks is backward and 'by straightforward business habits teach it efficiency and self-help on sound Liberal principles' (171). He wants to turn the dirty, squalid, and run-down Irish village of Rosscullen into a thriving garden city: 'I shall bring money here: I shall raise wages: I shall found public institutions: a library, a Polytechnic (undenominational, of course), a gymnasium, a cricket club, perhaps an art school' (171). In addition, he plans to attract tourists by building a golf course and a hotel. He also envisages offering motorboat trips on the river: 'There seems to be no question that the motor boat has come to stay. Well, look at your magnificent river there, going to waste' (168).

Though powerless to resist, Keegan is not impressed by Broadbent's materialist vision of prosperity for a poverty-stricken area or by his confident assertion that 'the world belongs to the efficient' (173). A nature lover, the spiritual Keegan typically responds to Broadbent's motorboat desecration of the river by closing his eyes in horrified incredulity and murmuring a line of poetry: 'Silent, O Moyle, be the roar of thy waters' (168). In a nostalgic gesture, Keegan asserts that all Ireland is 'holy ground' (172) and, in a lengthy diatribe that stands as the rhetorical climax of the play, he dwells percussively and ironically on the words 'efficient' and 'efficiently' (mentioned no less than twenty times) to denounce Broadbent's development scheme, predicting that it will wreak only havoc and desecration: 'For four wicked centuries the world has dreamed this foolish dream of efficiency; and the end is not yet. But the end will come' (174). In contrast to Broadbent's investment in efficiency, Keegan looks forward to the day when 'these islands,' presumably both England and Ireland, 'shall live by the quality of their men rather than by the abundance of their minerals' (173). If these characters were to be measured on how far they have creatively evolved towards the Superman ideal, the more spiritual Keegan would presumably be outpacing the intellectually plodding Broadbent. This supposition is reinforced in their last exchanges, when the two characters offer views of heaven as glimpsed in their dreams. In Broadbent's unimaginative dream, heaven was 'a sort of pale blue satin place, with all the pious old ladies in our congregation sitting as if they were at a service' (176); for Keegan, though, heaven manifests itself in a Blake-like mystical and apocalyptic vision of society both unified and transformed – a country, a commonwealth, a temple all at once, even a godhead: 'It is a godhead in which all life is human and all humanity divine' (177). Admitting that his dream is 'the dream of a madman' (177), Keegan himself confirms the play's repeated contrast with Broadbent's sanity.

There is no resolution at the end of the play. As in *Major Barbara,* where the world of Undershaft's munitions factory and that of the Salvation Army appear irreconcilable, there is no way out of the impasse pitting Keegan and Broadbent against each other: 'The world of facts and the world of dreams remain totally separate, the efficiency of the one vitiated by its lack of divine purpose, and the nobility of the other vitiated by its impotence' (Wisenthal 108). Intellectually limited and Philistine, Broadbent is temperamentally incapable of understanding Keegan's point of view: 'This lack of understanding makes Broadbent totally invulnerable to intellectual opposition,' and he is not at all disturbed by Keegan's 'trenchant, prophetic indictment of capitalist efficiency,' seeing no need to defend himself 'because he is not aware that he is being attacked' (91). Keegan's 'dream of a madman' will not stop Broadbent's projected developments from going ahead, and to that extent Broadbent comes across as the victor in the clash between the pragmatist and the visionary. But perhaps his triumph is empty in that it leaves him spiritually impoverished.

In keeping with his stock dialectical dramaturgy, Shaw sides overtly neither with Broadbent nor with Keegan; refusing to load the dice, he allows us to discern and judge the virtues and shortcomings of each protagonist. Yet in the preface to the play, he does tip his hand, suggesting that Broadbent's capitalist values may be 'out of date': 'I am persuaded that a modern nation that is satisfied with Broadbent is in a dream. Much as I like him, I object to be governed by him, or entangled in his political destiny' (15). It may appear that Shaw is contradictory or incoherent in his attitudes to efficiency. Yet if we look more closely, we find that Keegan's objections to Broadbent are not based on the utopian politics that are at issue in Shaw's prefaces and letters but on economic grounds. Efficiency is not condemned as such; it is only because Broadbent's efficiency is 'in the service of Mammon, mighty in mischief, skilful in ruin, heroic in destruction' (172) that Keegan condemns him. The target of Shaw's hostility is capitalism rather than political dictatorship. It is Keegan, then, who emerges as an early spokesman for a simple but crucial idea that runs throughout this book, the idea – which Shaw certainly shared in this play and others – that efficiency is not a good in itself: it is only a good if the ends it serves are good. However, the disjunction between economics and politics in *John Bull's Other Island* complicates this apparently simple idea by making it difficult to establish what constitutes the 'good' when Shaw seems rather contradictorily to be supporting the total-outcome model of efficiency in politics and the equilibrium model in economics.

H.G. Wells: The Elimination of Waste

In his 1941 essay 'Wells, Hitler and the World State,' George Orwell provides us with a witty and insightful retrospective on Wells: 'If one looks through nearly any book that he has written in the last forty years one finds the same idea constantly recurring: the supposed antithesis between the man of science who is working towards a planned World State and the reactionary who is trying to restore a disorderly past. In novels, Utopias, essays, films, pamphlets, the antithesis always crops up, always more or less the same. On the one side science, order, progress, internationalism, airplanes, steel, concrete, hygiene: on the other side war, nationalism, religion, monarchy, peasants, Greek professors, poets, horses' (163). In *Anticipations* (1902), his first non-fiction success, Wells appeals to efficiency to secure 'the material securities' of life, such as 'police and roads and maps and market rules'; he was especially keen on '"efficiency" and government,' but resented the stranglehold of 'busy little bureaucrats' and other vested interests imposing an 'order' at odds with 'the sheer power of naked reasonableness' (281–2). Illustrating the pitfalls of rational-choice models, he assumes that human beings consciously act on the basis of clearly established rational criteria; thus, in 'his desire for order and efficiency Wells represents the people of his utopia as being governed by reason to a greater extent than is usual, or perhaps even possible' (Haynes 120). Scientifically trained, Wells admired technological advances and put much trust in science for the betterment of mankind. A more efficient use of resources might hold the promise of greater material prosperity, while social engineering, especially eugenics, might improve the utopian prospect of a society consisting of rational social agents. At the same time, Wells seems more troubled than Shaw by scientific limitations and constraints on individual freedom or agency.

Where Shaw took no interest in efficiency experts, Wells admired Henry Ford and Thomas Edison for their technological innovations and approved of Taylor for putting the workplace on a scientific footing. Like Ford, Wells appreciated the enormous social impact of Edison's many technical inventions and improvements: 'The fruits of his resolution were to build up half-a-dozen new industries, to provide employment for millions, to extend the reach of civilization, and to enlarge the life of almost everyone who lives within it. Henry Ford says of him that he doubled the efficiency of modern industry – that it is due to Edison that America is the most prosperous country in the world' (Wells, *Work* 501). Along the same lines, he called Ford 'a very natural-minded mechanical genius' and appreciated his efficiency: He had the confidence to 'design an automobile as sound and good and cheap as could be done at that time, and to organize the mass production of his pattern with extraordinary energy and skill' (*Shape* 67). Ford is further credited with pushing 'economic expansion as if it were a necessity inherent in things, and never began to doubt continual progress until he was a man of over seventy' (67). What impressed Wells above all was Ford's totalizing desire for ever greater efficiency. Ford's 'system of production' is specifically praised for 'the elimination of waste': 'Not an inch of factory space is to be wasted, not a moment of time, not a fragment of scrap, not an ounce of physical strength nor of mental effort' (*Work* 510–11). The technical efficiency of Ford the mechanical genius begins to shade over into the ruthless social efficiency for which Ford Motor became infamous once Harry Bennett imposed his control on the workers. In his enthusiasm for standardization, Wells blinds himself to the dehumanization widely attributed to Ford's assembly line. Skating over its darker ideological implications, Wells dwells on the dynamic aspects of standardization: 'In fact, Ford's principle is that while every operation is standardized for the time being, no single operation, material or product is ever standardized in the sense that it is considered incapable of improvement' (512). For Wells, technological innovation is the mark of modernity as adventure. Discussing *A Modern Utopia* (1905), Mark Hillegas draws attention to the significance of the Crystal Palace in Wells's imagination by pointing out that the capital city of London is 'designed by the artist-engineer, who builds, with thought and steel, structures lighter than stone or brick can yield ... With its towering buildings, its moving ways, its domes of glass and great arches, it is ... a forecast of the modern architecture which ultimately traces its origins to the Crystal Palace of 1851 (a building Wells would have seen in its new form and location in South Kensington)' (74). Mesmerized by the wonders of technical efficiency, Wells disavows the dangers of standardization and routinization that worried Dostoevsky on his visit to the Crystal Palace.

Although Wells is known to have been suspicious of trained experts and administrators, he was, nevertheless, sympathetic to Taylor's attempt to bring order to the workplace. Noting that scientific management was being adopted by the National Institute of Industrial Psychology in London, he defended Taylor against workers who saw in his principles nothing but 'an attempt to put them under increased pressure and reduce employment,' arguing that Taylor, 'in his classical instances, seems always to have insisted that the workers concerned should have a substantial share in the economies effected in the form of increased wages' (*Work* vol. 1, 279). The basis for this optimistic view of Taylorization may well have been Well's contention that 'sixty percent of [Ford's] men earn more than [the minimum six dollars a day] under a system of rewards for efficient work' (*Work* vol. 2, 513).

Not surprisingly, Wells supported the Fabians in their campaign to replace traditional education with vocational training at the newly established London School of Economics. The Taylorized educational institution was to stress not Latin and Greek but science, technology, and sociology; the curriculum was to be based on a policy of 'Five E's,' that is, on efficiency, education, economics, equality, and empire. At a time when Britain proved highly resistant to Taylorization, Wells embraced the principle of efficiency first and foremost for Taylor's stress on method. Where others saw in Taylorism a vulgar Americanism, Wells was drawn to it as an antidote to the British tendency to elevate 'muddling through,' or pragmatic self-help, to an art form. In *The New Machiavelli* (1911), for instance, Wells attributes the disorder of society to haste and carelessness: 'Before men will discipline themselves to learn and plan, they must first see in a hundred convincing forms the folly and muddle that come from headlong, aimless and haphazard methods' (Haynes 115). Were his compatriots to realize the benefits of a clear method in all things, they would presumably be on the way to Wells's utopian society.

It is primarily as the enemy of waste that efficiency plays such an important role in Wells's thinking. As the 'natural and inevitable result of disorder,' waste became 'a perennial theme in his work – waste and inefficiency in domestic arrangements, the waste of natural resources, the wasted potential of men and women unable to make their full contribution to society ... waste in personal relationships, and in international relations, culminating in the immense waste of every kind involved in a world war' (Haynes 112–13). The novel most closely associated with this concern with waste is *Tono-Bungay* (1909), which 'surveys virtually the whole spectrum of waste in society' (113). Waste is a systemic condition that calls out for the systemic solution of efficient social planning. There are no individual villains on whom to pin the evils befalling individuals; it is 'the disorganization of society which prevents the "little man" from finding self-fulfillment' (117). As a character in *The New Machiavelli* puts it, it is 'muddle that gives us the visibly sprawling disorder of our cities and the industrial countryside, muddle that gives us the waste of life, the limitations, wretchedness and unemployment of the poor' (in Haynes, 115). Efficient technical and social engineering are thus touted as the panacea for the twin scourges of material want and the abrogation of human potential.

In spite of an often seemingly unremitting endorsement of efficiency to combat waste, Wells was at times troubled by the tension between the material benefits of efficiency and the social control of individuals it implied. Modernity as routine was not easily reconciled with modernity as adventure: 'In designing his utopian world-state, Wells was clearly torn between two divergent ideals – the desire for order and efficiency on the one hand and the desire to foster individual initiative on the other' (Haynes 118). Huxley's *Brave New World* satirizes precisely Wells's supposedly uncomplicated belief in an efficiently managed technological future, implicitly accusing him of sacrificing the individual to the system. Wells's *When the Sleeper Wakes* is thought to prefigure *Brave New World* in many ways, containing 'a remarkable series of technological anticipations of the Fordian world' (Firchow, *End* 65).[8] Yet, as Firchow stresses, Wells was 'not always a facile optimist, especially in his earlier books,' where he demonstrates how 'worlds controlled by technical ingenuity and moral ineptitude can go dangerously awry' (59). Wells certainly interpreted Huxley's novel as a personal attack, 'blustering about Huxley's "betrayal of the future"' (59).

Far from promoting science and efficiency without any thought to consequences, Wells was highly self-conscious of a paradox that neither he nor Huxley could resolve: 'How, in an age of science and technology, can the world achieve economic, social, political stability and efficiency and, at the same time, not dehumanize the individual by completely controlling him' (Hillegas 52). But, where Huxley decries all limitations on personal freedom, Wells is inclined to tolerate a high degree of social control in the interest of security and material prosperity. His concession to opponents of social engineering is an appeal to liberal-humanist morality to curb what he considers to be the excesses authorized by the logic of efficiency. Playing the liberal-humanist game, he is said to have realized, in Norman Nicholson's words, that 'science divorced from humanity and from the wisdom that sees beyond the first line of consequence may bring disaster on mankind' (Hillegas 17). But such concern for the 'wisdom' of 'humanity' seems all too often forgotten in Wells's drive for the elimination of waste.

As in the case of Shaw, the most controversial aspect of Wells's obsession with the efficient disposal of waste is his qualified support of eugenics. Unlike Shaw, Wells confronted the scientific claims of eugenicists and found them wanting. Theoretically in favour of improving humankind through procreation, Wells was not convinced that it was scientifically possible to breed selectively. In a letter to the *Daily Mail* (15? October 1913), he declared unambiguously, 'Not only do I not support eugenicists and the Eugenics Society, but ... I have written an entirely destructive criticism of their proposals' (in *Mankind in the Making*)' (*Correspondence* vol. 2, 348). As 'a scientist,' Wells was clearly interested in Mendelian experiments in heredity; as the editor of the *Correspondence* points out, Wells wrote 'a dozen articles or reviews on the subject of eugenics, the attempt to improve the stock by programmes of selective breeding' (349n). In another letter to the *Daily Chronicle* (16? December 1915), Wells dismissed eugenics as a 'sham science, with no stuff of any work behind it' (437).[9] In addition to his sceptical attitude to the science, comments the editor, Wells 'considered experimentation with humans unethical, and called for very careful scrutiny of all such experiments and their purported results' (349n). His scientific training thus made him even more opposed to 'stud farms' than Shaw. Yet his dismissal of the science does not mean that he was unsympathetic to the utopian desirability of selective breeding.

In *Mankind in the Making* (1903), Wells had earlier entered into a debate with Francis Galton, the pre-eminent scientist of eugenics, and

Victoria Woodhull Martin, the editor of the popular monthly the *Humanitarian,* who pontificated against 'the Rapid Multiplication of the Unfit' (*Mankind* 38–9). Distinguishing between theoretical desirability and practical impossibility, he approves of selective breeding as an efficient solution to the problem of overpopulation, but voices reservations about its scientific viability. Starting the chapter 'The Problem of the Birth Supply' with the concern that 'within the last minute seven new citizens were born in that great English-speaking community which is scattered under various flags and governments throughout the world' (34), Wells asks, 'What may be done individually or collectively to raise the standard and quality of the average birth?' (34). Calling this 'unending stream of babies' a 'supply of that raw material which is perpetually dumped upon our hands' (35), he affirms that 'to prevent the multiplication of people below a certain standard, and to encourage the multiplication of exceptionally superior people, was the only real and permanent way of mending the ills of the world' (39). But, after much reflection, he concludes that 'nothing of the sort can possibly be done except in the most marginal and tentative manner' (39). In 'The Problem of the Birth Supply' he considers both the 'method of selection' (positive eugenics) and the 'method of elimination' (51) (negative eugenics).

Not unlike Shaw, Wells locates the 'first difficulty' for positive eugenics in our inability to be clear on 'what points to breed for and what points to breed out' (40). Assuming that we do not want to create a 'homogenous race' but a society consisting of 'a rich interplay of free, strong, and varied personalities' (41), he cannot see how science could grapple with the problem of selective breeding, except on the most superficial and irrelevant level of predetermining physical traits like 'height, weight, presence of dark pigment in the hair, whiteness of skin' (44). It seems to Wells that the 'whole science of anthropology in its present state of evolution' is in no position to improve on the offspring produced through the 'sentimentalized affinities of young persons' (45). Conceding that we simply 'do not know enough' (50), he insists that selective breeding cannot be put on a scientific footing.

At the same time, Wells is not prepared to abandon procreation entirely to chance. Not unlike Shaw, he would like to encourage the right sort of people to mate. In *Anticipations* (1902), he clearly advocates, among other things, what sounds suspiciously like a utopian program of eugenics: 'The ethical system which will dominate the world-state will be shaped primarily to favour the procreation of what is fine and efficient and beautiful in humanity – beautiful and strong bodies, clear and powerful minds ... and to check the procreation of base and servile types, of fear-driven and cowardly souls, of all that is mean and ugly and bestial in the souls, bodies, or habits of men' (256–7). Although biological manipulation is scientifically out of the question, sexual attraction or love cannot be left to the whims of individuals: 'The New Republican, in his private life and in the exercise of his private influence, must do what seems to him best for the race; he must not beget children heedlessly and unwittingly' (*Mankind* 67–8). At the same time as Wells retreats from the eugenics laboratory, he remains open to social-engineering measures designed to optimize mating practices.

Although humanity may find it difficult to establish 'what is pre-eminently desirable in inheritance,' it may find it easier to isolate and define what is 'pre-eminently undesirable' (*Mankind* 52). Once again, Wells favours the theoretical tenets of negative eugenics, stressing that the citizens of the New Republic will have no pity for the 'multitude of contemptible and silly creatures, fear-driven and helpless and useless, unhappy or hatefully happy in the midst of squalid dishonour, feeble, ugly, inefficient' (*Anticipations* 257). But, unlike Hitler, Wells was not interested in improving humankind along race-based lines. Privileging the efficient and excluding the inefficient, his utopian society offers citizenship to 'white, black, red, or brown; the efficiency will be the test' (272). Countering anti-Semitism, he stresses that the state will treat the Jew 'as any other man' (272). At the same time, he would clearly not be impressed by our current emphasis on inclusivity and diversity. Independent of race or creed, all those who fail the efficiency test are discarded as waste: 'And for the rest – those swarms of black and brown and yellow people who do not come into the needs of efficiency? Well, the world is a world, not a charitable institution, and I take it they will have to go' (274). Like Shaw, Wells is prepared to sacrifice the inefficient to a utopian future dedicated to the promotion of the Nietzschean superman.

Whereas scientific breeding for superior offspring seemed unrealistic, the elimination of 'human waste' seemed well within the bounds of possibility. On the one hand, Wells cautions that science is not sufficiently advanced to establish with certainty how diseases, let alone personality traits, are passed on to future generations. On the other hand, though, his Malthusian concern with the problem of overpopulation tempts him to inveigh against the presumably inferior 'masses' that emerged with industrialization. Born in Bromley, Kent, in 1866,

Wells observed how the construction of the railway, together with the subsequent urban development to house a growing population, led to the ruin of the landscape: 'Anxiety about overpopulation, rooted in his childhood vision of woods and fields destroyed at Bromley, is the key to Wells's reading of modern history' (Carey 119). According to Carey, Wells is guilty of an elitist contempt for the 'useless': '"All over the world," [Wells] observes in *Anticipations,* "vicious helpless and pauper masses" have appeared, spreading as the railway systems have spread, and representing an integral part of the process of industrialization, like the waste product of an healthy organism. For these "great useless masses of people" he adopts the term "People of the Abyss," and he predicts that the nation that most resolutely picks over, educates, sterilizes, exports or poisons its "People of the Abyss" will be in the ascendant' (123). Yet he is also clearly disdainful of the 'queer flavour of absurdity and pretentiousness' (*Mankind* 39) that clings to the pages of the *Humanitarian* whenever solutions are proposed to avert the threat of 'the Rapid Multiplication of the Unfit.' In *The Time Machine* (1895), he was not even certain that the 'idea of a social paradise' was worth pursuing, since under 'conditions of perfect comfort and security' (30, 31) we would become physically and intellectually weak. It appears that Wells knows very well that the elimination of 'human waste' is scientifically a highly dubious proposition; however, he seems unable to divest himself entirely of the desire for a perfect garden free of weeds.

No matter what reservations he voices, Wells quite obviously hopes to improve the world by making everything from machines to human beings function more efficiently. As in the case of Shaw, though, he often adopts an ironic stance to distance himself from the more outrageous positions occupied by the Eugenic Society. It is never easy to tell if he really means what he says. This difficulty is exacerbated when he has characters expound views he himself may or may not entertain. In *When the Sleeper Wakes* (1899),[10] for instance, it is not clear if Wells himself supports the elimination of the 'inferior' that the dictator Ostrog encourages: 'The hope of mankind – what is it? That some day the Over-man may come, that some day the inferior, the weak and the bestial may be subdued or eliminated. Subdued if not eliminated. The world is no place for the bad, the stupid, the enervated. Their duty – it's a fine duty too! – is to die. The death of the failure! That is the path by which the beast rose to manhood, by which man goes on to higher things' (237–8). No matter how serious Wells may have been in his call for the elimination of the 'weak,' his proposals in both fiction and non-fiction strike today's readers as insufficiently sensitive to the rights of 'bare life' at issue in theoretical discourses. We need to remember that Wells, like Shaw, expressed opinions that found a receptive ear across Europe. It has even been rumoured that *Anticipations,* the non-fictional work most concerned with eugenics, may have been read by Hitler in German translation. Yet Wells, like Shaw, was not motivated by anti-Semitism or racism but by the utopian desire for an efficiently organized society. They consequently differ from Wyndham Lewis (1882–1957), whose openly fascist writings in the 1930s were motivated by anti-Semitism. What Wells's ideological investments may be anticipating is not the 'mad' Hitler raging against Jews, but the methodical Hitler making the trains to Auschwitz run on time.

While it is unlikely that either Wells or Shaw would have supported the actual extermination of human beings that they so callously seemed to advocate, their provocative proposals circulate as public statements to be used and abused. Although Shaw retreated from eugenic 'stud farms' and Wells opposed the scientific claims of selective breeding, both of them tolerated the elimination of the unfit or undesirable to speed up social progress. What they proposed in theory would undoubtedly have shocked them if put into practice. As Shaw's reaction to the Nazi camps shows, he could not bring himself to believe that human beings were capable of such cruelty. However, no matter what Shaw and Wells may have intended, their public pronouncements and literary output reflected and contributed to a cultural atmosphere obsessed with efficiency and rife with anti-liberal, quasi-totalitarian sentiments. While authors cannot be held personally responsible for the use others make of their writings, it is, nevertheless, important for readers to draw attention to suspect ideological investments. It may well be the critics' responsibility to warn against statements, no matter how ironic or unserious, that Hitler could have adopted as a blueprint for his own extermination policies. Whenever efficiency is promoted as a value that trumps all other considerations, readers ought to ask themselves whose interests are being served.

The broad appeal of eugenic arguments in British society extended even to a feminist and pacifist like Virginia Woolf. In *Modernism and Eugenics* Childs provides incontrovertible evidence for Woolf's familiarity with eugenics[11] and hence with the related concept of 'industrial efficiency'; in a diary entry, for instance, she describes medical experts as 'ineffective' by using 'neither the language of medicine nor the language of common prejudice, but rather the very language of Industrial

Efficiency' (Childs 26). According to Childs, Woolf's ironic stance towards the medical experts Holmes and Bradshaw expresses a rather ambivalent attitude towards eugenics; contrasting with her sympathetic treatment of the shell-shocked Septimus Warren Smith in *Mrs Dalloway* (1925), she concludes in a 1915 diary entry that people with mental disabilities 'should certainly be killed' (Childs 23). In *Three Guineas* (1938), Woolf later advocates that the state support women choosing motherhood as a profession, especially 'the daughters of educated men.' Encouraging women of 'the very class where births are desirable,' Woolf endorses a 'positive eugenics' designed to 'produce the "desirable" kind of future citizen who will help to create "peace and freedom for the whole world"' (Childs 23). Implicit in such eugenicist sympathies is an attitude favouring the elimination of waste in the interests of efficiency that is entirely consonant with the task Taylor assigns to the efficiency expert authorized to organize the workplace.

Dissenting Voices: Forster and Orwell

Although neither Shaw nor Wells was blind to potential downsides of the efficiency calculus, they are generally considered to be highly supportive of its progressive claims. But their vocal support for both technical and social efficiency provoked equally vocal dissent among intellectuals who vilified machines and the efficiency they promoted. Typifying such dissent, E.M. Forster's short story 'The Machine Stops' (1909) and George Orwell's sociological analysis *The Road to Wigan Pier* (1937),[12] cover the main themes and arguments mounted against the utopian societies imagined by Shaw and, more prominently, by Wells. Forster's short story has been called 'the first full-scale emergence of the twentieth-century anti-utopia,' the first of a series of '"admonitory satires" that includes Zamyatin's *We*, Huxley's *Brave New World*, and Orwell's *Nineteen Eighty-four*' (Hillegas 82).[13] A liberal humanist who argues that democracy, while not perfect, is 'less hateful than other contemporary forms of government' (Forster, 'Believe' 77), Forster expresses in 'The Machine Stops' some of the 'most important humanist fears about the machine' (Hillegas 89). Orwell joins Forster in his denigration of the machine civilization, but from a socialist perspective. In chapter 12 of *The Road to Wigan Pier*, the self-identified socialist worries that the justice agenda of socialism is being overshadowed by a worship of the machine that emphasizes efficient social-engineering projects that find their most sinister manifestation in Soviet Russia and Nazi

Germany. The efficient machine was for both authors the metaphor for a social future neither of them wished to embrace. Forster's dystopia concentrates on the impact of technology on human consciousness; it is a science-fiction story that predicts a civilization marred by a degraded natural world, alienated human beings, and an oppressively planned social order. With the exception of the ecological element, Orwell targets the same aspects of machine efficiency, but inflects his critique with an attempt to disentangle the real or assumed connections between socialism and progress, mostly mechanical progress. Even in areas where they diverge, they share similar attitudes to the efficiently organized civilization.

The most memorable parts of 'The Machine Stops' and *The Road to Wigan Pier* consist of lively and amusing depictions of the machines both authors dislike but also know are here to stay. In Forster's story, everything works by automated machinery: chairs, music, medical assistance, dressing, bathroom facilities, travel. It anticipates future discoveries in revolutionary technology, particularly in the areas of broadcasting (television and radio) and various forms of communication ('cinematophote' [video], video-conferencing, camcorders, computers).[14] These technologies dehumanize the inhabitants of the futuristic society. Virtual communication – talking on the phone, seeing a picture on a screen – is no substitute for the nuances and inflections of direct human contact. When the rebel, Kuno, wants to meet his conformist mother, Vashti, 'face to face,' he says to her: 'The Machine is much, but it is not everything. I see something like you in this plate, but I do not see you. I hear something like you through this telephone, but I do not hear you' ('Machine' 117). Having lost her sense of touch and being afraid of direct sunlight, Vashti is shown to be 'seized with the terrors of direct experience' (131, 122). 'Owing to the Machine,' comments the narrator, 'people never touched one another. The custom had become obsolete' (128). Machines have also thrown humanity into a world governed by speed; Kuno objects to this 'accelerated age' (119) in which his mother wants to limit her meeting with him to 'five minutes,' asking him to be 'quick' and rebuking him for being 'slow.' Rapid 'progress' in the use of technology has rendered the earth itself virtually unlivable. The ecological disaster is marked by the depletion of natural resources; in an amusing dig at writers like himself, Forster comments that the 'forests had been destroyed during the literature epoch for the purposes of making newspaper-pulp' (129–30). The atmosphere is so polluted that the inhabitants have to live underground. Once 'progress had come to mean the progress of the Machine,' humanity had caved in to 'its desire for comfort' and had thus 'over-reached itself': 'It had exploited the riches of nature too far. Quietly and complacently, it was sinking into decadence' (148–9). Forster anticipates the pact with the devil that sacrifices the long-term prospect of global ecological sustainability to the short-term self-interest of material comforts promised by the machine.

By asking himself 'whether there is *any* human activity which would not be maimed by the dominance of the machine' (*Road* 172), Orwell obviously shares Forster's misgivings. In *The Road to Wigan Pier* he laments that it is with 'sinister speed' (178) that the machine is able to impoverish human experience. A 'century of mechanization' has produced a 'frightful debauchery of taste' that manifests itself in the deterioration of the 'taste for decent food' under the impact of 'standardized' and 'factory-made' produce that is stored in 'tins' and may contain 'filthy chemical by-product' (179). Instead of enjoying a superior local apple, people prefer to eat 'a lump of highly-coloured cotton wool from America or Australia' (179). Targeting a world of substitutes and ersatz products, Forster similarly complained that the 'imponderable bloom of the grape was ignored by the manufacturers of artificial fruit' ('Machine' 118). For Orwell, such degradation is inescapably implicated in an unfortunate dialectical process: 'Mechanization leads to the decay of taste, the decay of taste leads to the demand for machine-made articles and hence to more mechanization, and so a vicious circle is established' (*Road* 180). For Orwell, then, 'the machine is here, and its corrupting effects are almost irresistible' (180). Butler's solution in *Erewhon,* to 'smash every machine invented after a certain date,' is no longer possible because we 'should also have to smash the habit of mind' (182) that feeds an addiction to efficiency.

Orwell traces this addiction to the assumption that the 'dull drudgery' of manual toil will be alleviated by labour-saving technology. Suspicious of the common notion that humanity's highest goal ought to be an increase in leisure, he argues that work ought to be savoured for the satisfaction the skilled craftsman gains from his activities. In *Down and Out in Paris and London* (1933), he approves even of efficiency as long as it remains associated with meaningful work: 'What keeps a hotel going is the fact that the employees take a genuine pride in their work, beastly and silly though it is. If a man idles, the others soon find him out, and conspire against him to get him sacked. Cooks, waiters and *plongeurs* differ greatly in outlook, but they are all alike in being proud of their efficiency' (*Down* 74–5). Aside from doubting that people would make good use of more leisure, he points out that what counts as 'work' for one person may signify 'play' for another: 'The labourer set free from digging may want to spend his leisure, or part of it, in playing the piano, while the professional pianist may be only too glad to get out and dig at the potato patch' (*Road* 173). Only the 'vulgarer hedonists' deny that 'life has got to be lived largely in terms of effort' (173). Nostalgic for the skill it took for a carpenter to make a table a 'hundred years ago' (174), Orwell points out that all that is left for him to do is 'buy all the parts of the table ready-made and only needing to be fitted together' (174). Since engaging in anachronistic skills in one's leisure time can never be more than an artificial substitute for 'the human need for effort and creation' (176), Orwell fears that the historical dialectic forecloses a return to the past.

Both Forster and Orwell single out the car and the airplane[15] as significant culprits in the shift from a humanly meaningful agricultural age to a hectic and disorienting industrial society. These technological inventions are blamed for the privileging of time over space that Forster links to efficiency. In his analysis of 'The Machine Stops,' Paul March-Russell draws attention to the nexus (efficiency-speed-control-totalitarianism) that we have already observed at work at the Ford Motor Company: 'The cultural orthodoxy of efficiency and advancement – that is to say, of *being seen to be* efficient and advanced – is predicated upon a notion of temporality: of being on time and in command of one's time-keeping ... To refuse to be on time but to enjoy its passing, as Kuno does, is to be automatically considered as degenerate. Only by the effective management of one's time can an individual ensure their [*sic*] true citizen status and, by extension, the organisation of the state' (60–1). Forster's obsession with speed and time management in 'The Machine Stops' is confirmed in a direct allusion to the Farman flight: 'To "keep pace with the sun," or even to outstrip it, had been the aim of the civilization preceding this. Racing aeroplanes had been built for this purpose, capable of enormous speed, and steered by the greatest intellects of the epoch' (127). The airplane threatened to colonize the air as the car had taken over the roads; it risked desecrating space that, Forster felt, should remain, as it was for the Greeks, the abode of the gods. One can only speculate on how horrified he would be by all the space junk now orbiting our planet.

For Orwell, the car and the airplane are symbolic of the 'tendency of mechanical progress ... to make life safe and soft' (*Road* 170). Although new inventions often start out as dangerous technologies, they are soon improved to increase humankind's craving for comfort and security. Mechanization is thus driven by a self-defeating dialectical logic that the history of the car and the airplane exemplify. Initially, the car was less safe than horse-drawn vehicles; the 'enormous toll of road deaths' (170) seemed to contradict the narrative of technological progress that the car represents. But engineers soon reduced the potential for accidents by seriously tackling 'our road-planning problem' (171). With better road conditions and more driving experience, operating a car became safe and easy. The same logic applies to the airplane, which 'does not at first sight appear to make life safer,' but, with the dedication of 'a million engineers,' will in time be 'made foolproof' (171). But every improvement tends to generate the need for yet more improvements. Every step forward sooner or later encounters an obstacle to be overcome; it follows that 'however high a standard of efficiency you have reached, there is always some greater difficulty ahead' (171). Even if we should ever reach a moment when there are no more problems to be solved, Orwell seriously doubts that this efficient outcome is desirable. Linking technical and social efficiency, he complains that the tendency is to make machines always 'more efficient, that is, more foolproof; hence the objective of mechanical progress is a foolproof world – which may or may not mean a world inhabited by fools' (171). In a final dialectical twist, he imagines that we would not be content with perfecting our own planet but would immediately 'start upon the task of reaching and colonizing another' (171). For Orwell, then, the desire for efficiency is always also a desire for the mastery of nature and political control over people.

But the machine is a Frankenstein monster poised to take control of its self-deluded masters. In 'The Machine Stops' Kuno expresses the popular anxiety that the machine will take on a life of its own: 'We created the Machine, to do our will, but we cannot make it do our will now. It has robbed us of the sense of space and of the sense of touch, it has blurred every human relation and narrowed down love to a carnal act, it has paralysed our bodies and our wills, and now it compels us to worship it. The Machine develops – but not on our lines' (140–1). As the unnamed narrator observes, the 'Machine was out of hand,' yet 'year by year it was served with increased efficiency and decreased intelligence' (148). Forster implies that devotion to the machine has compelled mankind to put a premium on instrumental rationality at the expense of reasoned contemplation. Orwell similarly fears that the 'process of mechanization has itself become a machine, a huge glittering vehicle whirling us we are not certain where, but probably towards the padded Wells-world and the brain in the bottle' (*Road* 182). For Forster, this 'padded Wells-world' is marked by a homogeneity that is being reinforced by the mobility made possible by both the car and the airplane. In the anti-utopian world of 'The Machine Stops,' all living spaces were 'exactly the same' and 'people were almost exactly alike all over the world' (131, 128). In spite of the advanced transportation system, 'few traveled in these days, for, thanks to the advance of science, the earth was exactly alike all over. Rapid intercourse, from which the previous civilization had hoped so much, had ended by defeating itself. What was the good of going to Pekin when it was just like Shrewsbury? Why return to Shrewsbury when it would be just like Pekin?' (124).[16] Now that we are able to drink Starbucks coffee in the Forbidden City in Beijing, it is becoming increasingly doubtful that travel is capable of broadening the mind.

In *The Road to Wigan Pier,* Orwell was above all concerned that socialism, having abandoned its commitment to justice and equality, was heading towards the efficient, standardized, 'foolproof' or 'padded' Wellsian world. Not surprisingly, then, he associates 'the idea of machine-production' (164) with a socially engineered 'world-system' that depends on 'some form of collectivism,' a globalized communications network, a 'centralized control' mechanism, and 'a certain uniformity of education' (165). As he sees it, the socialist world is 'always pictured as a completely mechanized, immensely organized world, depending on the machine as the civilizations of antiquity depend on the slave' (165). Although Orwell declares that many thinking people 'are not in love with machine-civilization,' he realizes that the machine is here to stay because 'no one draws water from the well when he can turn on the tap' (175). What he objects to is that socialism treats 'the idea of mechanical progress, not merely as a necessary development but as an end in itself' (165). Appalled by the enthusiasm for mechanical progress among socialists, Orwell complains that 'there will be no disorder, no loose ends, no wildernesses, no wild animals, no weeds, no disease, no poverty, no pain' in this utopian desire for 'an *ordered* world, an *efficient* world' (166). But it is precisely from this 'glittering Wells-world that sensitive minds recoil' (166). While there were earlier attacks on 'science and machinery' (166) – Jonathan Swift's *Gulliver's Travels,* Charles Dickens's *Hard Times,* Samuel Butler's *Erewhon* – these must be considered mere intellectual exercises. It is only 'in our own age, when mechanization has finally triumphed, that we can actually *feel* the tendency of the machine to make a fully human life impossible' (167). Suspicious of the claim that the machine will 'set humanity free' (169), he proposes to provide *reasoned* objections to the socialists' infatuation with machine civilization. By his own account, his arguments are so reasonable that they 'would be echoed by every person who is hostile to machine-civilization' (183).

To his great dismay, Orwell reports that the socialists envision 'an ever more rapid march of mechanical progress; machines to save work, machines to save thought, machines to save pain, hygiene, efficiency, organization, more hygiene, more efficiency, more organization, more machines – until finally you land up in the by now familiar Wellsian Utopia, aptly caricatured by Huxley in *Brave New World,* the paradise of little fat men' (169). Unlike Huxley, Wells assumed, even in his early pessimistic utopias, that technology could be engaged in the fight against class conflict. According to Orwell, Wells depicts in *When the Sleeper Wakes* a 'glittering, strangely sinister world in which the privileged classes live a life of shallow gutless hedonism, and the workers, reduced to a state of utter slavery and sub-human ignorance, toil like troglodytes in caverns underground' (177). The solution to the problem he identifies is to ensure that the products of the machine-civilization are 'shared out equally' (178). In Orwell's disdainful view, the 'only evil' Wells 'cares to imagine is inequality,' overlooking that 'the machine itself may be the enemy' (178). To add insult to injury, Wells returned in his later novels to 'optimism and a vision of humanity, "liberated" by the machine' (178). Wells, then, is a prime example of those socialists who failed to grasp that 'thinking people may be repelled by the *objective* towards which Socialism appears to be moving' (164). In other words, having lost sight of its commitment to justice, socialism has been sidetracked by a materialist agenda invested in the totalizing desire for always more efficient material outcomes: 'All mechanical progress is toward greater and greater efficiency; ultimately, therefore, towards a world in which *nothing goes wrong*' (169). Instead of striving to improve individuals by testing their mental and physical capacities, the world obsessed with mechanical progress will have the unfortunate 'tendency to eliminate disaster, pain, and difficulty' (170). In this perfect society, physical safety and social security would make us 'soft' and complacent, leaving no room for exercising distinctly human qualities like courage, 'loyalty, generosity, etc.' (170).[17] With a sideways glance at Huxley's *Brave New World,* Orwell suspects that 'the logical end of mechanical progress is to reduce the human being to something resembling a brain in a bottle' (176).

Orwell's critique of socialism is driven by his fear that fascism will triumph because those who hate mechanization assume that 'the Socialist is always in favour of mechanization, rationalization, modernization' (176), even though there is no necessary connection between social and technical progress. It is only if socialism reclaims the grounds of social justice that it can divest itself of the unfortunate 'nexus of thought' that aligns socialism with 'progress-machinery-Russia-tractor-hygiene-machinery-progress' (183). Preoccupied with progressive social-engineering schemes modelled on technical efficiency, the socialists 'have never made it sufficiently clear that the essential aims of Socialism are justice and liberty' (188). Since socialism is the only 'real enemy that Fascism has to face,' (188), he is concerned that socialism in England is no longer committed to 'revolution and the overthrow of tyrants' (188); instead, smelling of 'crankishness, machine-worship, and the stupid cult of Russia,' it propels conservative intellectuals suspicious of the 'materialist Utopia' (190, 188) and fond of 'tradition' as well as 'discipline' into the fascist camp. In 1937 Orwell was hardly in a position to recognize Hitler's investment in efficiency; he heard only the nostalgic yearnings the Nazis exploited in their rhetoric and pageantry.

6 'Criminal' Efficiency: Joseph Conrad's *Heart of Darkness*

Joseph Conrad's life (1857–1924) spanned the period of rapid transformation from an essentially agricultural or organic society to an industrial or mechanical one. Having spent a good part of his early adult life as a merchant seaman, he began writing fiction in 1895, publishing *Heart of Darkness* as a novel in 1902 (after it had started to be serialized in *Blackwood's Magazine* in 1899). Unlike most other writers of his time, he had first-hand experience of the working life. For the late-nineteenth-century seaman, work had not yet been reduced to the routinized drudgery of factory toil; it could still be imbued with a sense of romantic adventure and authentic male social bonding. In the isolated community of the ship, the skill of each seaman and his commitment to a common cause were essential for the safety of all. As Avrom Fleishman argued in his classic *Conrad's Politics: Community and Anarchy in the Fiction of Joseph Conrad* (1967), Conrad subscribed to a 'work ethic' that translates into 'subordination to authority, devotion to the given task, fidelity to comrades, identification with the mariner's tradition of service, acceptance of the difficulty of life within destructive nature, and the manifestation of effort and courage' (73). This work ethic is consonant with Conrad's embrace of the 'organicist tradition,' which Fleishman identifies with Burke's 'political conservatism' (55). In this tradition, society is not envisioned as a 'businesslike contract' but as 'a spiritual and cultural union' (57). The individual owes allegiance not to the state but to the nation or, more ominously, to ties of blood and soil captured in the romantic conception of the *Volk* later to be exploited by Hitler in his rhetoric of Aryan purity. In Conrad's novels we find a nostalgic longing for work as a meaningful activity contributing to a common good. Work is not just an activity but a 'calling'; it is 'a commitment to and identification with a role' (73). This organic notion of work was not, by definition, incompatible with the efficient execution of tasks. But for Conrad 'the value of work lies not in itself but in the performance of a social obligation' (73) that gives meaning to the individual's life. In contrast, the commitment to efficiency for its own sake is denounced as one of the evils of modernity that Conrad associates with such obvious culprits as progress, democracy, capitalism, and material interests. Since Conrad is not always clear on where he thinks purposeful work ends and efficiency for its own sake begins, his attitude to efficiency in his novels is often marked by ambivalence.

Even a passing reacquaintance with Conrad's fiction reveals an endorsement of devotion to work, an unmistakable dislike of inefficiency, and an uncertain attitude to efficiency for its own sake. In fiction narrated by sailors like Marlow, a man's worth is consistently measured by his dedication to the conscientious or 'efficient' execution of meaningful tasks. In 'Youth,' *The Shadow Line,* and 'The Secret Sharer,' for instance, we find officers who are not only committed to work but accept their social responsibility for the ship's crew. What thus irks Marlow above all is inefficiency. Complaining in *Chance* (1913) that 'no one seemed to take any proper pride in his work' (4), he typically expresses his disdain for a waiter's 'slovenly manner' (3), declaring that 'no ship navigated and sailed in the happy-go-lucky manner people conduct their business on shore would ever arrive in port.' He ascribes such 'universal inefficiency' to 'the want of responsibility and to a sense of security' (4) infecting a generation for whom life has become too easy.

Conrad's satirical treatment of smugly efficient policemen in *The Secret Agent* (1907) suggests his suspicion of those pursuing efficiency for its own sake. Congratulating himself that 'the foreign governments cannot complain of the inefficiency of our police' (185), the Assistant Commissioner boasts of the speed with which his men established the circumstances of a mysterious crime: 'Look at this outrage; a case specially difficult to trace inasmuch as it was a sham. In less than twelve hours we have established the identity of a man literally blown to shreds, have found the organizer of the attempt, and have had a glimpse of the inciter behind him' (185–6). But, as the reader knows, the external facts the Commissioner lists miss the complex truth of Stevie's death. The efficient policemen are no longer in touch with the community and therefore misread the broader social circumstances of the novel's central tragedy.

In *Lord Jim* (1900), though, Marlow is less certain in his judgment of

Jim's failure to act as an efficient and morally responsible seaman. He sympathizes with this flawed protagonist, whose self-reflexive search for existential meaning distinguishes him from an outwardly efficient but inwardly hollow seaman such as Captain Brierly. Devotion to work is here suspect in that it allows a man to escape self-knowledge. Finding excuses for Jim's shortcomings, Marlow seeks to reintegrate the anguished outcast within the organic community. Since Jim's redemption at the end of the novel comes at the expense of the very community he had rescued to earn Marlow's approval, *Lord Jim* illustrates a highly conflicted attitude to efficient social planning when it is divorced from the cultural or organic roots of the native community. Broadly speaking, then, we find in Conrad a deep investment in work as an organic ideal, an unmistakable disdain for inefficiency, and a highly ambivalent attitude towards efficiency.

While ambivalence to efficiency is thus pervasive in Conrad's fiction, it manifests itself in a particularly complex and sustained manner in *Heart of Darkness*, a novel that severely tests Prince Albert's naive belief that the prosperity Britain celebrated at the Great Exhibition of 1851 would promote peace through commerce and lead to the 'realization of the unity of mankind' (in Ffrench, 51). The conditions Marlow encounters as he travels up the Congo to extricate Kurtz from the jungle constitute an unmistakable indictment of the colonial enterprise that has been the topic of much critical debate.[1] However, as has often been pointed out, Conrad primarily targets the particularly appalling colonial administration of King Leopold of Belgium. Looking at a colour-coded map of Africa, Marlow notes with approval that there 'was a vast amount of red' (the colour symbolizing Great Britain), indicating that 'some real work is done in there' (*Heart* 25). As Stephen Ross points out, the difference between the two imperial powers can be traced to specifically political decisions:

> For example, though the East India Company had been granted monopoly trading rights to India in 1600 and became a ruling enterprise in 1757, its rapacious activities were restrained by the British parliament's Regulating Act in 1773 and it was eventually dissolved by Parliament in 1858. By contrast, King Leopold of Belgium's governance of the Congo in the late nineteenth and early twentieth centuries virtually erased the distinction between government and commerce. In the place of governmental restraint over commercial activity, Leopold established commercial interests as the de facto principle of governance in the Congo, and either did not

> regulate commercial activity or else promoted it through governmental policy. (10)

Marlow's journey into 'the yellow' (*Heart* 25) or Belgian Congo thus constitutes a confrontation with 'the transition beginning to take place at the turn of the twentieth century as commercial interests started to supersede governmental authority' (Ross 10). Marlow's reactions to the inefficient methods he decries on his visit to the Company Stations and to the unsound but highly efficient method he encounters at Kurtz's Inner Station display a corporate logic that Ross convincingly identifies not with the imperial ambitions of nation states but with the global capitalism of Empire theorized in Michael Hardt and Antonio Negri's influential study *Empire* (2000).[2] An awareness of the slippery distinctions between inefficient, properly efficient, and improperly efficient behaviours in *Heart of Darkness* thus contributes to a recent debate on the economic underpinnings of colonialism in Chris Bongie's *Exotic Memories* (1991), Christopher GoGwilt's *The Invention of the West* (1995), and, most recently, Stephen Ross's *Conrad and Empire* (2004). This focus on global capitalism does not seek to invalidate trends in Conrad criticism taking their cue from Chinua Achebe's initially shocking declaration, in his seminal article 'An Image of Africa' (1977), that Conrad is 'a thoroughgoing racist' (257), a position that has been either more subtly argued or seriously disputed by Conradians already mentioned as well as by Edward Said, Frances Singh, Andrea White, and Linda Dryden.[3]

What if Kurtz wanted to satisfy, not his lust for unrestrained savagery, but his totalizing desire for unrestrained efficiency? In a letter dated 31 December 1898 to William Blackwood, the editor of *Blackwood's Magazine,* Conrad famously describes the novel in terms that foreground a concern with efficiency by emphasizing its opposite: 'The title I am thinking of is "*The Heart of Darkness*" but the narrative is not gloomy. The criminality of inefficiency and pure selfishness when tackling the civilizing work in Africa is a justifiable idea' (*Collected Letters* 139–40). Written at a time when the novel was only just taking shape, this letter frames the implicit binary opposition efficiency/inefficiency within the juridico-ethical category of 'criminality.' For a long time, it had generally been assumed that 'inefficiency' is here meant to be roughly equivalent to colonial 'exploitation.' The more recent distinction between 'good' and 'bad' colonial practices suggests that the tag 'criminal' attaches to the unnecessary suffering caused by inefficient colonial administrations. Moreover, tracing changes in Conrad's conception of his story, Michael Levenson convincingly argues that the 'criminality of inefficiency' applies only to the first part of the novel. He consequently concludes that 'no one would dream of saying that the subject of the work is "inefficiency"' (395). For him, the emphasis in part one on the appalling conditions in the Congo turns in part two into 'a drama of officialdom' (395) and finally ends with psychological and ethical preoccupations. What Levenson fails to pursue is the persistence of the 'criminality' criterion as the novel transforms itself into a drama of officialdom and psychopathology. It is finally the ideology of efficiency as such that is on trial in *Heart of Darkness.* The 'drama of officialdom' plays out a pointed confrontation with the violence implicit in both 'sound' and 'unsound' manifestations of efficiency.

In *Heart of Darkness* various characters act in different ways as carriers of the ideology of efficiency. There are the inefficient and unprincipled colonial administrators or 'pilgrims,' who make no pretense of following a higher calling and who are guilty of wasting natural and human resources. Then there is Kurtz, who lets his insatiable desire for more and more ivory override his perhaps initially laudable moral intentions. Finally, there is Marlow, who, finding himself caught between 'two nightmares,' seeks an uneasy refuge in the devotion to the efficient performance of work. The 'pilgrims' are dull because they have succumbed to a routinized life devoid of any meaningful purpose. Kurtz is infinitely more fascinating because, in his case, aim and method are at cross-purposes. Although Marlow condemns the pilgrims out of hand, he remains ambivalent towards Kurtz, whose failure to uphold his principles strikes him as tragic, while Kurtz's 'unsound' method makes him shudder with fascinated abhorrence. Where Marlow looks for a balance between an efficient method and moral aims, the pilgrims fail on both counts, while Kurtz becomes guilty of pursuing efficiency for its own sake. Throughout *Heart of Darkness,* efficiency is at issue both on the level of the personal behaviour of characters and on the level of the socio-economic power shift from nation state to global corporation.

Conrad's comment to William Blackwood draws attention to his opposition to the botched colonial effort in the Belgian Congo. If the colonizers proved to be efficient administrators, then 'the civilizing work in Africa' could take its proper British course. As long as the colonizers were 'emissaries of light' serving a noble end, the efficient execution of the 'civilizing process' would presumably be desirable. Moral aims justify the application of efficient means just as efficient means are only tolerable if they support moral aims. This means-ends approach implies that Conrad understands social efficiency to function according to the total-output model symbolized by the machine. In the first instance, the novel indicts the undesirable consequences of an inefficiently run colonial enterprise through the unflattering portrayal of the white administrators he disdainfully refers to as 'the pilgrims.' Disembarking at the first station, Marlow steps 'into the gloomy circle of some Inferno,' a 'place where some of the helpers had withdrawn to die' (*Darkness* 34, 35). These 'moribund shapes' (35) filled Marlow with pity for the victims of colonialism and indignant anger at the colonizers. Confronted by a picture reminding him of 'a massacre or a pestilence,' he 'stood horror-struck' as he watched one of 'these creatures' going off 'on all-fours towards the river to drink' (36). But his pity is overlaid with his disgust with the inefficient management of the station, which this scene of death illustrates. In the final analysis, Marlow explains the inhuman suffering he witnesses, the 'pain, abandonment, and despair' (35) of these 'creatures,' as a waste of human resources: 'Brought from all the recesses of the coast in all the legality of time contracts, lost in uncongenial surroundings, fed on unfamiliar food, they sickened, became inefficient, and were then allowed to crawl away and rest' (35). The dying 'black shadows' are above all an indictment of the pilgrims as slovenly, careless, rapacious, and wasteful managers. Their dehumanization already draws attention to the indifference of Empire to the claims of individuals, social formations, and geographic or cultural specificities.

The pilgrims not only waste human labour power but prove to be so disorganized that Marlow is left waiting interminably for rivets he needs to fix his boat. Although he later suspects that the arrival of the rivets was impeded so as to delay his journey up the river, the incident underscores the general lack of order he observes in the administration of the colonial stations he visits. Singling out the manager of the Central Station for special opprobrium, Marlow spells out what he dislikes most about the colonizers:

> He was a common trader, from his youth up employed in these parts – nothing more. He was obeyed, yet he inspired neither love nor fear, nor even respect. He inspired uneasiness. That was it! Uneasiness. Not a definite mistrust – just uneasiness – nothing more. You have no idea how effective such a ... a ... faculty can be. He had no genius for organizing, for initiative, or for order even. That was evident in such things as the deplorable state of the station. He had no learning, and no intelligence. His position had come to him – why? Perhaps because he was never ill. (41–2)

Typical of other colonial administrators, the manager 'originated nothing,' but 'he could keep the routine going' (42). As a man who had 'nothing within him' (42), his behaviour was dictated by routine problem solving rather than by an ambition to initiate anything new. He was an exponent of the 'flabby, pretending, weak-eyed devil of rapacious and pitiless folly' (34) that Marlow blamed for the inhuman dying of the black workers. Although Marlow could be said to be responding 'appropriately' to the misery of the black slaves, 'his major criticism of Belgian rule is leveled against the inefficiency of the administration, the laziness of the "pilgrims," the sporadic blasting for the railway, the lack of straw for bricks and bolts for the steamboat, the waste of pipe in a gully, the general lack of devotion to the task' (Fleishman 105). Although 'criminal' *inefficiency* is perhaps not the overriding theme of the whole novel, it is clearly targeted in Marlow's descriptions of the Outer and Central Stations as inefficient bureaucracies.

It is usually assumed that the disreputable performances of the colonizers – the 'pilgrims' as well as Kurtz – is made possible by a loss of institutional safeguards that keep civilization from reverting to Hobbes's state of war. But the 'pilgrims' justify their behaviour by appealing to the very administrative apparatus that is meant to prevent their rapaciousness. No matter how inefficiently the administrators carry out their tasks, they exemplify the psychopathology already implicit in Taylor's principles of scientific management, Whyte's 'organization man,' and Weber's bureaucrat. Marlow's unfavourable depiction of 'inefficient clerks, disaffected functionaries, envious subordinates, and defensive superiors' (Levenson, 'Value' 395) is entirely compatible with Max Weber's analysis of bureaucracy and the reification of modern consciousness it entailed. Weber's contention that bureaucracy as a 'general system of legal domination' constructs the individual as 'a small cog in a ceaselessly moving mechanism which prescribes to him an essentially fixed route of march' (Weber, 'Bureaucracy' 228; in Levenson, 'Value' 396) is entirely compatible with Ford's assembly line, which reduces the individual to a 'cog in the machine,' and with Taylor's principles of management, which subordinate Whyte's 'organization man' to the rational system. Whether we associate Conrad's 'pilgrims' as representatives of a newly emergent social figure with Weber, Ford, or Taylor, the crucial point remains that Marlow is appalled by the bureaucratic sensibility of which the Manager is the most exemplary figure. The 'pilgrims' hypocritically conceal 'the banality of their evil' under an 'institutional formalism' (Levenson, 'Value' 396) that Marlow condemns for its petty criminality and hypocrisy. The veneer of bureaucratic order and efficiency conceals the 'violently anti-institutional facts' (396) that Marlow condemns as he witnesses the senseless suffering in the 'grove of death.' Appealing to the Company's institutional apparatus to justify their callous indifference to the collateral damage they cause, the 'pilgrims' embody a corporate logic that breaks down the traditional opposition between 'civilization' and 'barbarism.'

The only notable exception to this spectacle of slovenliness and inefficiency in part one is the Company's chief accountant, a 'white man' who, wearing a 'high starched collar, white cuffs, a light alpaca jacket, snowy trousers, a clear silk necktie, and varnished boots,' dresses with 'unexpected elegance' (*Heart* 36). This ostentatiously efficient figure embodies the Taylorized bureaucratic ideal in its most extreme manifestation; he is the defender of civilization even under the most unpropitious circumstances. Marlow can't quite decide how he feels about such dedication to the efficient life. While he admits that the accountant's appearance was 'certainly that of a hairdresser's dummy,' he 'respected the fellow' for keeping up his appearance 'in the great demoralization of the land' (36). In one of Conrad's implicitly racist moments, he has Marlow praise the Accountant for having taught one of the reluctant native women to iron his clothes. Most importantly, though, the Accountant 'was devoted to his books, which were in apple-pie order' (37). But Marlow undercuts his enthusiasm for the efficient Accountant by suggesting that he displays signs of the kind of psychological disassociation increasingly necessitated in those serving global capitalism. When a sick agent was temporarily put into the Accountant's office, he 'exhibited a gentle annoyance. "The groans of this sick person," he said, "distract my attention. And without that it is extremely difficult to guard against clerical errors in this climate"' (37). The Accountant is both admirable in the efficient execution of his duties and despicable for his insensitivity to the suffering of another human being. Inured to the groans of a white man in his office, the accountant is not a likely source for the alleviation of the suffering Marlow had witnessed in the grove of death. In his farewell from the Central Station, Marlow summarizes his experience of colonialism by juxtaposing three scenes: 'In the steady buzz of flies the homeward-bound agent was lying flushed and insensible; the other, bent over his books, was making correct entries of perfectly correct transactions; and fifty feet below the doorstep I could see the still tree-tops of the grove of death' (38). To the only efficient colonial agent Conrad attaches an absurd unreality and unreflecting dehumanization that anticipates Hannah Arendt's memorable judgment of Eichmann as the emblematic representative of the 'banality of evil' that allowed the atrocities of the Third Reich to unfold virtually unchecked.

Marlow finds it relatively easy to condemn the pilgrims as 'flabby' devils who lack both an efficient method and a moral purpose. His reactions to Kurtz are far more complex. According to prevalent critical opinion, Kurtz's adoption of an 'unsound method' exhibits a return of savage instincts that fortunately tend to remain repressed in modern, 'civilized' subjects. Marlow clearly suggests that Kurtz has in fact 'gone native,' a condition in which 'the whites become more savage than the "savages"' (Fleishman 90). Troubling the boundary between 'civilized' and 'barbaric' behaviour, Kurtz here reinforces a point that Conrad had already raised in the parallel between the colonization of Africa and the Roman conquest of Britain at the beginning of the novel. Addressing on board the *Nellie* in London a director of companies, a lawyer, and an accountant, Marlow recalls for these representatives of modern civilization the temptations faced by the 'civilized' Romans who were sent to administer 'barbarian' Britain during the historical 'darkness' that, from a longer temporal perspective, was 'here yesterday' (*Heart* 19). He imagines 'a decent young citizen in a toga' who, having perhaps been tethered to 'some inland post,' would have felt 'the savagery, the utter savagery' (19) that surrounded him. Having to 'live in the midst of the incomprehensible, which is also detestable,' the Roman would have been susceptible to the 'fascination of the abomination' (20) to which Kurtz will surrender in the story Marlow is about to narrate. Unlike the pilgrims, these Romans did not disguise the fact that they used 'brute force' to conquer Britain; they did not dress up 'robbery with violence' (20) as the philanthropic gift of Western civilization to the African continent.

Although Kurtz is a conqueror using 'brute force,' he represents for Marlow 'a principle of opposition to social machinery' (Levenson, 'Value' 397) that both Conrad and Weber sought in the 'organic' concept of the charismatic leader. When Marlow sides with Kurtz rather than the other colonial administrators, he expresses a hostility to the efficiency calculus that he shares with most other modernists discussed in this study (Forster, Ford Madox Ford, Lawrence, Huxley). Charisma escapes the rational bureaucratic organization of society precisely because it is 'the "gift" that is invested in a leader whose authority depends on neither tradition nor law, who indeed overturns every traditional and legal norm in the name of a personal calling acknowledged by an entire society' (398). Like Hobbes's sovereign or Taylor's expert, the charismatic leader stands over and above the system he governs; as we have seen in chapter 4, Weber stresses that the charismatic leader is bound neither by tradition nor bureaucracy. For Fleishman, charisma would constitute anarchy, whereas for Derrida it would indicate a suspect nostalgia for self-presence. Kurtz's voice, in particular, is emblematic of the immediacy Rousseau celebrated in his privileging of nature. The charismatic voice that commands the tribe over which Kurtz presides is associated with the violence perpetrated by the colonizer as conqueror or 'robber.' But it is also an eloquent voice that mesmerizes Marlow through the rhetorical power with which Kurtz expounds on the 'idea' of bringing the 'lighted torch' (*Heart* 46) of civilization into the darkness of Africa. According to the brickmaker, Kurtz came to the Congo as 'an emissary of pity, and science, and progress,' a prominent member of 'the gang of virtue' whose 'higher intelligence, wide sympathies, a singleness of purpose' made them representatives of the 'guidance of the cause intrusted to us by Europe' (47). Marlow is thus willing to excuse Kurtz's later violence on the basis of his initial idealism: 'The conquest of the earth, which mostly means the taking it away from those who have a different complexion or slightly flatter noses than ourselves, is not a pretty thing when you look into it too much. What redeems it is the idea only. An idea at the back of it; not a sentimental pretence but an idea; and an unselfish belief in the idea – something you can set up, and bow down before, and offer sacrifice to' (20). This passage already hints at the deterioration of this noble idea into the temptation to which Kurtz succumbs when he took 'a high seat amongst the devils of the land' and let himself be 'adored' (81, 93) by his tribe. Charisma has replaced rational planning. According to Marlow's favourite explanation, Kurtz 'goes native' because he lacks the inner restraint or moral strength to remain faithful to his ideals. But the story he narrates reveals itself as a more complex moral tale: 'Through its own strenuous logic *Heart of Darkness* pursues the representation of bureaucracy until it becomes the representation of a monstrous passion' (Levenson, 'Value' 401). What then is the nature of this 'monstrous passion'?

There is certainly evidence in *Heart of Darkness* that contradicts Marlow's suggestion that 'the whisper' of the wilderness 'had proved irresistibly fascinating' (95) to a man who was insufficiently armed by the protective layers of civilization to resist its dark lure. Could it be that civilization itself had had a hand in Kurtz's degeneration? What if we consider Kurtz's 'unsound method' to be the most 'criminally' efficient means for maximizing his output of ivory? Although Kurtz may have convinced himself that he was engaged in a civilizing mission, he was from the start serving the economic interests of his colonial masters. His main purpose for going to the Congo was 'to make his fortune' (Ross 42) so as to make himself acceptable as a suitor to the family of his Intended. Behind the 'immense plans' (*Heart* 106) – his intention to 'exert a power for good practically unbounded' (83) – hides the 'vulgar economic principle' (Ross 42) to remedy the poverty standing in the way of his sexual desire and social aspirations. The report commissioned by the 'International Society for the Suppression of Savage Customs' (*Heart* 83) was thus a mere sideshow to the main business of extracting as much ivory from the Congo as possible. For Ross, Kurtz makes 'a Faustian deal with the Company and enters the realm of capitalist desire and production in which there is no such thing as satisfaction, but only the perpetuation of desire' (43). In the final analysis, though, it may be neither the profit motive nor the satisfaction of libidinal desires that doom Kurtz but his 'criminal' obsession with efficiency for its own sake.

Although Kurtz's civilizing 'idea' was first of all corrupted by the profit motive, it later transpires that the ivory he hoards constitutes a fetish to be located outside the circulation of capital. It is certainly the case that ivory is a privileged signifier or fetish in the novel; it is 'a commodity that is of the essence of Africa and vital to European profit margins' (Ross 34). Beyond its status as a commodity rather than a material good, ivory is 'the means by which the company controls commerce' (35) and 'gives identities, establishes purposes, assigns destinies' (Levenson, 'Value' 395). The first time Marlow hears Kurtz's name, it is in the context of his colonial and capitalist status as 'a first-class agent' who 'sends in as much ivory as all the others put together' (*Heart* 37). Being 'at present in charge of a trading-post, a very important one, in the true ivory-country' (37), Kurtz is the object of much envious speculation among the pilgrims; they fear that he will outperform them and be 'a somebody in the Administration before long' (38). Assuring Marlow that 'Mr Kurtz was the best agent he had, an exceptional man, of the greatest importance to the Company' (43), the Station Manager then proceeded to delay Marlow's mission to retrieve the sick agent. Kurtz's unsound method did not meet with moral opprobrium on the part of the Manager, but only with anxiety about the competitive commercial advantage it entailed.

As the story unfolds, we discover that the Manager need not have bothered to obstruct Kurtz's return to Company headquarters or 'civilization.' When Marlow reaches the Inner Station, Kurtz refuses to leave with him, in effect repeating an earlier scene in which he turned his canoe around instead of accompanying a load of ivory to the Central Station, where he was also expected to buy supplies. Ross correctly reads this retreat into the wilderness as a breaking away from the Company in order to establish 'his independence at the Inner Station' (60). Attributing this drive for independence to frustrated libidinal desires, Ross draws on Zizek's association between 'fetishism' in Freud's sexual economy and Marx's commodity logic to argue that 'Kurtz is foremost a victim of the impingement of Imperial economic imperatives into the libidinal life of the subject' (42). The profit motive thus appears as a displaced manifestation of sexual frustration. Africa offers Kurtz 'sexual license' (44) in the form of his African mistress and 'frees' him from subservience to the Company. Emphasizing the profit motive, Ross sets Kurtz up as a competitor within the capitalist economy: 'Further, by refusing to restock his provisions from the Company's stores, Kurtz maximizes the profits of his station at the expense of the company – he all but declares himself an independent supplier of ivory to the Company rather than its agent' (60). From a different perspective, though, the profit motive can be seen to have been superseded by an obsession with efficiency that constitutes a nostalgic refusal of multinational capitalism. As Ross comments, the geopolitical space is so saturated by global capitalism that Kurtz could not possibly operate independently. Since the Company holds a monopoly, he would have to sell his ivory to it. His 'independence' is not only illusory but reinforces the Company's position in that 'the use of free agents can in fact enhance corporate margins' (61). In addition to constituting a challenge to the bottom line and to the Company's geopolitical control, Kurtz's retreat into the wilderness may well signal an obsession with efficiency for its own sake, an obsession common to both the amoral 'barbarism' of the premodern charismatic hierarchy and the 'barbarism' of the postmodern decentred corporate hegemony.

Conrad's *Heart of Darkness* is marked by the cultural crisis that Fredric Jameson locates in an economic shift from an emphasis on production to an emphasis on consumption. Weber's analysis of the type of Puritan personality ideally suited to the early stage of capitalism thus no longer adequately accounts for the type of personality required by third-stage or multinational capitalism. It could be said that Kurtz reverts to an early stage of capitalism premised on the accumulation of capital needed for the expansion of the market. In his arrogation of autonomy, he further exemplifies the principle of the 'free' market rather than the global market of monopolies and trusts. In this rejection of consumer capitalism, he hoards ivory in its material form rather than to circulate it in its artificial or mediated form for exchange in a money economy. When Marlow finally catches up to Kurtz, he discovers large stockpiles of ivory that this 'rogue' agent had accumulated by leading his adoring tribe to raid neighbouring villages. In his obsession with ivory, Kurtz even threatened to shoot his harmless Russian disciple unless the latter surrendered the small amount of 'ivory' (92) he had collected. Having sealed himself off from civilization and hence from markets, he collects ivory for its own sake rather than for the profit it would fetch if traded in the money economy. He has lost interest in the pursuit of profit because there are no goods to be consumed in the tribal community he is no longer willing to leave. Marlow recounts how he had piled up the ivory on the boat so that the dying Kurtz could 'see and enjoy' it as long as possible. 'You should have heard him say, "My ivory"' (81), recalls Marlow. In this scene, Marlow equates the wilderness with the ivory coveted by the Company: 'The wilderness had patted him on the head, and, behold, it was like a ball – an ivory ball; it had caressed him, and – lo! – he had withered; it had taken him, loved him, embraced him, got into his veins, consumed his flesh, and sealed his soul to its own by the inconceivable ceremonies of some devilish initiation' (81). Although his insatiable desire for ivory was initially stimulated by the profit motive, he regressed from a focus on the abstract form of ivory as a commodity to an obsession with it as a material fetish. What was deemed his 'unsound method' was highly efficient in the maximizing of the output of ivory he thereby achieved. In one sense, then, the 'unsound method' is analogous to global capitalism in that it thrives on the violent exploitation of native populations. In another sense, it represents a fantastical distortion of the nostalgic retreat from capitalist modernity into nature so typical of high modernists like Lawrence or Forster. But it also constitutes a radical challenge to multinational capitalism in that in the raids on native villages and in the gratification of his various lusts, Kurtz refuses the injunction to consume. When Marlow sides with Kurtz, he does so primarily because he appreciates him as a site of resistance to modernity as routine. As a charismatic leader he represents the principle of modernity as adventure. His unsound method has no need of organization. Although the heads on the stakes at Kurtz's compound speak of a violence at least equivalent to that perpetrated by the Company on the shadowy figures in the grove of death, Marlow seems to be saying that Kurtz was no bureaucrat but embodied the spirit of modernity as adventure.

As the agent of a colonial corporation, Kurtz embraces the 'freedom' from social constraints that corporate capitalism today enjoys in an increasingly deregulated global environment. Kurtz's moral regression is symbolized not only by the cannibalism he condones but also by the unrestrained greed for ivory he exhibits to the very end of his days. Symbolically, then, capitalism reproduces the barbaric savagery of the cannibalism Kurtz had set out to eradicate. What 'redeems' Kurtz in Marlow's eyes is his honesty; unlike the pilgrims, he does not hide his exploitation of the native population behind smokescreens of mystification, and, again unlike them, he acknowledges his moral degeneration. Marlow thus sides with Kurtz because he is destroyed by the moral sense of which he cannot divest himself in the very process of rendering it obsolete. Kurtz stands out for the moral insight he has gained into himself after having cut himself loose from the social contract laws that govern civic society but not capitalist corporations. He is the Faustian striver who, having been tested under extreme conditions, made a 'bargain for his soul with the devil' (*Heart* 82). In his defence of Kurtz, Marlow contends that the men listening to him on board the *Nellie* are protected from both temptation and self-knowledge: 'Here you all are, each moored with two good addresses, like a hulk with two anchors, a butcher round one corner, a policeman round another, excellent appetites, and temperature normal – you hear – normal from year's end to year's end' (80). The unidentified narrator stresses that Marlow's audience consists of the 'Director of Companies,' the 'Lawyer,' and the 'Accountant' (15, 16), figures clearly complicit with the dehumanization in the colonies that Kurtz first witnessed and then perpetuated. Whereas these accomplices of colonial capitalism are able to conceal from themselves the violence they make possible, Kurtz has no such saving illusions.

Marlow's own perspective on efficiency makes him highly disdainful of the pilgrims' inefficiency as well as anxiously critical of Kurtz's relentless pursuit of efficient outcomes. In the proleptic opening pages of *Heart of Darkness*, Marlow concludes his musings on the savagery of the Roman conquerors with an affirmation of efficiency: 'What saves us is efficiency – the devotion to efficiency' (20).[4] Although Marlow tends to approve of the organic conception of work rather than efficiency, in this metaphorically significant passage he refrains from making this distinction. Since Marlow's own work furthers the aims of global capitalism, it seems that any efficient execution of a task makes personal satisfaction possible. In both its organic and its mechanical conception, 'devotion to efficiency' is paradoxically positioned as both a connection to the inner self and as an escape from it. On the one hand, Marlow claims that it is through work that he learns who he really is: 'I don't like work – no man does – but I like what is in the work, – the chance to find yourself. Your own reality – for yourself, not for others – what no other man can ever know. They can only see the mere show, and never can tell what it really means' (52). In an affirmation of the Puritan ethic, Marlow sees in the execution of tasks a test or confirmation of an individual's moral fibre and meaningful existence. Throughout *Heart of Darkness* he praises those who show even the slightest dedication to work. As we have already seen, no matter how ambivalent he may feel about the Accountant, he is clearly impressed by an uncompromising commitment to work in the midst of waste and chaos. In his evaluation of the colonial enterprise, he often judges a man's[5] worth by his attitude to work. Coming upon the abandoned abode of a white trader, he finds a book – 'An Inquiry into some Points of Seamanship' (65) – that speaks to the owner's concern with the proper way of doing his job. Noticing recent margin notes in cipher, he includes even the Russian trader, who 'looked like a harlequin' (87), in his list of admirable men. Even among the 'pilgrims' he approves of 'a boiler-maker by trade' who is 'a good worker' (53). Not unlike Henry Ford, Marlow insists that he preferred to 'chum' with the 'mechanics that were in that station, whom the other pilgrims naturally despised' (53). It is on similar grounds that Marlow senses a kinship with the native fireman who has been trained to perform his work against what Conrad imagines to be his 'savage' instincts. Achebe has rightly criticized the implicit racism in Marlow's supposedly kindly attitude to the fireman. Even if we could be certain that Marlow does not treat the fireman as racially inferior, he would still not see him as a human being but only as an instrumental cog serving the needs of the steamboat. In spite of their faults, the characters who are committed to the efficient performance of work are favourably compared with the corrupt and wasteful company men who exploit others without contributing any productive work of their own.

On the other hand, though, work is identified as an escape from the deeper self-knowledge that both dooms and ultimately redeems Kurtz. Marlow is intent on probing the depths of Kurtz's moral degeneration because his 'unsound' pursuit of efficiency constitutes an affront to his own sense of work as a humanly meaningful activity. Kurtz's undoubted devotion to the work of extracting ivory is marked by a capitalist disregard for the collateral damage he inflicts on both the land and its inhabitants that is at odds with Marlow's own sense of the satisfaction the craftsman derives from his toil. The novel owes much of its dramatic tension to the exploration of characters who need to negotiate the clash between a residual pre-industrial and an emergent capitalist ideology. Where Kurtz was forced to confront his lack of 'innate strength' (*Heart* 82), Marlow shies away from such terrible self-knowledge: 'True, he had made that last stride, he had stepped over the edge, while I had been permitted to draw back my hesitating foot' (113). What saves Marlow from following Kurtz is precisely his devotion to efficient work. Steering his steamer up the river into the heart of darkness, he remains busy keeping it going and navigating it safely: 'When you have to attend to things of that sort, to the mere incidents of the surface, the reality – the reality, I tell you – fades. The inner truth is hidden – luckily, luckily' (60). It was not 'fine sentiments' that stopped him from going 'ashore for a howl and a dance,' but the fact that he 'had no time' while putting 'bandages on those leaky steam-pipes' and circumventing 'those snags' (63). As Marlow concludes, 'There was surface-truth enough in these things to save a wiser man' (63). The social norms provide Marlow's audience on the *Nellie* with an illusory self-understanding that protects them from probing too deeply into their own culpability. Unlike those who are 'too dull even to know [they] are being assaulted by the powers of darkness' (82), Kurtz self-consciously divested himself of 'the surface-truth' of the efficiently organized social order that Marlow ultimately also endorses.

When Kurtz first set out for Africa, he did so under the illusion that he acted as a representative of 'modernity as adventure.' But the colonial administration of the Congo epitomized a particularly degraded form of 'modernity as routine.' Being a 'gifted' or charismatic figure, he rejected not only the inefficient pilgrims but also the very logic of routinized effort embodied in the highly efficient Accountant. Joining other modernist protagonists, he retreated from modern civilization into the 'natural world' or wilderness. Not unlike Huxley, Conrad does not romanticize 'nature' or the 'primitive' as is the case in the novels of Lawrence, Forster, or Ford Madox Ford. Instead of endorsing Rousseau's myth of the spontaneous cooperation of 'free' individuals, Kurtz encounters and incites Hobbes's conception of the state of nature as a war of all against all. Although Kurtz's intrusion into the wilderness undoubtedly disrupted existing tribal social formations, he is not guilty of introducing 'violence' into a previously 'innocent' social space. His situation is comparable to that of Lévi-Strauss in Derrida's analysis of the ethnographer's incursion into the territory of the Nambikwara. In his desire to distance himself from the ethnocentrism he wishes to combat, Lévi-Strauss stages an elaborate *mea culpa* in which he blames himself for having introduced 'violence' to a people he considered to have been previously untouched by it. Accusing Lévi-Strauss of a Rousseauistic nostalgia for metaphysical presence, Derrida meticulously deconstructs the ethnographer's discourse in order to illustrate that there never has been a historical moment before 'violence.' Instead of indulging in the myth of Rousseau's 'noble savage,' Conrad confronts the disavowed increase in violence made possible when multinational corporations are allowed to ravage natural and social worlds unchecked. If Kurtz's 'violence' is not the eruption of a repressed atavism but the logical extension of the efficiency calculus central to global capitalism, then the Empire of multinational corporations leaves no sites of resistance as it insinuates itself into every corner of the globe's surface. Kurtz's openly violent conquest is for Marlow preferable to the concealed violence of global capitalism; Kurtz's attempt to reclaim 'modernity as adventure' is an anachronistic gesture or parodic repetition of a historical moment that has long since been defeated by an increasingly rapacious and soulless 'modernity as routine.'

There are some uncanny parallels between Kurtz and Henry Ford. Where Kurtz wanted to improve the lives of Africans, Ford sought to alleviate the hard toil of farmers. Throughout his life, Ford also affirmed his sympathies for the workers he employed, preferring their company to that of high society. Similarly, Kurtz did not want to leave the tribe he exploited. Yet both men became infected with the desire to be more efficient than anyone else in the output of cars or ivory. Ford cared no more about profits than Kurtz did about the money his ivory would fetch. Ultimately, then, neither of them was driven by either moral motivations or material rewards; both were obsessed with the most efficient methods at hand to maximize an output to which they were essentially indifferent. It is this devotion to efficiency that dehumanizes both the real man and the fictional character. As we have seen, against his own instincts, Ford allowed Harry Bennett to institute methods of intimidation designed to abuse the very men Ford had claimed to respect. Perhaps the most bitter irony of Kurtz's fate is the note he added to the incomplete report on the suppression of savage customs: 'Exterminate all the brutes' (*Heart* 84). Ford and Kurtz illustrate the psychopathology of those who divorce their efficient methods from social or moral aims. If we scratch below the surface of the rhetoric of progress, the industrialist and the colonial agent employ unsound methods that make them the corporate-capitalist equivalents of the Roman conquerors of Britain. At the same time, both express nostalgia for the 'simple life' of the tribe or the farming community. Illustrating the conflicted cultural logic arising with the emergence of late capitalism, they exemplify the psychopathological disassociation of personality that permits them to disavow their moral degeneration. It is not that Kurtz represses a secret; he knows that what he does is morally reprehensible but he nevertheless persists in his behaviour.

It is interesting to note that the preoccupation with efficiency in *Heart of Darkness* discussed in this chapter surfaces in Francis Ford Coppola's reworking of Conrad's novel in his acclaimed Vietnam War movie *Apocalypse Now* (1979). In his analysis of the film's adaptation of the novel, Simon During foregrounds Coppola's radical extension of Conrad's still hesitant acknowledgment of efficiency as a pervasive social force. The film's Willard repeats the journey Marlow undertakes in the novel as he travels up a river in an attempt to reclaim Kurtz from the heart of darkness. Where the novel represents Kurtz's rhetorical power as a voice without content, the film fills in this void with quotes from Eliot and Frazer and Weston as well as 'a Nietzschean tirade on greatness as the capacity to bear the suffering of others' (During 454). This content introduces a 'historical incongruity' in that Coppola indicts the Vietnam War by falling back into a 'mere monumentalization of modernism' (454). In other words, Kurtz in *Apocalypse Now* is the product of 'a liberal arts education' that does not 'add up to charisma,' in spite of assurances that Brando-Kurtz is to be regarded as a 'genius' (454). The film's Kurtz stands out not for his moral insights or existential suffering but for his devotion to efficiency: 'His true distinction in the film's own terms is his efficiency, his refusal to play the hypocritical game of army bureaucrats. But in having him killed they do not play their own game either – so there is no final difference here. Ultimately, efficiency rules everywhere. The values of honour, truth and work for work's sake, which Conrad upholds as he reveals their limits, have disappeared along with the autonomous subject and work of art' (454). During's reading of *Apocalypse Now* brings into sharp focus the cultural anxiety with the emergence of efficiency as a dominant ideology that in *Heart of Darkness* remains tentative and shadowy.

7 Efficient Management: D.H. Lawrence's *Women in Love*

The son of a miner, D.H. Lawrence had witnessed Taylorized rationalization processes in the mining industry that had reverberated throughout the social fabric at the turn of the century. A novel ostensibly concerned with the sexual experiences of two sisters, *Women in Love* (1920) is marked by deep cultural anxieties that find their most explicit exposition in chapter 17, 'The Industrial Magnate.' Taking over his father's coal mines, Gerald Crich revolutionizes the family business by changing the process of production from a patriarchal to a modern scientific model indebted to an understanding of the Taylor expert wielding his stopwatch. Focused on the emergence of entrepreneurial capitalism, this chapter constitutes an almost self-contained historical essay, which seems to interrupt the main narrative concentration on the courtships of Ursula/Rupert and Gudrun/Gerald. The appearance of this historical digression in the centre of the novel alerts us to the possibility that we are dealing with more than Gerald's family background. The critical consensus has indeed stressed that 'The Industrial Magnate' occupies a 'special place within the novel'; it is 'a narrative within a narrative' that 'assumes a privileged role within the text as a tale of origins, a governing fiction by means of which we might interpret the historical crisis which is the torment of these characters' (Knapp 154). Crucial to this historical crisis is Gerald's emphasis on making the mines more efficient; as Hubert Zapf has shown, Frederick Winslow Taylor was at the very least a diffuse source for the mechanical social system that Lawrence extrapolates from the scientific industrial model and compares unfavourably with earlier organic alternatives. Haunted by Taylor, this central chapter thus invites an analysis that takes into account not only Lawrence's unequivocal hostility to technical efficiency but also the contradictory logic implicit in the novel's treatment of social efficiency.

Like other modernist literary figures, Lawrence manifests highly conflicted attitudes towards modernity. Having known poverty as a child, he escaped from his working-class background into the professional rank of the schoolteacher and eventually into the higher socio-cultural spheres opened up to the acclaimed writer. Initially sympathetic to socialist aspirations, he later embraced a cultural conservatism hostile not only to the left but also to bourgeois liberalism and democratic institutions. Fearing the unpredictability of worker unrest, he sought refuge in the paternalistic fantasy of a strong leader capable of imposing order on an ignorant mob. In a letter to Bertrand Russell, he admonished him to 'drop all your democracy' because 'there must be an aristocracy of people who have wisdom, and there must be a Ruler: a Kaiser: no Presidents and democrats' (Lawrence, *Letters* 364; in Daly, 23). The state is for him 'a vulgar institution,' while 'life in itself is an affair of aristocrats' (*Letters* 254; in Daly, 22). The social unrest epitomized by the miners' strikes he had witnessed had for him culminated in the carnage of the First World War. His first reaction to such unrest was the 'experimental utopian colony ("Rananim"), to be peopled by the literati, professionals, and aristocrats who now formed his regular circle' (Daly 21). Throughout his life, Lawrence fell back on various versions of utopian communities to be established in such places as Florida, Australia, New Mexico, Mexico, and Tuscany. If he could prevail on decent and highly educated people to congregate in a socially isolated space, they would presumably spontaneously form an organic community. It is this kind of utopian vision that Huxley satirized in *Brave New World* when a group of Alphas were left to organize themselves on the Island of Cyprus. In the satire, these most intelligent beings immediately descended into chaos and petitioned the World Controller to let them rejoin the totalitarian social order. When Lawrence had to admit to himself that the anarchic or 'free-floating paradise' of 'Rananim' would not materialize, he compensated for this failure by opting instead for the charismatic principle of the 'absolute *Dictator*' (Daly 21, 23). As in the case of Wells's and Shaw's qualified approval of eugenics, Lawrence resorts to dictatorship as a compensatory abstract solution to real or intractable social contradictions. In a rather typical vacillation, he told Russell in yet another letter: 'I don't want tyrants. But I don't believe in democratic control' (*Letters* 1981, 370–1; in Daly, 23). Stressing 'manifold social and political contradictions' (Daly 31) in Lawrence's sociopolitical thinking,

Daly concurs with the predominant view, expressed by Delany, that the novelist's 'conversion to belief in "a kaiser" was just one of a series of intellectual somersaults' (Delany, *Nightmare* 121) marking his writings; as Delany cautions us, it is 'only by forcing the facts [that] we can fit these changes into a coherent pattern of intellectual or emotional development' (121–2). No matter how incoherent his search for sociopolitical alternatives may have been, Lawrence leaves little doubt about his hostility to the 'efficient' modern life most graphically embodied by Gerald's rationalization of the coal mines.

The flashpoint for Lawrence's fictional dramatization of the detrimental consequences of Taylorization in the coal-mining industry is the historical miners' lockout of 1893. While some critics are preoccupied with the historical accuracy of Lawrence's depiction of the miner's strike (Knapp, Daly), Holderness stresses above all the 'three distinct stages' in the 'development of the mining industry' (203), an industry that is in Lawrence always symbolic of industrial capitalism as such. There is 'an initial "*laissez-faire*" stage of primary accumulation,' followed by 'a "paternalistic" stage, in which the sufficiently wealthy capitalist becomes philanthropic,' and finally 'the stage presided over by Gerald Crich, in which the Victorian paternalist and philanthropic system is transformed into a modernized, mechanized system' (203). For this critic, this last stage is 'still "entrepreneurial" – decisively dominated by the will and activity of a single man' (203). However, 'The Industrial Magnate' provides no examples of 'primary accumulation'; the anarchistic logic of laissez-faire systems finds expression in the aristocratic tradition represented by the mother, Christiana Crich. Thomas Crich is always already a representative of the 'paternalistic' stage that is being challenged by his son's 'Americanization' of the production process. But, in the figure of Gerald, Lawrence also points beyond entrepreneurial capitalism to its transformation into a corporate stage in which the charismatic figure of the entrepreneur will be replaced by the faceless 'manager.' In 'The Industrial Magnate,' Lawrence rails against industrial capitalism as a threat to liberal-bourgeois individualism, which he locates not in parliamentary democracy but in a laissez-faire political anarchism predicated on the spontaneous cooperation of a cultural elite. Crucial to this central chapter of *Women in Love* is the emergence of efficiency as an ideology that Lawrence feared as a juggernaut intent on the destruction of nature, meaningful individual life, and organic social bonds. Although Lawrence is brilliant in his depiction of the desire for domination and violence implicit in Gerald's investment in efficiency, he overlooks the possibility that his alternative social models remain complicit with capitalism. As Holderness argues, Lawrence's 'pleading for individual freedom' reinforces capitalism in its 'transformation' from 'the entrepreneurial system of capitalism' to the later stage of 'monopoly capitalism' (195). More recently, John Marx has convincingly demonstrated that Rupert's supposedly nostalgic yearnings for more organic social bonds are in fact implicated in Britain's imperial ambitions; his elite 'connoisseurship' of 'things' like tropical objects and antiquities is representative of the consumer stage of global capitalism, which constitutes an advance over Gerald's production stage of entrepreneurial capitalism.

In 'The Industrial Magnate,' though, the condemnation of efficiency is, as Holderness claims, predominantly a concentrated confrontation with industrial capitalism as an emergent mode of *production* emblematic of the entrepreneurial stage rather than as an incitement to *consumption* typical of the corporate stage. The critical disagreement over the historical accuracy of the miners' lockout at stake in 'The Industrial Magnate' allows us to gauge the ideological mystifications to which Lawrence is driven by his inability to deal with real social conflicts. James F. Knapp, for instance, comments on the 'interpretive clarity which this wonderfully coherent narrative of historical change seems to offer the reader' (154); for him, Lawrence is to be praised for capturing the 'degradation of industrial labour' (159) under the impact of Taylorization and for the 'historically accurate' depiction of 'the miners' assent' to the 'scientific rationalization of industry' (159, 167). In contrast, Macdonald Daly foregrounds historical inaccuracies that he convincingly attributes to Lawrence's ideological need to reconfigure historical events in the metaphysical register of cultural apocalypse. For him 'The Industrial Magnate' is not a 'historically accurate version' of 'the industrial climate of the 1890s and early 1900s' (25); on the contrary, Lawrence superimposes on this historical moment an apocalyptic vision of cultural decline more typical of the period of social unrest after the First World War, a period during which the novel was in fact composed. Daly painstakingly demonstrates that Lawrence projects later moments of worker unrest back onto 'the Midlands mining lock-out of 1893' (26), converting the relatively successful rebellion of the miners into a defeat. The dispute of 1893 thus 'fostered a solidarity of a radically different complexion from the submissive unity of instrumentality which Lawrence tries to establish as its upshot' (Daly 29). The miners have to be defeated and transformed into hapless cogs in the efficient machinery of production so as to allow Lawrence to condemn Taylorism unequivocally as a scourge of cultured civilization. The conflict between the mother's aristocratic values, the father's patriarchal style of management, and the son's new industrial system is analysed in terms that carry the mark of Lawrence's virulent opposition to Fordism and Taylorism from a nostalgic conservative position that unwittingly reinforces the very capitalism it seeks to resist. A prototype of the 'visible' style of management characteristic of entrepreneurial capitalism, Gerald seems to be held personally responsible for his actions and punished for his obsession with efficiency. Yet, as we will see, his inner emptiness and inability to enter into meaningful relationships with others take on their fatal dimensions only after his entrepreneurial role had been exhausted and had made way for the later stage of corporate capitalism.

The Taylorization of industry portrayed in 'The Industrial Magnate' holds the key not only to an understanding of the anxieties and contradictions mobilized by the emergence of efficiency for its own sake but also to the struggle for control over others concealed by the appeal to a supposedly ideologically 'neutral' scientific model. In the novel's historical genesis, the transition from the father's 'patriarchal' to the son's 'modern' capitalism arises out of the mother's adhesion to an earlier aristocratic tradition of landed property. Given Lawrence's own social aspirations, it may strike us as surprising that Christiana Crich should be portrayed in highly unfavourable terms. The novel's hostility to the aristocratic mother could, perhaps, be explained by Lawrence's gendered attitudes to women or by the working-class man's disavowed resentment towards an elite he simultaneously envies. More to the point, the unflattering treatment of Christiana may be structurally necessary to convey the aggressive stance of a social order finding itself increasingly in a defensive position. In her hysterical reaction to her husband's entrepreneurial capitalism, she may even constitute a displaced expression of Lawrence's own impotence in the face of an apparently unstoppable social investment in efficiency. Described as 'a pure anarchist, a pure aristocrat' (Lawrence, *Women* 292), she is representative of a loosely organized caste system subject only to the will of powerful individuals whose privileges are based on accidents of birth. Anarchy is here aligned with individual 'freedom' from responsibility or capricious whim. Not surprisingly, then, she abhors her husband's sense of stewardship of the miners he employs. Unwilling to see the lower classes as human beings, she is as dehumanizing in her attitude as Gerald will prove in his drive for productivity.

In contrast, Thomas Crich adopts a 'patriarchal' form of leadership that conceals from itself the violence it perpetrates. A 'large employer of labour,' he resembles the early Henry Ford in that 'he had never lost this from his heart, that in Christ he was one with his workmen.' He was a 'philanthropist' whose generosity to others irked his wife who, 'like a bird of prey,' pounced on 'her husband's soft, half-appealing kindness to everybody.' In her eyes, he seemed to be 'some subtle funeral bird, feeding on the miseries of the people.' It is no wonder that he is said to be living in 'dread' of his wife, 'the destroyer.' Recoiling from 'this world of creeping democracy,' she effectively deconstructs his philanthropy as the self-gratification and self-love it really was: 'He liked hearing appeals to his charity.' In a relationship marked by 'utter interdestruction,' she offered intense passive resistance to the husband who 'had subdued her.' The pre-industrial age, symbolized by Christiana, has been supplanted by Thomas's 'heroic' or emergent phase of industrial capitalism; however, at this historical moment, the residual ideology vested in her is still capable of bleeding 'the vitality' out of the husband. Although reduced to an insane and hysterical woman who has 'no connection with the world' (*Women* 286–90), she nevertheless voices an impotent anger at the new order that compares favourably with the willing submission Lawrence will attribute to the miners who bend to Gerald's will.

In his depiction of the intergenerational conflict between father and son, Lawrence quite obviously romanticizes Thomas Crich and demonizes Gerald; ideologically, he thereby privileges what he assumes to be an organic connection between generous master and grateful servant that is being threatened by the indifference of the Taylorized system to the social needs of human beings. Although couched in Oedipal terms, Gerald's reactions to his father are inflected by the socio-economic transformation the two figures represent. When the dying Thomas Crich appeals to his son for compassion, Gerald feels a 'poignant pity' that is simultaneously 'shadowed by contempt and unadmitted enmity' (*Women* 291) not only for the father but also for the outmoded values he represents. According to Roger Ebbatson, Lawrence here dramatizes Max Weber's contention that, 'by the turn of the century, capitalism was so highly developed that it no longer required the foundation of ascetic Protestantism, but had attained a secular value-system of its own which he named "economic rationality"' (99). The 'self-contained' (*Women* 292) Gerald has nothing but disdain for a 'master' who seeks acceptance from the 'servants.' In Hegelian terms,

Thomas thrives on a mimetic rivalry that binds master and slave together in a dialectical formation. The master depends for his sense of identity on the acknowledgment of the servant whose desire to replace him confirms his own position of superiority. It would thus not be in the master's interest to kill the servant; where Thomas refuses to commit the 'Hegelian murder' that would also have doomed him, Gerald will have no such compunction. While Gerald is able to displace Thomas's patriarchal system, the son's suicide proves a cultural degeneration that Lawrence can only express in apocalyptic terms.

It is not difficult to realize that Lawrence's sympathies lie entirely with Thomas Crich. As Daly contends, the novel exaggerates both the extent of the generosity of conscience-stricken owners and their philanthropic motives. In the owners' support of the miners, 'there was a great deal of sheer business perspicacity' (Daly 28); the handouts in the novel were in reality loans to be repaid once the miners had been tided over the hard times. When we are told that Thomas was always thinking 'only of the men' (*Women* 297), we are treated to a romanticized version of the supposedly organic relationship between owners and miners. During the lockout, he is shown to have valiantly upheld an unpopular stand against other owners, falling in line with them only under duress. He 'had lived and striven with his fellow owners to benefit the men every time' (297), and he had never turned hungry men and women from his door, providing them with leftovers from his own table. Lawrence ultimately absolves Thomas from personal blame, showing that both the miners and the other owners eventually drove him to lock out his men. In the first place, the miners are guilty of greed and ingratitude. Since 'man is never satisfied,' the miners passed from 'gratitude to their owners' (297) to complaints and demands for more. In the second place, Thomas was exonerated from culpability by the distress he felt when he was forced by fellow owners to close down the mines: 'He, the father, the patriarch, was forced to deny the means of life to his sons, his people' (297). It is important for the reader to understand that the 'men were not against *him*' but against 'the masters' (298) who had forced his hand. Thomas is thus presented as a tragic figure unable to understand that his personal moral code had been superseded by the logic of a new social order. Although tinged with ironic condescension, the representation of Thomas's allegiance to an outmoded feudal form of patronage acknowledges the irrevocable passing of a nobler age. At the same time, the strike forced the father to acknowledge his complicity in the money economy, robbing him of the comforting illusion that he 'wanted his industry to be run on love.' What was 'breaking his heart' was not the hostility of the miners but the solidarity with the owners the strike imposed on him. Trying to counteract the damage he was inflicting as a capitalist, he was 'giving away hundreds of pounds in charity' (298–9). Incapable of separating his personal moral code from his public position, he was destroyed by a schizophrenic contradiction he could not resolve: 'He wanted to be a pure Christian, one and equal with all men. He even wanted to give away all he had, to the poor. – Yet he was a great promoter of industry, and he knew perfectly that he must keep his goods and keep his authority' (299–300). The striking miners forced him to confront his own moral emptiness and self-serving hypocrisy.

In stark contrast with his father, Gerald not only acknowledged his power position but relished the thought of bending both nature and men to his will. Inheriting the mine, he was determined 'to put the great industry in order' (300). To this end, he adopted a gospel of efficiency quite obviously indebted to Frederick Winslow Taylor. Juxtaposing passages from Taylor's *Principles of Scientific Management* with passages from *Women in Love,* Zapf illustrates convincingly that the 'emphasis on efficiency, rationalization, and science' (130) in the novel must have been directly influenced by Taylor. Maintaining that Taylorism had been 'highly influential' in Britain at the time 'when *Women in Love* was written' (130), Zapf contends that Lawrence need not necessarily have read the *Principles of Scientific Management* to have been aware of their application in the coal mines in which his father had been employed. In his view, Gerald proceeds to modernize the mine by 'rationalizing the administration,' 'improving and standardizing [the] production tools,' introducing 'new technology and machinery,' transferring 'the control of the working process from the workers themselves' to a 'group of scientifically trained experts,' 'subdividing the working process into its smallest elements,' 'increasing the speed of production,' 'giving absolute sway to the scientific principle on which the whole system must be based,' 'replacing the inherited class conflict' by the 'functional cooperation' between class antagonists, and 'interpreting the new system as a great ideological step in the progress of human civilization' (133–6). Associating Taylorism with mechanical operations, Lawrence sees it above all as a threat to the organic principle he values in social relations. In 'The Industrial Magnate,' Gerald's commitment to efficiency is dehumanizing and disruptive of meaningful social bonds. Reducing Taylor's Pareto-efficient utopian vision to its assembly-line degeneration at Ford Motor, Lawrence exhibits an antipathy to Taylorism as well as Fordism that is rather typical of the British resistance to scientific management that Merkle identifies in *Management and Ideology.* Gerald resolves the detrimental impact of strikes on economic productivity not by soliciting the cooperation of workers, as Taylor advises, but by brutally imposing his own will as Henry Ford allowed Harry Bennett to do at Ford Motor. The power asymmetry typical of his father's patriarchal style of management carries over into Gerald's Taylorized substitution. The class conflict is resolved through the suppression of the workers who consent to their oppression. Beyond the reinforcement of organic over mechanical social organization, 'The Industrial Magnate' illustrates that 'the abstract system' is allowed to take precedence over the 'individual person' (Zapf 137) because both owners and workers have internalized the totalizing desire of the efficiency calculus.

Already a man of great fortune, Gerald was not driven by the lure of money but by the prospect of controlling his natural and social environments. His commitment to industrial efficiency is inextricably linked to a totalizing desire for sociopolitical mastery. In his early years he sought to increase the efficiency of the mine for the sake of self-satisfaction: 'The profit was merely the condition of victory, but the victory itself lay in the feat achieved' (*Women* 296). He was a Faustian striver who put all his efforts into building a modern edifice he might not want to inhabit once it was completed. The dominant metaphor for this modern edifice is the machine, which is contrasted with organic images of nature throughout *Women in Love;* the novel's dramatic tension is unmistakably structured around an ideological opposition between the character pairs Gerald/Gudrun (machine efficiency) and Rupert/Ursula (natural spontaneity). The mine is for Gerald a microcosm of society as a whole. Committed to 'the great social productive machine,' he 'abandoned the whole democratic-equality problem as a problem of silliness' (300). The father's liberal aspirations are dismissed in favour of a totalized system predicated on the rigorous integration of socially isolated atoms that far exceeds the mother's feudal authoritarianism. The emphasis in 'The Industrial Magnate' is neither on the amorality of the profit motive nor on the social inequalities on which the capitalist system thrives, but on the threat that the efficiency calculus poses to the cultural or spiritual needs of 'free' individuals and to the spontaneous social bonds Lawrence sought to 'engineer' in utopian communities like 'Rananim.'

In 'The Industrial Magnate,' Lawrence explores the ideological implications of Taylorization by focusing most specifically on the psychological drama unfolding as Gerald internalizes the efficiency calculus.

Once in charge of the mine, he intended to impose his will first on 'the resistant Matter of the underground; then the instruments of its subjugation, instruments human and metallic; and finally his own pure will, his own mind' (301). This trajectory traces in an uncanny parallel the stages in Henry Ford's shift from technical to social efficiency and ultimately to the reification of his own consciousness. Gerald begins his 'great reform' by replacing miners with machines: 'New machinery was brought from America' and the control over the machines 'was taken out of the hands of the miners' (304). In a description that directly references Taylor's principles of scientific management, Lawrence explains: 'Everything was run on the most accurate and delicate scientific method, educated and expert men were in control everywhere, the miners were reduced to mere mechanical instruments' (304). Not only do the miners have to work harder than ever, but the work was also 'terrible and heartbreaking in its mindlessness' (304). Although seeing himself as 'the God of the machine' in control of 'a great and perfect machine, a system of activity of pure order, pure mechanical repetition, repetition ad infinitum, hence eternal and infinite' (301), Gerald understands that even he himself is only a mere instrument in a larger design: 'He himself happened to be a controlling, central part, the masses of men were the parts variously controlled. This was merely as it happened' (300). He is able to act inhumanly because his commitment to efficiency has resulted in the reification and instrumentalization of his own consciousness. His own dehumanization is the price he is willing to pay for the social-engineering experiment that is meant to replace his father's soft hypocrisy with his own cold acceptance of the 'practical world order.' Explaining that Gerald was filled 'with an almost religious exaltation' by the idea of 'this inhuman principle in the mechanism he wanted to construct' the narrator provides one of the most accurate articulations of the ideology of efficiency anywhere in literature: 'And for this fight with matter, one must have perfect instruments in perfect organization, a mechanism so subtle and harmonious in its workings that it represents the single mind of man, and by its relentless repetition of given movement, will accomplish a purpose irresistibly, inhumanly' (301). Although this formulation could be read as Gerald's Nietzschean will to power, it in fact reiterates the machine-god's submission to the system. It is not that the system serves the 'single mind of man'; rather, the system functions with single-minded concentration. Gerald himself is simply a replaceable aspect of the totalized system that depends for its perfection on the efficient deployment of efficient instruments. The 'purpose' this system will accomplish so 'irresistibly' and 'inhumanly' is its eternal self-reproduction; the drive for efficiency has no purpose beyond itself. Once Gerald has totalized the system, he no longer has a purpose in life; as Ursula recognizes, 'he'll have to die soon, when he's made every possible improvement, and there will be nothing more to improve' (99).

While efficiency for its own sake has no meaningful purpose, it has dire and destructive consequences for those who submit to its logic. Lawrence is not concerned with collateral damage to the environment or the workers slaving in the mines; his concern is with the damage sustained by (conservative or high) culture as individuals internalize industrial society's obsession with efficiency. On the most obvious level, the miners 'became more and more mechanized' and, more importantly, they 'were satisfied to belong to the great and wonderful machine, even whilst it destroyed them' (304). Critics have complained that Lawrence has no Marxist faith in the proletariat as a subject of history. In his attitude to the miners, he exhibits what Holderness identifies as an ideology that Marx and Engels called 'critical-utopian socialism,' which 'can "see the class antagonisms" in bourgeois society, but which does not recognize the proletariat as capable of any "historical initiative"' (196). Expanding on this line of argument, Daly offers historical evidence showing 'how the dispute of 1893 fostered a solidarity of a radically different complexion from the submissive unity of instrumentality which Lawrence tries to establish as its upshot' (29). It is for ideological reasons that 'he must write as if the miners had lost rather than won the 1893 dispute' (29). Although in 1893 Lawrence was only a boy, he had been close to the events on which the novel's historical trauma is directly based; he must have changed the historical record knowingly. Daly surmises that Lawrence excluded 'socially transformational events' so as to 'achieve "a perfect, complete, summation of the epoch," a finished, framed, and hung picture of a doomed civilization' (Daly 30). One might want to add that this apocalyptic doom is specifically traced to the Taylorization not only of industry but of the social fabric as a whole. The miners are shown to be mesmerized by the invention of machines and the efficient new order they make possible. There is no explanation for the 'fatal satisfaction' with which the miners eventually accepted the changes Gerald introduced. In psychoanalytic terms, the miners could be said to embrace the death drive or Thanatos: 'There was a new world, a new order, strict, terrible, inhuman, but satisfying in its very destructiveness' (*Women* 304). This self-destructive urge anticipates Gerald's own submission to Thanatos in the Tyrolean Alps.

Lawrence's reactions to changes in the mining industry take on metaphysical dimensions that make more than one critic claim that 'Lawrence is primarily concerned with spiritual change' (Worthen 99). The treatment of the miners' strike is thus 'more concerned with myth than with history'; Lawrence is 'talking about souls, not about working selves' (99). Yet the spiritual crisis is firmly located in the economic sphere. While Lawrence's reactions to the strike may be more hysterical than historical, the novel nevertheless captures the loud wail against Taylor's infamous assertion 'In the past the man has been first. In the future the system must be first' (Kanigel 19). Like Huxley, Lawrence objected to the threat the system posed to the 'free' and unique individual. Most disturbingly, the 'free' individual consented to his or her victimization. Without the participation of the miners, 'Gerald could never have done what he did' (*Women* 304). Without needing to take recourse to Henry Ford's carrot of the $5 day or the stick wielded by Harry Bennett, Gerald was 'giving them what they wanted,' namely, their 'participation in a great and perfect system that subjected life to pure mathematical principles.' Their allegiance was not to a charismatic leader but to a system that was radically egalitarian in that it was equally indifferent to masters and servants. The men were thus 'not important to [Gerald], save as instruments, nor he to them, save as a supreme instrument of control.' If the miners sacrificed freedom so as to embrace a totalized order, they did so not because they craved security or happiness but because they were psychopathologically addicted to the 'terrible purity' (305) of the new industrial order.

Since master and servant subscribe to the same 'terrible purity,' the system may not be socially *just* but it is destructively *egalitarian*. The system's total integration of all social agents robs both master and servants of individual agency. The efficient system binds individuals to the 'mechanical' totality at the same time as it cuts them adrift by severing their 'organic' social bonds. The new order is thus paradoxically both rigorously organized and disturbingly chaotic:

> This was a sort of freedom, the sort they really wanted. It was the first great step in undoing, the first great phase of chaos, the substitution of the mechanical principle for the organic, the destruction of the organic purpose, the organic unity, and the subordination of every organic unit to the great mechanical purpose. It was pure organic disintegration and

> pure mechanical organisation. This is the first and finest state of chaos. (305)

This chaos should not be explained away as the individual's inner or spiritual disintegration as he or she internalizes the demands of the external system. If chaos is the experience of being a flexible or infinitely interchangeable participant in a system indifferent to such human needs as, for instance, being rooted in a community, then 'the state of chaos' is indeed indispensable to 'pure mechanical organization' and difficult to distinguish from it.

Unlike Kurtz, Gerald does not rely on charisma to impose his will on the miners; he owes his dominance to his initiative in the entrepreneurial installation of an efficient system of organization. Yet *Women in Love* consistently portrays Gerald as a man who violently masters both animals (nature) and human beings (culture). In highly memorable scenes, he cruelly seeks to subdue his sister's rabbit and to tame the Arabian mare spooked by an oncoming train. Where Ursula affirms that the mare is an autonomous 'living creature,' Gerald counters that the 'mare is there for my use' (200). It is on the basis of this utilitarian calculus that he justifies his cruel treatment of the animal: 'And if your will isn't master, then the horse is master of you' (201). It is his expectation of how the mare will act that compels his preemptive aggressive stance. In terms of game theory, he consistently allows his calculation of how others will behave to influence his own conduct. It is his fear of being 'suckered' into a typical Prisoner's Dilemma scenario that makes him choose individually and collectively self-defeating options. The Freudian emphasis on unconscious desires in the novel tends to overshadow the intersubjective dynamic at stake in Gerald's homoerotic relationship with Rupert and his heterosexual one with Gudrun. It is a dynamic that one recent critic identifies with Lawrence's general exploration of 'a drive to argue, to force a reaction,' a drive at work in confrontations between characters as well as in the 'coercive voice' Lawrence uses to engage with 'a listener' in order to 'get a reaction' (Fernald 193).

In the aptly named chapter 'Gladiatorial,' the nude wrestling match between Gerald and Rupert is marked simultaneously by the violence of male rivalry and by bonds of homosexual attraction. Beyond the psychoanalytical level, Gerald's rejection of Rupert's offer of love is more broadly speaking indicative of his investment in an efficiently run social order. In reminiscence of Lawrence's hope for a utopian 'Rananim,' Rupert advocates a Hegelian identity thesis predicated on the sublation of opposites that significantly does not submerge one term of the opposition within the other, but maintains their integrity within a state of dynamic tension. In a conversation with Ursula, he specifies that he wants 'a strange conjunction' with her that is precisely not a 'meeting and mingling' but 'an equilibrium, a pure balance of two single beings; – as the stars balance each other' (210). He thereby endorses a Pareto-efficient model of society in which each player agrees that cooperation constitutes his or her optimal choice. The short-term self-interest of each individual is in perfect alignment with the long-term interests of the community. In this kind of equilibrium, players make best responses to each other without seeking an advantage at the expense of others. In Rupert's terms, this equilibrium model is 'a lovely state of free-proud singleness' (332) rather than an erasure or submersion of individual differences. Rupert's reliance on 'love' as the 'life centre' (109) of both individuals and communities is a mystified version of a perfect equilibrium emerging spontaneously when players make best responses to each other. The organic community is thus the desired outcome of a laissez-faire or 'free' market ideology of personal and political decisions. For Gerald, though, social life does not arise spontaneously, but is 'artificially held together by the social mechanism' (109). As we have seen, Taylor undermines his Pareto-efficient aspirations by introducing an external authority figure to ensure the cooperation on which his equilibrium model is premised. Gerald has to refuse Rupert's offer because, like Taylor, he privileges 'the system' over 'the man'; he realizes that he is no longer 'free' to make personal choices.

Where Rupert nostalgically yearns for an 'autonomy' no longer available to the modern social subject, Gerald acquiesces in a new social order that Lawrence deems to be a 'doomed' civilization. In his intersubjective relations, Rupert expects the reciprocity and trust appropriate for a social equilibrium achieved through cooperation. In contrast, Gerald's responses to others are defensive; he protects himself from being 'suckered' by competitors. It is to his credit that he recognizes his own subservience to the system. Once he had achieved the transformation from an organic to a mechanical order, the system worked so much by itself that he himself became a dispensable aspect of the totalized order: 'The whole system was now so perfect that Gerald was hardly necessary any more.' From Rupert's or Lawrence's romantic conservative perspective, Gerald embodies Dostoevsky's modern engineer who is doomed to create an inhuman, uninhabitable edifice; his 'substitution of the mechanical principle for the organic' (305) left him without inner resources so that he 'was afraid that one day he would break down and be a purely meaningless bubble lapping round a darkness' (306). The novel leaves it unclear whether Gerald was psychologically predisposed to embrace an indifferent system or whether he let himself be corrupted by the system. In either case, he is not an autonomous master but a master who depends for his sense of self on the acknowledgment of another. Instead of acting independently, he is entangled in intersubjective dynamics that shape his personality. In his relationship with Rupert, for instance, the narrator stresses that Gerald was 'held unconsciously by the other man' (110). With respect to the miners, the narrator even contends that 'Gerald could never have done what he did' unless he thought that 'it was what they wanted' (304). Here Lawrence's historical myth-making serves to illustrate once again that the conditions he deplores have their source in a systemic shift and should not be dismissed merely as the actions of a culpable individual. If Gerald deserves our sympathy, it is because he is as much victim as victimizer.

Rupert's Hegelian sublation of opposites is the mirror image of laissez-faire capitalism; just as bonds of love form a unified social community through spontaneous cooperation, so the 'free' market is thought to create a balance between forces through spontaneous competition. In contrast, Gerald and Gudrun are not content to let Adam Smith's 'invisible hand' of the market regulate their intersubjective economy; on the contrary, they aspire to occupy a monopolistic position more representative of the 'visible hand' of managerial capitalism. In this late phase of capitalism, faceless managers coordinate the economic activities of a corporation without having any stake in the ownership or value of the goods they circulate. As manager of the mine, Gerald occupies a hybrid historical moment; like Henry Ford, he manages a family firm in which the link between ownership and management has not yet been *formally* severed. Rejecting the self-regulating mechanism of laissez-faire capitalism (or Rupert's utopian ideal), he tries to reconstitute in the sexual arena the loss of personal control he suffered in the economic sphere. In his relationship with Gudrun, he falls back on his father's patriarchal authority structure, which he had transformed into a decentred managerial model. However, his participation in the Taylorization of the social fabric has blocked access to the father's organic conception of social relations. In his desire to master Gudrun, he seeks to occupy a monopolistic position that anticipates the competitive struggle of corporations for global control. Stripped of his father's ideological mystifications, Gerald is alienated both from himself and from his society. It is thus in an effort to reaffirm his self-identity that he turns to Gudrun for acknowledgment. In psychoanalytic terms, he illustrates Lacan's logic of intersubjective relations, which have their source in Alexandre Kojève's rereading of Hegel's Master-Slave dialectic. Since the unconscious is for Lacan always already social, the conflicted sexual relations between Gerald and Gudrun should not be explained away as responses to repressed unconscious drives or instincts. If the unconscious is marked by the logic of social relations, then the failure of these lovers to establish a mutually satisfactory relationship hinges on the intersubjective desire of each to control the other. If they are from the beginning locked in combat, it is because Gerald is afraid of being trumped by Gudrun, on whose acknowledgment his own sense of identity as master depends. In return, Gudrun knows she has to resist and challenge his demands in order to preserve her own sense of self. Unable either to accommodate each other's needs or to walk away from each other, they are frozen in a repetitive competitive struggle, with each of them selecting the 'lose-lose' scenario that game theory schematizes as 'I win/you lose.' In their monopolistic aspirations for control, their relationship is tragically doomed to end in 'Hegelian murder.'

The turbulent story of the sexual attraction and repulsion keeping Gerald and Gudrun bound to each other is punctuated by scenes highlighting a repetitive pattern of their violent struggle for mastery over each other. Much of the novel's interest clearly revolves around the question of who will be master and who will be mastered. The combative stage is set from the first moment Gerald confesses his love for Gudrun. It is highly significant that the declaration of love was preceded by a blow she had just struck across his face, a blow the novel immediately elevates to 'a race to the bottom' doomed to end in outright victory for one and outright defeat for the other: '"You have struck the first blow," he said ... "And I shall strike the last," she retorted involuntarily' (237). Given that 'both felt the subterranean desire to let go, to fling away everything, and lapse into a sheer unrestraint, brutal and licentious' (367), it would be in each lover's self-interest to cooperate by allowing the other to act on this desire. In other words, all the ingredients are present for each 'player' to get what he or she wants. But such 'win-win' scenarios are consistently refused by an equally strong desire to control the other. Throughout the novel, they keep changing positions within an asymmetrical gender framework. In one scene, it is Gerald who felt 'liberated and perfect, strong, heroic,' while Gudrun acknowledged that she 'died a little death' after her 'heart fainted' as she felt 'herself taken.' At the same time as she 'quailed under [his arm's] powerful close grasp,' she already exulted that her 'fingers had him under their power' (412, 416). In a later scene, he visits her in order to renew himself after the stress of his father's funeral. Seeking 'vindication' and 'relief,' he reduces her to an instrument facilitating the reassertion of his self-identity: 'Into her he poured all his pent-up darkness and corrosive death, and he was whole again.' Contaminated by the 'terrible frictional violence of death' he brought with him, she 'had no power at this crisis to resist.' In Lawrence's often-analysed suspect gender economy, it is Gudrun who is said to submit in an 'ecstasy of subjection' while Gerald slept 'the sleep of complete exhaustion and restoration' (430, 431). Yet it is Gerald who admits his dependence on Gudrun while she reaffirms her autonomy. 'If there weren't you in the world,' he confesses, 'then *I* shouldn't be in the world, either' (429). In contrast, she eventually withdraws from him by reminding herself that 'one must preserve oneself' (435). It may appear as if Gudrun acted more selfishly than Gerald. However, it is because she knows that he expects her to complete him that she reacts by frustrating an expectation she recognizes as a threat to her own expectations; she understands the asymmetrical nature of a position that falsely presents itself as reciprocal.

In later scenes, the expectations each lover attributes to the other intensify the violence with which they meet each other's demands. Gudrun experiences Gerald's demand for love as an aggression that she must repulse by withdrawing her love from him. Her refusal then makes him redouble his demand. In Lacanian terms, they exhibit the logic of desire that is generated when the child addresses to the mother a demand for love that neither she nor her substitutes throughout the subject's life can ever satisfy. This absolute demand for love thus constitutes a hole that the subject seeks to fill with always unsatisfactory substitutes. The asymmetry between his expectation that she submit to his desire for mastery and her expectation that he acknowledge her as his equal inevitably leads to misreadings and misrecognitions. On the one hand, a slave to his desire for her acknowledgment of his superiority, Gerald felt 'dominated by the constant passion, that was like a doom upon him' (492). Feeling that he would be destroyed 'if he were not fulfilled,' he thus violently imposes himself on her, causing her to move 'convulsively, recoiling away from him.' At this moment of crisis, he is said to feel 'strong as winter,' whereas she is described as 'lost' and 'without hope of understanding, only submitting' (493, 494). On the other hand, his masterful self-assertion makes her form a 'deep resolve' to 'combat him' because she clearly realizes that 'one of them must triumph over the other' (506). For the rest of the novel, both characters consider their options in the 'murderous' terms of radical victory or radical defeat.

In the novel's suspect gender economy, Gerald is consistently portrayed as the initiator of campaigns in the sexual conquest, leaving Gudrun to resist by withholding and withdrawing what he desires. But both lovers acknowledge their intersubjective dependence on each other. In one scene, she fears falling 'down at his feet, grovelling at his feet, and letting him destroy her' (508). But once this moment of 'terrible panic' has passed, she realized that she once again 'had the whip hand over him' (509). In a later scene, he despairingly concedes that his will to power is driven by his need for her acknowledgment of his existence: 'But then, to have no claim upon her, he must stand by himself, in sheer nothingness.' Contemplating his alternatives, he concludes that he can either 'give in, and fawn to her' or 'finally, he might kill her' (543). Living in 'fear of his power over her, which she must always counterfoil' (540), she nevertheless submits once again to his 'passion' which 'was awful to her, tense and ghastly, and impersonal, like a destruction, ultimate. She felt it would kill her. She was being killed.' In this see-saw battle, then, 'sometimes it was he who seemed strongest, whilst she was almost gone, creeping near the earth like a spent wind; sometimes it was the reverse' (542). Where his desire can be seen as a compensatory gesture for the nothingness he experienced once the routine world he had initiated left him not knowing 'what to do' (305), her resistance can be similarly located in his surrender to the instrumental logic of efficiency. Late in the novel, she seriously considers marrying him: 'She would marry him, he would go into Parliament in the Conservative interest, he would clear up the great muddle of labour and industry.' Realizing that he 'was a pure, inhuman, almost superhuman instrument,' she frustrates the reader's expectation by admitting that his 'instrumentality appealed ... strongly to her.' It was only when she asked herself 'the ironical question: "What for?"' that she recoils from the idea of 'Shortlands with its meaningless distinction, the meaningless crowd of the Criches' (511). It is not only that his power to control her frightens her, but also that his 'faculty of making order out of confusion' (510) served no meaningful purpose. What she rejects in Gerald is his commitment to efficiency for its own sake.

It is therefore ironic that Gudrun endorses Loerke's concept of 'art for art's sake,' an aesthetic theory that mirrors in its self-conscious abjuration of purpose the very meaninglessness for which she condemns Gerald. Loerke's transposition of Gerald's lack of purpose into the aesthetic arena suggests that Gudrun's final triumph over Gerald may well be a pyrrhic victory. In contrast to Gerald's vitality, Loerke is for Gudrun 'the very stuff of the underworld of life,' for Rupert he is a 'wizard rat,' and for Ursula he is 'indescribably inferior, false, a vulgarism.' In freeing herself from Gerald, Gudrun could be said to follow the 'little obscene monster of the darkness' towards 'a vacuum' (522–3) that condemns her to a living death. What ends the intersubjective dialectic binding Gerald and Gudrun to each other is thus the zero-sum game of the 'Hegelian murder.' Before Gerald's physical death in the Tyrolean Alps, both of them had harboured murderous thoughts. Feeling tortured by her, he had at one time said to himself: 'If only I could kill her – I would be free' (540). But just as he thought of killing her, so she had dreamed of killing him: 'And because she was in his power, she hated him with a power that she wondered did not kill him. In her will she killed him as he stood, effaced him' (553). As the narrator comments at one point, in this intersubjective combat 'there is only repetition possible, or the going apart of the two protagonists, or the subjugating of the one will to the other, or death' (550). Since they cannot be 'strangers' (555), separation is not a possibility, and since neither will submit to the other, death is the only way out. It is only after he had almost strangled her that he seeks his own death in the cold snow. In their intensely competitive struggle for mastery, they have embarked on a race to the bottom by consistently selecting Pareto-inefficient options.

In 'The Industrial Magnate,' Lawrence foregrounds the logic of industrial capitalism, which he condemns for causing the dehumanization of the individual and the degeneration of culture and society. In *The Modernist Novel and the Decline of Empire* (2005), John Marx delineates a connection between 'industrial innovation' and 'Britain's decline' (151) that takes on its fullest significance within the eugenics debate of the day: 'For the concept of degeneration, Lawrence was indebted to a model laid out not only by Wells but also by Thomas Huxley, who himself borrowed heavily from the father of evolutionary thought' (151). Gudrun's choice of the 'industrial capitalist' illustrates a doubt Darwin expressed in his later work *The Descent of Man,* that natural sexual selection may not ensure 'species health' (152): 'The moment when females select the most agreeable partners gives [Darwin] pause to worry that in certain circumstances their choices will not be for the best' (151). Such doubt leads Thomas Huxley to favour what he describes as 'a "horticultural process" governed by an intrusive gardener, a eugenicist who "restricts multiplication ... and ... attempts to modify the conditions, in such a manner as to bring about the survival of those forms which most nearly approach the standard of the useful, or the beautiful, which he has in mind"' (in Marx, 152). In short, as long as 'female taste is involved in sexual selection' (152), there is no guarantee that reproduction proceeds along optimal lines. By choosing first Gerald and then Loerke, Gudrun exhibits a 'habit of sexual selection' that is 'the surest indication that industrialization positions humanity on the slippery slope to degeneration' (155). The optimal efficiency Gerald achieves in industrial production is thus inversely proportional to the cultural decline this potentially vital character falls victim to as he fails to enter into meaningful intersubjective relationships. Taylor's technical efficiency in the coal mines is once again unproblematically conflated with his failed socially engineered utopia. On the one hand, Lawrence is hostile to industrialization as an efficient machine; on the other hand, he dreams of a society so perfectly engineered that no weeds dare grow. Since the same totalizing desire is operative in both technical and social efficiency, Lawrence has to resort to the laissez-faire principle of the spontaneously organic community which is not only a nostalgic fantasy but is logically entirely compatible with the very capitalism he condemns along with industrialization. On the one hand, Lawrence's privileging of the individual freedom symbolized by Rupert's opposition to Gerald is entirely compatible with both the laissez-faire stage of capitalism and with the ideology of the unique individual of 'patriarchal' entrepreneurial capitalism. On the other hand, the values Lawrence opposes to the efficiency calculus are politically suspect, especially in view of the conflation of the lost organic community and eugenic cleansing in Hitler's rhetoric. Lawrence's apparently diametrically opposed alternatives of the free-floating paradise of 'Rananim' and the charismatic leadership principle do in fact share a questionable resistance to the privileging of rational control basic to both Taylor's efficient industrial model and the rule of law in liberal democracy. Yet what Lawrence immediately, if perhaps unwittingly, grasped was not only that efficiency is an ideology but also that it was historically deeply implicated in the shift from visible human-centred forms of control to invisible decentred disciplinary structures. As commentators consistently point out, *Women in Love* is unconvincing in its depiction of the 'organic' alternative represented by Rupert and Ursula to the 'mechanistic' system represented by Gerald and Gudrun. Although Lawrence was an astute analyst of the dehumanizing impact of efficiency, he implicitly capitulated to its impending triumph by shifting to the register of apocalypse in his opposition to its ideology.

In the first instance, then, 'The Industrial Magnate' indicts the dehumanizing impact of Taylor's privileging of the 'system' over the 'man'; on this level, the logic of efficiency is treated synchronically. But Lawrence also situates Gerald's industrial innovations within the history of changing modes of production; on this diachronic level, efficiency is linked to successive phases of capitalist expansion. Finally, efficiency is implicated in the utopian longings that Lawrence encodes in characters who either find qualified success or unqualified failure in their relationships with others. On this level, Taylor's Pareto-efficient aspirations are at stake in that the intersubjective dynamics in the novel are marked by Prisoner's Dilemma scenarios. In *Women in Love,* Lawrence thus draws attention to the broader social implications of the industrial innovations he pointedly foregrounds in 'The Industrial Magnate.'

8 Efficiency and Perverse Outcomes: Ford Madox Ford's *The Good Soldier*

Ford Madox Ford's *The Good Soldier* (1915) appears at first sight to be a highly unlikely candidate for inclusion in a study of efficiency. A fine example of 'literary impressionism,' the novel has predominantly been appreciated for the epistemological crisis it signals through the narrator's legendary unreliability. Alluding to the sense-making mechanisms of storytelling, the narrator John Dowell is tentative in his assertions, often revises what earlier seemed to be factual information, and castigates himself for having allowed himself to be deceived by others. Moreover, as the critics keep pointing out, Dowell deceives himself in ways that he himself fails to see or to acknowledge. *The Good Soldier* is a 'text' in the current theoretical understanding of an unstable construct always open to reinscription. Samuel Hynes is quite typical of the epistemological emphasis in the criticism of the 1960s when he describes Dowell as being 'a limited, fallible man, but the novel is not a study of his particular limitations; it is rather a study of the difficulties which man's nature and the world's put in the way of his will to know' (230). Literary impressionism is here seen as a mode of representation that not only grapples with but also performs the rapid transformations that 'the world' had undergone since about the 1890s. Narrative experimentation in literary modernism was thus a response to the 'far-reaching reorganization of spatio-temporal experience brought about by changes in transport, energy, urban planning, communication and media, the "taylorizing" of labor in the factory, and the mass slaughter of modern mechanized warfare' (Britzolakis 1). Recent criticism tends to link Dowell's individual psychological trauma with the experience of a society faced with the 'shock' of urban modernity and global capitalism. *The Good Soldier* is now considered to be 'a story about the ephemeral, private, and anxiously guarded interiority of the bourgeois psychological subject' seeking to 'secure the border of the psychological interior against the swift urban rush of mobility' in a 'world of contingent international finance capital' (Mickalites 288). Dowell's unreliability as a narrator is no longer simply an epistemological problem; his inability to know dramatizes 'the traumatic evacuation of meaning from accepted categories' (Henstra 184) once Britain's self-understanding as a civilizing force was thrown into crisis by the 'heart of darkness' at the core of capitalist and colonial expansionism. After decades of psychological and metaphysical probing, *The Good Soldier* is beginning to be analysed as a symptom of the cognitive disorientation engendered by modernity's rapid transformation of society's material conditions.

In '*The Good Soldier* and Capital's Interiority Complex,' Carey J. Mickalites stresses the 'inextricability of modernist psychology from material capitalist culture,' arguing that the 'perverse logic of the commodity form' (296, 297) restructures traditional conceptions of the relationship between subject and object. As the emphasis in capitalism shifts from production to consumption, there emerges a new 'problematic relation between synthesis and dissociation – or unity and fragmentation in traditional literary terms – [that] is indicative of the way capital flux both informs and challenges modernist psychic interiority' (297). Referring to the novel's main character, this critic associates Edward Ashburnham's 'possession of women' and 'territorial expansion'; it marks 'a continuous expenditure of desire' which then 'meets with the "rational" management of capital, the historical absorption of all surplus' (298). It is within this exploration of late capitalism and psychic interiority that a consideration of efficiency both reinforces and reconfigures the terms of an emerging debate focused on *The Good Soldier* as a critique of late capitalism (Mickalites), British imperialism (Sarah Henstra), and patriarchy (Karen Hoffman). In *The Good Soldier,* efficiency does not exactly advertise itself as a major concern. However, as Anthony P. Monta has shown, efficiency is more obviously a concern in Ford's war tetralogy, especially in *Parade's End*[1] which 'engages with the discourse of "national efficiency," a government reform movement which expresses itself most noticeably during the decade before the first world war' (41). An examination of the novel's 'metaphorical texture' reveals an unmistakable tension 'between images and values associated with mechanism and abstract calculation and those aligned with physical vitality and intimacy' (42). Drawing on the enthusiasm for efficiency on the part of the Fabians and the Co-Efficients (George

Bernard Shaw and the Webbs), Monta demonstrates that machines and mechanisms were used 'as tropes for the state,' especially by Arnold White in his polemical work *Efficiency and Empire* (1901). The 'image of the state as a machine' (45) was in the first decade of the new century effortlessly extended to the discourse of imperialism. Uncovering two versions of 'administrative idealism' (46) in *Parade's End,* Monta argues that Tietjens eventually rejects the mechanical model of efficiency endorsed by Macmaster by nostalgically retreating to rural Groby, the romanticized family estate. Treating efficiency as an evil of modernity, Ford explicitly targets in this later novel what remained a more muted but also more broadly conceived concern in *The Good Soldier. Parade's End* thus 'retreats from *The Good Soldier*'s radically destabilizing treatment of identity to a late-Victorian view of Englishness' (Henstra 193) in order to reaffirm, in less complex terms than *The Good Soldier,* the hero's reintegration into a pre-capitalist social order.

For 'A Tale of Passion' so obviously concerned with concealed passions and misunderstandings, *The Good Soldier* seems almost inordinately preoccupied with questions of money and conflicting conceptions of efficiency. Aligning himself with other literary modernists hostile to modernity, Ford equates a commitment to efficiency with the threat that standardization was generally thought to pose to the cherished notion of the unique human being. It is not that all devotion to efficiency is absolutely vilified. As long as efficiency remains a means to an end, it may even receive unqualified approval. But once the efficiency calculus becomes severed from a commitment to some public good, it is thought to contribute to possessive individualism and the reification of consciousness under late capitalism. Distinguishing between Edward Ashburnham's efficient stewardship of those dependent on him and Leonora Ashburnham's coldly self-interested drive for efficient outcomes, the novel dramatizes a conflict between competing social models that privilege either an organic or a mechanical view of social relations. Along these lines, *The Good Soldier* seems merely to reiterate standard reservations by cultural conservatives about the detrimental impact of social modernity on the individual human subject. But Ford also provides us with an illustration of the potential and limitations of Pareto efficiency in his depiction of the intersubjective relationships of the characters in *The Good Soldier.* Anticipating the terms of game theory, Ford dramatizes how the behaviour of his characters constitutes a response to the expectations each of them projects onto others and how these are reflected back to them. Although Dowell concludes that the story's tragic outcome could be attributed to Edward being a sentimentalist, Leonora a cold guardian of her 'property,' Florence a manipulative sensualist, and he himself a timid fool who sought to protect himself by disavowing what he saw, he also acknowledges systemic material conditions that exceed such psychological categories.

The shift in critical emphasis from psychological analyses of characters' individual motivations to a focus on the representation of the material conditions informing the practices of social agents has resulted in the disruption of the opposition between interiority and exteriority that has underpinned distinctions between surface and depth, authenticity and inauthenticity, and appearance and reality in earlier critical discourses. The 'bourgeois psychological subject' is no longer endowed with an 'interiority' that is not always already social or ideological. Since Althusser's Marxist extrapolations from Lacan's rereading of Freud, critical commentary has by and large acknowledged that 'interiority' and 'exteriority' are mutually reinforcing categories. In his influential essay 'The Mirror Stage as Formative of the *I* Function,' Lacan famously contends that the individual's sense of self does not originate in some core essence, but is produced through specular identifications with others who act as models and obstacles to its desires. Instead of conceiving of the unconscious as the site of uniquely personal instincts that are being repressed by social prohibitions, Lacan argues that the unconscious comes into being only at the moment the subject enters into consciousness or language and is consequently always already social. The individual is not so much *relating* to its social conditions as it is *immersed* in them; in other words, the individual no longer expresses an interiority that precedes material conditions but is spoken by these. Drawing on Lacan's specular model, Althusser draws out the ideological implications of the socially constructed subject by arguing that 'it is not their real conditions of existence, their real world, that "men" "represent to themselves" in ideology, but above all it is their relation to those conditions of existence which is represented to them there' (Althusser 164). Althusser's formulation breaks with both the Romantic tradition of the unique individual and with the Hegelian identity thesis predicated on the coincidence of subject and object. The subject receives its sense of self from the 'instructions' it receives from the society in which it functions; these 'instructions' do not issue from a centre of power, but are to be located in overdetermined ideological apparatuses. Although the subject responds to some ideological 'hailings' and rejects others, such choices do not take place outside the specular logic of ideology. In *The Good Soldier,* the narrator seeks to probe the depth of psychological motivations at the same time as he foregrounds the material conditions that *speak* the subjects of his tale.

The calamitous outcome of the 'tale of passion' Dowell narrates can, to no small extent, be traced to the failure of a collectivity to function as efficiently as it could have. *The Good Soldier* anticipates a similar failure in *Parade's End:* 'If Tietjens represents England's best administrative mind trying to maintain a productive balance between devotion to abstract efficiency and sensitivity to local needs, Ford's plot stages this balance's betrayal and disintegration' (Monta 47). Although Monta does not distinguish between technical and social efficiency, the balance in question alludes to the Pareto-efficient utopia of a noise-free system in which everyone is perfectly integrated into the totality. The social agents of this system bring socio-economic assumptions to the 'game' that structure their behaviour in ways they can neither completely understand nor control. Although the four characters of the opening minuet seem to line up along national and sexual lines, they also exhibit different attitudes towards capitalist modes of production and the efficiency calculus specific to each. Ford's precise socio-economic location of each character alerts us to the ideological 'obviousnesses' structuring their understanding of self and other. More specifically, the 'new money' of the American couple John and Florence Dowell is most obviously contrasted with the 'old money' of the English couple Edward and Leonora Ashburnham. On a more complex level, the four characters embody different forms of economic activity. In the first instance, Edward and Dowell line up as men of property whose wives endorse a speculative capitalism that threatens the stability of the landed gentry with the flux and contingency of market relations. But even these alignments break down as Edward and Dowell exhibit different views of property just as Leonora and Florence occupy different positions within the economy of market exchange.

Dowell's relationship to property is highly mediated and abstract; living as he does off rent rather than the produce of the land, his property has no more material significance for him than a stock certificate. Unlike the 'mostly professional people' in his family, Dowell was a man of property who had weathered 'the financial panic in 1907 or thereabouts' (Ford 154) that had impoverished his relatives. Owning 'mostly real estate in the old-fashioned part of the city,' he was an absentee landlord who did just enough to ensure that 'the houses were in good repair and the doors kept properly painted' (154). Employing others to work on his buildings, he neither produces nor uses what he owns. Disarmingly quoting Florence's aunts who described him as 'the laziest man in Philadelphia' (15), he stresses his parasitical exploitation of others' labour and need. His relationship to the real estate he owns is so abstract that he has no emotional ties to a geographical location; on the contrary, his material assets provide him with the financial means and leisure to become 'a wanderer upon the face of public resorts' in Europe: 'I had no attachments, no accumulations' (21). In a compensatory gesture, he identifies with Edward, a man deeply rooted in Branshaw Teleragh, the family estate. At the end of the novel, he declares that he loved Edward 'because he was just myself' (253), a claim that fails to convince since he lacks the very character traits he ascribes to the man he admires. Yet, as a man of property, he is in fact the American equivalent of the British landowning classes. What he can neither buy nor imitate is Edward's claim to an aristocratic title that allows even the 'dispossessed aristocrat' to retain 'his nobility, which has a "spiritual existence"' (Michaels 93). Such title circulates outside the economy of capitalist production and consumption; it cannot be lost or appropriated. Financial troubles can no more deprive Edward of a stable identity than financial security can guarantee Dowell a firm hold on his self. Aspiring to *be* Edward, he tries 'to define himself in terms of the gentry's wealth by emphasizing how long his family had held its land' (Hoffman 40). His only emotional ties are to virtual land, to the historical title to a farm once owned by the Dowells of Philadelphia, one of the city's 'old English families' (Ford 5): 'I carry about me, indeed – as if it were the only thing that invisibly anchored me to any spot upon the globe – the title deeds of my farm, which once covered several blocks between Chestnut and Walnut Streets' (5). Having initially been 'legally' expropriated from 'an Indian chief' (5), the land is no longer a farm but has been transformed into blocks of commercial property. The title deeds are pieces of paper that have lost any material connection to the land they designate; they indicate 'a nostalgia for the time when land had not yet been transformed into a commodity' (Michaels 94). Where English gentry seem to be rooted in the land they own, American ownership of property is inescapably implicated in the capitalist logic of exchange. Having been transformed into blocks of commercial property, the family farm constitutes a nostalgic fantasy of belonging that is belied by Dowell's extraction of surplus value from real estate. Concealing from himself the source of his wealth in the market, he seeks to associate the luxury of leisure and laziness with the aristocrat's hereditary and hence inalienable claim to land and identity. Constructing himself as neither a producer nor a consumer, Dowell places himself outside and above the utilitarian logic of the efficiency calculus.

Instead of relying on real estate, Florence resorts to the tested method of marriage to ensure both the security of wealth and an identity based on social standing. With the possible exception of Jimmy, she used sex to transform 'new money' into 'old money.' A husband was for her a commodity she was 'coldly and calmly determined' (80) to acquire according to precise specifications: 'She wanted to marry a gentleman of leisure; she wanted a European establishment. She wanted her husband to have an English accent, an income of fifty thousand dollars a year from real estate and no ambitions to increase that income' (79). Marrying Dowell is part of her efficiently planned campaign to 'get to Fordingbridge and be a county lady in the home of her ancestors' (90). Money is for her as incidental as it is for Dowell; 'the fascinations of Wall Street' (80) have no hold on her imagination. Abstract capital is for her something not to be increased but to be exchanged for the materiality of feudal property. Her design first on Fordingbridge and then Branshaw Teleragh indicates a desire to escape from the uncertainties of the market she had witnessed as a child. Where Dowell was used to the relative solidity of 'old-fashioned' property, Florence had been brought up by Uncle John Hurlbird, the owner of a factory 'which in our queer American way, would change its functions almost from year to year. For nine months or so it would manufacture buttons out of bone. Then it would suddenly produce brass buttons for coachmen's liveries. Then it would take a turn at embossed tin lids for candy boxes' (17). The materiality of buttons or tin lids is secondary to their virtual status as commodities to be exchanged. In spite of her hostility to capitalism, she has internalized the logic of commodification. Having married Dowell in order to buy the ancestral home with his money, she seizes the chance of exchanging him for Edward, whose ownership of Branshaw offers a quicker route to taking 'her place in the ranks of English county society' (80) through marriage. Objectifying Dowell and Edward, she sees them not as objects of love or passion but as goods to be acquired, exchanged, and discarded at will. It is not only in her travels that she is a 'kind of consumer-tourist abroad' who 'has "the seeing eye" that encounters each new vista as though checking it off her list' (Henstra 185). But in the process of commodifying others, she effectively reifies her own consciousness. In Dowell's final judgment, she was 'a personality of paper' who 'represented a real human being with a heart, with feelings, with sympathies, and with emotions only as a bank note represents a certain quantity of gold' (Ford 121). Ironically, then, the woman who disdained Wall Street in favour of an identity anchored in inalienable property rights is here pointedly identified with the circulation of abstract money in the market. The bank note metaphor indicts her not only for her superficiality but also for the reifying logic she exemplifies in her pursuit of the aspiration she shares with Dowell of becoming English gentry. If Edward epitomizes for Dowell a nobility of spirit beyond capitalist acquisitiveness, then Florence demeans herself by foregrounding the vulgar utilitarianism of market efficiency that Dowell seeks to disavow as he nevertheless thrives on it.

Being anglophiles, the Americans are sympathetic outsiders whose perspective on Englishness allows Ford to offer 'subtle though limited critiques of the landed elite, patriarchy, and imperialism' (Hoffman 33). It is in the treatment of Edward and Leonora Ashburnham that the ideology of efficiency is directly targeted. Both of these 'good people' are in their different ways highly efficient social agents. Yet their perspectives on efficiency are radically incommensurate. Their inability to connect should be ascribed not only to purely psychological factors but also to their inability to read each other's ideological investments. Where Edward considers efficiency to be a means to a noble end, Leonora approaches it as an end in itself. Contrary to what one might expect, Ford does not impugn the Americans for contaminating British culture with an investment in efficiency popularly associated with Ford's assembly line and Taylor's stopwatch. Although both Dowell and Florence cannot entirely escape from socio-economic determinants, they seek to position themselves outside capitalist market forces. They do not function as a model to be emulated or rejected but rather as an outside perspective from which to evaluate the crisis of cultural consciousness enacted by the British couple. It is thus in Ford's portrayal of the Ashburnhams that the efficiency calculus acts as a noticeable point of reference. As we will see, the conflict between Edward and Leonora foregrounds Ford's astute realization that the efficiency calculus is more of a threat to the community than it is to the individual.

In Dowell's eyes, Edward Ashburnham exemplifies efficiency in its most positive manifestations. Described as 'a hard-working, sentimental, and efficient professional man,' Edward is consistently praised for being 'the cleanest sort of chap; an excellent magistrate, a first-rate soldier,' and, above all, 'one of the best landlords, so they said, in Hampshire, England' (152, 11). Yet he was not some hidebound landowner contented with riding to hounds in his local fiefdom; he was up to date 'on soldiering, keen on mathematics, on land-surveying, on politics, and, by a queer warp of his mind, on literature' (137). Most significantly, perhaps, he was prepared to participate in the new money economy. Having invested a 'good deal' of money 'in rails,' he at one point had advised the astonished Dowell 'to buy Caledonian Deferred, since they were due to rise' (144, 26). As Dowell discovers, the stock did in fact rise, making him wonder where Edward 'got the knowledge' about stocks that 'seemed to drop out of the blue sky' (26). It appears that Edward was far more attuned to capitalist investment strategies than Dowell allows in his construction of him as a member of the British gentry who enjoy a 'natural and non-performative' (180) state of being. At the same time as Edward is prepared to take advantage of market speculation, he remains deeply committed to 'the feudal theory of an overlord doing his best by his dependents, the dependents meanwhile doing their best for the overlord' (146). He assumes that the feudal order naturally creates a Pareto-efficient balance, a balance Taylor sought to reproduce artificially by mobilizing the expertise of a supposedly disinterested external 'social engineer.' In her analysis of the novel's conception of 'Englishness,' Henstra helpfully draws on Ford's *The Spirit of the People* (1906) in which the 'English race' is being thanked for 'its evolution of a rule of thumb system by which men may live together in large masses' (in Henstra, 179). For her, the 'descriptor "rule of thumb" – authority accrued over time and through reiteration instead of formally codified law' is of interest to her because it 'points to the peculiar relationship of Englishness to language in general' (179). But this descriptor also references Taylor's scientific-management model, which was intended to replace the old 'rule-of-thumb method' (Taylor 36) with a new rational approach. Ford's pride in English pragmatism is typical of attitudes Merkle identifies as sites of resistance to Taylorism in Britain. The cultural investment in a stable class structure privileged the classically educated gentleman over the technically trained professional. Since the gentleman was committed to the liberal values inculcated at Oxford and Cambridge, he would 'naturally' accept responsibility for the welfare of those less fortunate. The English tradition of rugged individualism, pragmatic self-help, and social paternalism had no use for 'questioning or making explicit the rule of thumb whose truth is apt to collapse under scrutiny' (Henstra 179). However, as Henstra demonstrates from a different perspective, *The Good Soldier* in effect deconstructs 'the comfortable rule-of-thumb system perfected by the English' as consisting of nothing more than 'ugly compromises and cruel renunciations' (193). As we will see, it is as if Ford had been compelled to demystify the very Englishness he so nostalgically sought to defend.

As the representative of a spontaneous Pareto-efficient social order, Edward considers it his responsibility to act as an efficient landlord at home and an efficient officer abroad in order to ensure the welfare of both his tenants and his soldiers. Not unlike Gerald Crich's father in Lawrence's *Women in Love,* he 'managed his estates with a mad generosity towards his tenants' and 'subscribed much too much to things connected with his mess' (143, 141). In his management of others, he embraces a paternalistic model of social efficiency rather than the total-output model of technical efficiency. The novel clearly privileges Edward's notion of feudal stewardship. In a dispute over the best response to a financial crisis, Edward's insistence on protecting his tenants is shown not only to be socially responsible but to produce a more efficient technical outcome. When Leonora proposes that they replace long-term tenants with cheaper labour imported from Scotland, Edward refuses to 'turn out people who've been earning money for us for centuries' (145). Although motivated by his class sense of having 'responsibilities' (145) for his tenants, his feudal stance turns out to be economically advantageous. As the land steward realizes, Edward ensured the long-term prosperity of the land by resisting the temptation of short-term gains represented by the Scottish farmers, whose practices 'just skinned your fields and let them go down and down' (145). In retrospect, even Leonora had to acknowledge that 'Edward was following out a more far-seeing policy in nursing his really very good tenants over a bad period' (143–4). In terms of game theory, his devotion to tradition and duty leads to a win-win outcome; accepting that 'everyone had to feel the pinch, landlord as well as tenants' (145), he exemplifies a cooperative strategy that results in everybody being better off. As the embodiment of a Pareto-efficient ideal, Edward is presented as a man beyond reproach. But, as we will see, this ideal construction conceals a power asymmetry that belies the appearance of social justice in this supposedly win-win scenario.

In contrast to Edward's feudal form of efficiency, Leonora pursues efficiency gains for her own purposes rather than for the communal good. No other character in the novel is so consistently associated with a commitment to efficiency. Unlike the other rather helpless female characters, she is described as capable and pragmatic, traits best summed up by the comment that she 'drove with efficiency and precision' (205). In her outward appearance and manner, she strikes a note of perfection making her seem coldly inhuman: 'Leonora was extraordinarily fair and so extraordinarily the real thing that she seemed to be too good to be true. You don't, I mean, as a rule, get it all so superlatively together. To be the county family, to look the county family, to be so appropriately and perfectly wealthy; to be so perfect in manner – even just to the saving touch of insolence that seems to be necessary' (8–9). A smoothly integrated totality, she is almost uncanny in her rather mechanical exactitude: her evening dress is 'too clearly cut, there was no ruffling' and 'her shoulders were too classical' (32). In his considerate judgment, Dowell surmises that her shoulders would be 'slightly cold' to his lips but 'not icily, not without a touch of human heat, but, as they say of baths, with the chill off' (32). Unlike Florence, she is not presented as a 'personality of paper' but as a signifier coinciding with its signified. From the moment he met her, Edward 'admired her for her truthfulness, for her cleanness of mind, and the clean-runness of her limbs, for her efficiency, for the fairness of her skin, for the gold of her hair, for her religion, for her sense of duty' (140). Outer poise is here linked to commendable character traits. But this studied perfection is an affront to Edward's 'rule-of-thumb' pragmatism; in her purposive efficiency, she implicitly disdains the role of submissive female essential to Edward's patriarchal paternalism. He could admire but not love Leonora because 'she was never mournful; what really made him feel good in life was to comfort somebody who would be darkly and mysteriously mournful' (140). Where Edward is the remnant of a residual feudal ideology, she is emblematic of the emergent ideology of bourgeois capitalism.

The 'saddest story' Dowell narrates centres on the inter-destructive dynamic structuring the relationships of the 'good people' he claims to love and admire. As the critics were quick to point out, the narrator seeks to explain what happened by resorting to the key metaphors of a 'rotten' apple and 'a minuet de la cour' to express his bafflement at the unravelling of his 'tranquil life' (7, 6). The first metaphor alludes to Dowell's sense of having been fooled by taking deceptive appearances at face value:

> If for nine years I have possessed a goodly apple that is rotten at the core and discover its rottenness only in nine years and six months less four days, isn't it true to say that for nine years I possessed a goodly apple? So it may well be with Edward Ashburnham, with Leonora his wife, and with poor dear Florence. And, if you come to think of it, isn't it a little odd

> that the physical rottenness of at least two pillars of our four-square house never presented itself to my mind as a menace to its security? (7).

The 'rotten apple' analogy signals above all Dowell's epistemological crisis; mistaking appearance for reality, he loses confidence in his ability to know. But this gap between exterior surface and interior depth also operates in psychological and moral registers. The rotten 'core' of the apple is indicative of the diseased modern psyche and society's moral decrepitude that Dowell's paranoid search seeks to unveil. In contrast with this surface/depth paradigm, the minuet metaphor functions spatially in that it foregrounds the degeneration of an initial ideal state of equilibrium into an unanticipated and undesirable rearrangement. The stately 'minuet de la cour' with which Dowell's narrative opens is likened to 'an extraordinarily safe castle' (6). Both the 'apple' and the 'minuet' are thus configured in terms of a sense of security under threat. As Dowell puts it, 'that long, tranquil life, which was just stepping a minuet, vanished in four crashing days at the end of nine years and six weeks' (6). It had gradually dawned on him that it 'wasn't a minuet that we stepped; it was a prison – a prison full of screaming hysterics' (7). The Pareto-efficient ideal parading as a safe castle had always already been inhabited by imprisoned hysterics. In Dowell's imagination, the German resort town of Nauheim is emblematic of a social totality that offers him 'refuge,' 'permancence,' and stability' (6), a physical setting distinguished by 'carefully swept steps,' and 'carefully arranged trees in tubs upon the carefully arranged gravel whilst carefully arranged people walked past in carefully calculated gaiety, at the carefully calculated hour' (21). The novel's reigning irony is that the 'four-square coterie' (5) was from the beginning an elaborate disguise for the hysterical drama that had been in full swing by the time the couples had first met in Nauheim. Edward had already betrayed Leonora with several women just as Florence had already carried on an affair with Jimmy. The minuet thus functions as a nostalgic longing for a perfect world that had never in fact existed. It also suggests that the underlying hysteria ought to be located in the private psychic interiority of each dancer. As a spatial metaphor, though, the minuet draws attention to the strategic decisions that structure intersubjective spaces.

The opening minuet proves a highly unstable castle in that Edward and Leonora misread each other's expectations because they do not share the same 'priors.' We may recall that a system functions in Pareto-efficient terms when players act on the assumption that they are all 'instrumentally rational,' that they agree on what rationality means, that they 'hold common priors,' and that they 'know the rules of the game' (Hargreaves and Varoufakis 6). In Dowell's astute assessment, Edward and Leonora behaved according to different prior assumptions; they are locked in combat because 'whereas his traditions were entirely collective, his wife was a sheer individualist' (Ford 146). Although Leonora aims to gain Edward's love, her efficient management of both his finances and his love affairs has the unintended effect of appalling him. Where she sees her behaviour as supportive of her husband's feudal ambitions, he interprets her efficiency as a totalizing desire for mastery over both his estates and his heart. In the novel's narrative economy, it is the efficient who flourish and who are condemned for doing so. But Leonora was not born the cold and calculating woman who seeks to master her husband. On close examination, it is always Edward's behaviour that is shown to be responsible for her self-defeating decisions. In the first instance, the marriage between Edward and Leonora had been arranged by their parents. Although Edward had some choice among the daughters or 'the goods that [Colonel Powys] was marketing,' Dowell doubts 'that there was ever any question of love from Edward to her' (138, 139). Unfortunately for the supposedly cold and calculating Leonora, Edward aroused in her a love and passion that drove her later strategic decisions. As Henstra points out, 'the young, Irish Catholic Leonora' could be seen as 'Edward's victim in his acquisitive, imperial attitude towards courtship' (191) for 'it seems that, calmly and without any quickening of the pulse, he just carried the girl off, there being no opposition' (Ford 140). In this scene, it is he who is coldly efficient while she is passionately in love. Moreover, her interest in his financial affairs was motivated by her discovery that he was being blackmailed by the husband of a former mistress. In response to the economic threat posed by his sexual indiscretions, she 'had had such lessons in the art of business from her attorney' (56) that she was thereafter armed to deal expeditiously with subsequent emergencies. When she discovered his dalliance with the Grand Duke's mistress, Leonora 'had her plan as clearly drawn up as was ever that of General Trochu for keeping the Prussians out of Paris in 1870' (56). Edward's irrational actions compel Leonora to devise efficient strategies to ward off the financial threat to the feudal social position underpinning his sense of self.

Dowell's estimation of Leonora diminishes as she increasingly asserts her right to rationally 'manage' the husband who defensively entrenches himself in the 'rule-of-thumb' ideology of a lost feudal order.

For Hoffman, Dowell's 'fear of women's rising power' compels him to depict 'Leonora as transgressive and threatening,' especially in her 'disturbing' attempt to wrest 'the control of property and finances from Edward' (42). From a feminist perspective, it is to protect 'his sense of women's place in British and American patriarchy' that Dowell tends to begin by acting as the 'gallant protector' of women only to 'gradually move toward attributing more power and blame' (42) to them. But the narrator's consistent sympathies for Edward and increasing hostility to Leonora are not only indicative of his gender politics but also of unacknowledged economic convictions. In his narration, Edward and Leonora exhibit diametrically opposed tendencies in both their economic and sexual dealings. Edward's collectivist traditionalism makes him a protectionist, whereas Leonora's individualist ideology marks her as a laissez-faire bourgeois capitalist.

In spite of flirtations with stock-market speculation, Edward embodies the traditional values of a class intent on safeguarding its rights not only to hereditary property but also to colonial expansion. As has been pointed out, in his description of Edward's serial infidelities Dowell 'employs a metaphor of colonial expansion' (Henstra 190) that compares his desire 'to possess ever more women' with Britain's craving 'to possess ever more colonies' (Hoffman 37). In his appraisal of Leonora and Florence early in the story, Edward's possessive gaze reminds Dowell of a later incident when he contemplated 'the sunny fields of Branshaw' and said: '"All this is my land!"' (Ford 29). For Hoffman, 'this account explicitly presents the gentry patriarch in terms of raw greed,' establishing a 'common acquisitiveness' (36) in the attitudes of feudal paternalism and patriarchy. Responding to the same incident, Henstra similarly identifies Edward's expression 'of pride, of satisfaction, of the possessor' as evidence of his 'explore-and-conquer attitude' (191). Such territorial acquisitiveness exemplifies the logic of technical efficiency; it is predicated on a totalizing desire for optimal outcomes. We can now see that the 'social responsibility' that had prompted the Pareto-efficient solution to the short-term economic distress of his tenants illustrates not the cooperation implicit in social-justice models but the asymmetrical structure of paternalistic control. When Dowell repeatedly calls Edward a 'sentimentalist,' he puts his finger on the 'mawkish' aspect of his romanticized self-image. According to the dictionary, 'sentimental' refers not only to 'showing tenderness, emotion, delicate feeling' but also denotes being 'affectedly or superficially emotional; pretending but lacking true depth of feeling; maudlin; mawkish'

(Webster's). His paternalistic sentimentalism allows him to disavow the violence he perpetrates both as the embodiment of 'Englishness' and as the self-appointed protector of women.

The sad story of *The Good Soldier* unfolds as Leonora seeks to defend her husband's protectionist traditionalism with efficient measures that make her the ideological carrier of the very bourgeois capitalism he resists. What the novel dramatizes is a moment of crisis when the dialectical tension between a residual and an emergent ideology imposes itself on the cultural consciousness. At the centre of this drama stands Branshaw Teleragh, the Ashburnham family estate to which the main characters want to lay claim in one form or another. By the end of the narrative, this symbol of the age of property will have been transformed into an object of exchange in a market economy that makes it possible for Dowell simply to buy it from Leonora. The fate of Branshaw illustrates that capitalism opportunistically moves into a feudal space that had already been undermined from within. Edward is symbolic of an exhausted feudal ideology that has become complicit with its own demise. Dowell attributes Edward's sexual infidelities, which lead him to incur the debts that Leonora then proceeds to combat with rigorous efficiency, to his friend's chivalric instincts. In the Kylsite case, Dowell believes Edward who claims that his motives were misunderstood by a court that did not acknowledge the chivalric impulse of his class to comfort a servant girl:

> He assured me that, before that case came on and was wrangled about by counsel with all the sorts of dirty-mindedness that counsel in that sort of case can impute, he had not had the least idea that he was capable of being unfaithful to Leonora. But, in the midst of that tumult – he says that it came suddenly into his head whilst he was in the witness box – in the midst of those august ceremonies of the law there came suddenly into his mind the recollection of the softness of the girl's body as he had pressed her to him. And, from that moment, that girl appeared desirable to him – and Leonora completely unattractive. (156–7)

According to historical stereotype, the paternal concern for servant girls had in fact often enough ended with their seduction. But the reader is inclined to share Dowell's trust in Edward's innocent motives. Insisting that he had not experienced any sexual feelings for the girl while he was comforting her, Edward describes a psychological process that Freud calls *Nachträglichkeit* (belatedness). It was only after he had been faced with the prohibition of sex that his sexual feelings for the servant girl were retroactively aroused. The law prohibiting sexual advances to young girls in a train created the very transgression it was meant to prevent; exemplifying a logic Michel Foucault has so brilliantly analysed, the Kylsite incident suggests that the law creates the very desire it was set up to punish.

The logic foregrounded in the Kylsite incident resurfaces in the ways in which Edward and Leonora generate perverse outcomes by misreading what each expects the other to be expecting. Both of them make entirely rational decisions that nevertheless result in suboptimal outcomes. Faced with Edward's debts, Leonora energetically pursued practical solutions to avert the looming financial crisis threatening Branshaw Teleragh. The blackmailing costs of her husband's sexual indiscretions made Leonora decide to live on his officer's pay in India and to eliminate all waste at Branshaw. Since her husband's taking 'upon his own shoulders the burden of his troop, of his regiment, of his estate, and of half his county' struck her as a sort of 'madness' and 'megalomania' (149), she considers her attempts to cure him of these diseases to constitute a perfectly rational response to his irrational excesses. Not unlike Henry Ford, she set herself clearly articulated efficiency outcomes. According to her calculations, 'properly worked and without rebates to the tenants and keeping up schools and things, the Branshaw estate should have brought in about five thousand a year when Edward had it ... Edward's excesses with the Spanish lady had reduced its value to about three – as the maximum figure, without reductions. Leonora wanted to get it back to five' (166). To achieve this figure, she instigated economy measures designed to squeeze out as much profit as possible from Branshaw; she therefore 'put the rents back at their old figures, discharged the drunkards from their homes, and sent all the societies notices that they were to expect no more subscriptions' (59). However, while she saw herself engaged in pragmatically resolving a utilitarian problem, he considered her rational actions to be destructive of the public good he considered it his duty to serve. In the process of riding to Edward's financial rescue, she made herself and her attorney the trustees of Branshaw Teleragh 'and there was an end of Edward as the good landlord and father of his people' (167). Although Leonora is a Catholic, her dedication to thrift makes her exemplary of the Puritan personality that Max Weber singled out for being serviceable to capitalism. She is the ascetic accumulator of capital who decries her husband's generosity to others as a deplorable form of waste. But from his phallocentric and patriarchal position, he found it particularly 'intolerable' that his wife should be 'worrying about his managing of the estates' (146). Her interference was all the more galling as her decidedly capitalist approach was incommensurable with his own feudal tradition.

In her pursuit of technically efficient outcomes, Leonora succumbs perhaps inevitably and undoubtedly inadvertently to the same temptation of social engineering and control that marked the Ford Motor Company. As Dowell comments, the need to develop 'her purposeful efficiency' by the age of twenty-three made her 'perhaps have a desire for mastery' (141). Caught up in the technical project of revitalizing Branshaw Teleragh, she failed to recognize that Edward 'began to perceive a hardness and determination in his wife's character' that made him 'regard her as being not only physically and mentally cold, but even as being actually wicked and mean' (146, 149). Committed to efficiency, she could thus not help herself from 'beginning to try to rule with a rod of iron' the husband she 'loved passionately' (149). To her detriment, Leonora employs the resources of instrumental rationality to gain the love of a husband, an 'irrational' end beyond her utilitarian paradigm. Her propensity for 'mastery' impelled her to set herself up as an instrument of surveillance. Describing herself as 'watching him as a fierce cat watches an unconscious pigeon in a roadway,' she was so adept at keeping an eye on Edward and Nancy that the astonished Dowell comments: 'I don't know how she managed it, but, for all the time they were at Nauheim, she contrived never to let those two be alone together, except in broad daylight, in very crowded places' (130). Throughout the novel, the characters watch each other watching others. As in the case of Ford Motor, Leonora's tactics of surveillance eventually transmogrify into overt terrorism. As Henstra points out, the 'collective commitment to the game' of English conventions relies heavily 'upon repression,' and hence the 'mastery of the spoken word' (188, 189) of which Leonora is the prime embodiment. Once she realizes that Edward is hopelessly in love with Nancy, silent surveillance alone is no longer effective. In a memorable scene, Leonora 'finally gives in to the temptation to talk,' a torrent of words that has 'a force akin to physical violence' (189). Once the verbal floodgates had been opened, Leonora, Edward, and Nancy 'had talked and talked' until Nancy, 'a silent, a no doubt agonized figure, like a spectre,' suddenly offered 'herself to him – to save his reason!' (Ford 201). Terrorized by Leonora's tactics, Nancy finally agreed to leave for India, thereby precipitating the final catastrophe of Edward's suicide and her own madness.

But Leonora's surveillance extended to first 'seeing' and then managing the accounts that Edward considers to be protected from the probing eyes of a dutiful wife by the social conventions he takes for granted. Her transgression of the code that guarantees his feudal identity is almost as potent a source of miscommunication as his sexual adventures. Unable to understand that the estate was for him a way of life rather than a source of income, she could not see that the efficiencies she introduced were destructive of the very values embodied for him in Branshaw. Where she 'thought that he cared very much about the expenditure of an income of five thousand a year and the fact that she had done so much for him would rouse in him some affection for her,' he hated her precisely for her efficient management of his affairs since he 'imagined that no man can satisfactorily accomplish his life's work without loyal and whole-hearted co-operation of the woman he lives with' (175, 146). Instead of applauding her 'great achievement' of having extracted profit from the land and from investments to pay off debts and make them 'as well off as they had been before the Dolciquita had acted the locust' (175), Edward met her display of the figures with stony silence. Her expectations of his expectations were as miscalculated as were his of hers. To her consternation, he responded to the prospect of being set up 'as the Lord of Branshaw again – as a sort of dummy lord, in swaddling clothes' with nothing but distaste and hatred for her, 'even if, by accident, an act of hers were kind' (176, 177). Dowell's narrative suggests that Leonora's investment in technical efficiency either created or at the very least exacerbated a desire for the efficient management of both Edward's financial and sexual affairs, a desire that leads to the final catastrophe.

Leonora's capitalist commitment to maximizing short-term efficiency gains is an affront to Edward's sentimental nostalgia for a communal world focused on long-term stability. Her attention to 'the minutest items of his expenditure' (194) mathematizes what is for him an organic social community; he sees her as a coldly calculating efficiency expert and himself as a caring father figure. In this scenario, he accepts his social responsibilities while she selfishly satisfies her own needs. But from her perspective, he is ruled by irrational sexual instincts, and thereby inconsiderately and self-interestedly jeopardizes the survival of Branshaw. In contrast, she selflessly sacrifices herself to ensure his patriarchal privileges. In a dialectical spiral, husband and wife generate the very reactions they seek to forestall. Leonora's interventions in Edward's financial and sexual affairs are consistently traced to sexual adventures on his part that are in turn shown to have been generated by her financial decisions to safeguard the feudal order to which he appeals in the Kylsite case to rationalize subsequent sexual indiscretions. It is often to console himself for Leonora's affronts to his paternalistic values that he seeks comfort in the company of other women. Dowell informs us, for instance, that 'it was old Mr Mumford – the farmer who did not pay his rent – that threw Edward into Mrs. Basil's arms' (169). By attempting to stem Edward's feudal largesse to his tenants, Leonora in fact exacerbates the Ashburnhams' financial difficulties in that this affair opens the way for Colonel Basil to blackmail Edward. And it is only because Maisie Maidan is not intent on extracting money from her husband that Leonora has the disastrous idea of taking the young woman to Nauheim. In short, Edward's sexual infidelities are attributed to an initial chivalrous gesture nurtured by the feudal ideology that will from then on necessitate Leonora's management of his estate and mistresses. It is out of a sense of chivalric duty that he pays extravagant sums for the sexual favours of the Dolciquita, just as it is out of feudal obligation that he is overly generous to tenants on his estate and soldiers in his regiment. Ultimately siding with Edward, Dowell seems to accept that Edward's behaviour was not only appropriate but reasonable. But his narrative also justifies Leonora's reactions to her husband's sexual indiscretions as being entirely normal and hence reasonable. Yet her very normal efforts to combat his infidelities tend to further incite him to transgress her 'laws' of economic management. The mutual interdependence of love and money finds its culmination in Edward's suicide and Nancy's madness. Although Dowell attributes the suicide primarily to grief over Nancy's telegram from Brindisi, he nevertheless considers the possibility of another precipitating event: '[Leonora] threatened to take his banking account away from him again. I guess that made him cut his throat. He might have stuck it out otherwise – but the thought that he had lost his Nancy and that, in addition, there was nothing left for him but a dreary, dreary succession of days in which he could be of no public service ... Well, it finished him' (195). The conflict between Edward and Leonora thus hinges on their differing perspectives on efficiency: for him, efficiency was to serve the public good whereas for her it was to assist the individual in the capitalist accumulation of wealth.

Dowell's story conveys not only why the initial minuet's stable totality has always already been an illusion, but also how collectively self-defeating choices lead to the dissolution of social relationships. If for 'nine years and six weeks' (Ford 6) Dowell inhabited a 'brave new world' in which everybody seemed to have what they wanted and nobody wanted what they did not have, he ends his tale of passion on a note that reverses this initial Pareto-efficient ideal: 'Not one of us has got what he really wanted. Leonora wanted Edward, and she has got Rodney Bayham, a pleasant enough sort of sheep. Florence wanted Branshaw, and it is I who have bought it from Leonora. I didn't really want it; what I wanted mostly was to cease being a nurse-attendant. Well, I am a nurse-attendant. Edward wanted Nancy Rufford and I have got her. Only she is mad. It is a queer and fantastic world. Why can't people have what they want? The things were all there to content everybody; yet everybody has the wrong thing' (237). As Dowell insists, the 'things were all there' to satisfy the desires of all the 'dancers' in this minuet. The calamitous outcome is not entirely attributable to the inner life of individual characters. Although it may seem that Edward simply had a more passionate nature than Leonora, their actions are shown to be largely conditioned by intersubjective expectations formed by social conventions they have internalized. It is not that either Leonora or Edward could have modified their behaviour to achieve a different outcome. Dowell's commentary on the sad story he has just narrated makes it clear that the tragic ending was predetermined by the place each character had always already been assigned in the final configuration of the minuet. Insisting repeatedly that all the actors in this drama were 'good people,' the narrator tries to explain what compelled them to behave so 'badly': 'For I ask myself unceasingly, my mind going round and round in the weary, baffled space of pain – what should these people have done?' (233). Even when the 'end was perfectly plain to each of them' (233), they continued along a path that would lead to Edward's suicide and to the girl's madness. In Dowell's narrative, Edward and Nancy, as well as the other characters, passively perform the fate prescribed for them.

The Good Soldier is thus less a story of the individual heart than of the ideologically conflicted material conditions of a society in transition. More specifically, Ford's 'tale of passion' is also a tale of conflicting notions of efficiency. In the course of the novel, Edward's sentimental affirmation of a Pareto-efficient feudal order falls victim to Leonora's instrumentally rational efficiency. The efficient Leonora is finally condemned for being the ideological carrier of an increasingly standardized bourgeois-capitalist society that 'can only exist if the normal, if the virtuous, and the slightly deceitful flourish, and if the passionate, the headstrong, and the too-truthful are condemned to suicide and madness' (253). In his final judgment of Leonora, the narrator dwells almost obsessively on her privileging of efficiency. At the end of the novel, when Edward and Nancy painfully await their impending separation, 'only Leonora, active, persistent, instinct with her cold passion of energy, was "doing things"' (233). With her energetic commitment to efficiency, she strides across the ruins of the feudal order so as to create a new life with Rodney Bayham, a husband described as an 'economical person of so normal a figure that he can get quite a large proportion of clothes ready-made' (255). The destruction of Edward and Nancy is then not the outcome of the psychological predispositions or moral choices of the characters, but merely the human price to be paid for the triumph of social modernity: 'What then, should they have done? It worked out in the extinction of two very splendid personalities – for Edward and the girl *were* splendid personalities, in order that a third personality, more normal, should have, after a long period of trouble, a quiet, comfortable, good time' (233). In Dowell's sadly ironic assessment, the events he described demonstrate that 'it worked out for the greatest good of the body politic. Conventions and traditions I suppose work blindly but surely for the preservation of the normal type; for the extinction of proud, resolute, and unusual individuals' (238). If Dowell's rather lofty assessment of Edward and Nancy seems contradicted by a narrative that has, after all, depicted Edward as rather typical of his class, it is to indicate that he can only be described as an 'unusual' individual after his code of conduct has in effect already been superseded by different social norms. It is not as a person but as a social type that he faces extinction. Leonora is thus indicted for embodying the standardized and routinized mass society from which Dowell chooses to withdraw: 'So Edward and Nancy found themselves steam-rolled out and Leonora survives, the perfectly normal type, married to a man who is rather like a rabbit' (238). Seen as the inaugurator of a new but diminished social order, she 'will shortly become a mother of a perfectly normal, virtuous, slightly deceitful son or daughter' (252). In spite of the fact that the harmonious minuet opening the story had never actually existed, Dowell blames its dissolution on society's need 'to breed, like rabbits,' a society of which Leonora is the representative who inhabits not Branshaw Teleragh but a 'modern mansion, replete with every convenience and dominated by a quite respectable and eminently economical master of the house' (254, 252). The 'economical master' who fits into standardized clothes is indeed the appropriate husband for Leonora, who has consistently acted to prove that she 'desired to avoid waste' (240). In short, the 'brave new world' that Dowell so clearly abhors will be that of masses of standardized men and women reproducing themselves with the speed of rabbits. *The Good Soldier* thus demonstrates that the efficiency calculus has infiltrated the whole social fabric; it has so reified individual consciousnesses that 'good people' who act rationally seem doomed to precipitate perverse outcomes.

9 Efficiency and Its Alternatives: E.M. Forster's *Howards End*

In its nostalgic celebration of traditional English life, *Howards End* indirectly instals efficiency as the primary threat to Forster's devotion to his often acclaimed 'ideas of decency, humaneness, the civilized private life in which the disparities of the human condition might be resolved by honesty and good will' (Gordon 89). This devotion implies a typically British attachment to the rule-of-thumb approach to life that prevented easy acceptance of Taylor's principles of scientific management in manufacturing and public administration. Where the genteel Schlegel sisters represent the residual ideology of liberal humanism, the enterprising Wilcoxes stand for the consolidation and spread of an increasingly globalized capitalism. Like other modernists, Forster wanted to 'reconcile liberalism's commitment to the life of the spirit' with the 'competing tugs of power and property' (Born 141). This desire has produced readings of the novel insisting that the final scene affirms Forster's liberal-humanist belief in the reconciliation of the warring Schlegels and Wilcoxes. Such comforting interpretations are now being challenged by analyses that shift their focus away from relations between the sexes to problematize the synthesis school of criticism centred 'around love, marriage, mating and begetting' (Gordon 95). Situating characters within their socio-economic specificity, J. Hillis Miller identifies 'as central themes the redistribution of money and property through marriage' (469); Peter Widdowson stresses a dominant concern 'with money and its relationship to the life of liberal-humanist values' (63); Mary Ellis Gibson insists on 'the novel's importance as a critique of cosmopolitanism' (106); Daniel Born uncovers a preoccupation 'with the business of real estate' (141); Elizabeth Outka intervenes in 'recent critical debates on nostalgia and commerce' (331); and Henry S. Turner addresses the 'attempt to reconcile the ubiquity of the 'New Economy' with the 'Old Morality' (328). This shift in critical focus tends to indicate, usually only in passing, that the Wilcoxes are presented as figures of efficiency. The narrator is said to give 'the Wilcoxes their due for the efficient exercise of patriarchal power' (Miller 475), just as the Schlegel sisters 'have a certain respect for Wilcox efficiency' (Widdowson 82). But the 'organizing power of the Wilcoxes' (Delany 'Islands' 286) is more often than not considered to be destructive of the spiritual values embodied by the Schlegel sisters; they are dominant men who are generally indicted by both Forster and the critics for their 'narrow, bloodless efficiency' (White 52). Although the efficiency calculus seems to play a supporting role in Forster's depiction of Britain's rush to modernize, it suffuses the social fabric of *Howards End* and reconfigures our understanding of the novel's hostility to the 'evils of modernity' in ways not previously analysed.

Efficiency rears its ugly head in the disfigurement of the English landscape that Forster associates primarily with the motorcar, a technical innovation graphically symbolizing everything he seems to despise. *Howards End* could be said to confirm the contention that the positive as well as the negative aspects of the Ford Motor Company changed our world in significant ways. With the development of the low-cost Model T, the car made it possible for motorists to enjoy 'a brand-new freedom – the freedom to go anywhere they wanted, anytime' (Brinkley 118). Making the car available to the masses, Ford inspired 'the automobile vacation' and 'camping trips' (123). As Bill Ford Jr comments, 'prior to the Model T, most people never traveled more than twenty miles from home in an entire lifetime' (125). The car also had the 'capacity to render rural life more bearable' (127), alleviating the isolation of farmers that caused 'depression and mental illness' (128) and 'modernizing farming practices' (127). Moreover, women discovered 'an exhilarating sense of independence that could be had behind the wheel of an automobile' (119). But this new mobility was also feared for uprooting people and cutting them off from long-standing traditions and practices. In *The Magnificent Ambersons* (1918), Booth Tarkington cautions that motorists 'with all their speed forward ... may be a step backward in civilization – that is, in spiritual civilization' (in Brinkley, 115). For him, 'men's minds are going to be changed in subtle ways because of automobiles' (115). In *Howards End,* the car is treated as the triumph of technical efficiency that Forster targets for its destructive effect on both the natural and the social world. What needs to be highlighted in Forster's depiction of the impact of efficiency on society is that he resorts to categories – speed and flux, compartmentalization, fragmentation, standardization, homogenization – closely associated with the assembly-line production process at the Ford Motor Company.

But the socially transformative power of the efficiency calculus manifests itself not only in the direct disfigurement of natural landscape and individual consciousness, but also in competing conceptions of social organization. Margaret's liberal-humanist maxim 'Only connect!' (*Howards* 174) assumes a Pareto-efficient social model of cooperation capable of reconciling the individual's spiritual 'inner life' and efficient 'outer life' as well as healing society's class divisions. The opposition is once again between an organic 'good totality,' affirming the spontaneous harmony of all its willing parts, and a mechanical 'bad totality,' predicated on the artificial integration of resistant fragments. It is not that the organic model is inimical to efficient organization; rather, in the mechanical model efficiency is linked to social control. The personal relationships in *Howards End* are consequently examined for the social logic they imply. Margaret's emphasis on 'connection' and 'proportion' is thus beset by the logic of the Prisoner's Dilemma infecting all equilibrium models. Instead of unproblematically accepting the principle of connection as the guarantee for a harmonious liberal-humanist social order, a perspective informed by the logic of efficiency reinforces what Turner identifies as Forster's recognition that 'every system has a cost, requires an expenditure, a leftover, an outside' (340) and must consequently be examined for its dark underside.

The focus on efficiency in this chapter on Forster's *Howards End* seeks to complicate the 'exact specification of the characters' social placement' (Miller 470) and intervene in current debates on limitations of the liberal-humanist imagination and its complicity with the very ideology it seeks to resist. Creeping into the consciousness of individuals and infecting the social fabric as a whole, the efficiency calculus disrupts the categories on which the novel's oppositions build, undermines its privileged values, and fails to resolve the increasingly glaring contradictions arising with the dominance of global capitalism. This scrutiny of Forster's ideological critique of efficiency establishes *Howards End* as a far more astute document of a society in transition than appears in the critical tendency to dismiss the novel as a deplorable instance of its author's nostalgic yearnings and anemic idealism.

The Motorcar: Speed and Hurry

Howards End opens with a car bearing Mrs Munt to Howards End on a mission to save Helen from Paul Wilcox and ends, more or less, with another car taking Henry and Charles once again to remove Helen from Howards End. In both instances, the car is pressed into service against Helen's wishes and initiates a calamity. The speed with which Mrs Munt reaches her destination complicates Helen's retreat from the Wilcoxes just as the speed with which Henry and Charles later arrive at Howards End allows for the fateful encounter between Leonard and Charles. Associated with precipitated deeds, speed is decried as the enemy of contemplation and carefully considered actions. Margaret objects to the car not only because it is a danger to children and chickens, but because it is symbolically identified with Henry, a man who 'was always moving and causing others to move' (*Howards* 309). At home in the new world of speed and mobility, Henry is consistently appreciated for his ability to manage weddings, search for houses, and run a colonial business empire. In line with her doubts about Henry's inner life, Margaret dislikes riding in cars, objecting to the sensation of speed blurring and hence obliterating individual aspects of the landscape. In contrast with her investment in wholeness as a harmonious connection between self-identical parts, the car strikes her as a representative of modern efficiency destined to eliminate distinctions and differences in a totalizing drive threatening to homogenize the natural as well as the social world. Although Forster makes no mention of either the rationalization of the production process or the expansion of the consumer society symbolized by the Ford Motor Company, the narrator dwells on the impact of the car by foregrounding precisely such negative aspects of Fordism and Taylorism as an obsession with speed and flux, the compartmentalization of unified processes, the fragmentation and atomization of whole social areas, the standardization of previously individualized practices, and the pervasive homogenization of formerly distinctive cultural spaces. In 'E.M. Forster and the Motor Car' (2000), Andrew Thacker similarly demonstrates that the car serves as 'an ambivalent symbol in *Howards End*' (49); connecting geographical locations, this new mode of transportation is for him closely associated with Margaret's privileging of 'connection' in her articulations of philosophical wisdom. 'Obsessed with the notion of the nomadic basis of modernist identities,' Forster treats the car as an index of the changes in 'the experience of space' (39) so central to the novel's symbolic economy. Thacker's article provides excellent historical details about the anxieties the car generated in England, anxieties that reflect Forster's concerns about 'the psychic consequences of spatial disturbance brought about by travel' (46) and point to his more general objections to the 'new' civilization. While Thacker provides a thorough exploration of resistance to the car in *Howards End,* he does not contextualize it as a pre-eminent manifestation of a new commitment to efficency.

As a 'profound symbol of technological modernization' (Thacker 39) or a 'new instrument of disruption' (Weissman 433) at the disposal of the Wilcoxes, the 'automobile' is a constant presence in the novel, inviting readers to make sense of 'this new machine' that means a 'new mode of transportation, new economic fact, and, emblematically, new attitude toward the earth' (433). Above all else, the car expresses Forster's dismay at the way in which 'the fast unconnected urban time is speeding into the country' (Outka 337). Margaret's resistance to the car tells us much about the broadly reifying impact that this new technology has on human consciousness. She is shown to detest a 'motor-drive' (*Howards* 184) because she finds that the scenery 'heaved and merged like porridge' (185). The solidity of space is abolished by the speed of time, making it difficult for her to orient herself in the world and to connect with it. After her arrival at Howards End, she had to recapture 'the sense of space which the motor had tried to rob from her' (188). Similarly, on the return trip to London, she once again lost 'the sense of space; once more trees, houses, people, animals, hills merged and heaved into one dirtiness' (191). It was only after having spent a pleasant evening at home that she felt the 'sense of flux, which had haunted her all the year,' disappear for a while. Reorienting herself in her own home, she 'forgot the luggage and the motor-cars, and the hurrying men who know so much and connect so little.' For her, 'the sense of space' is in fact 'the basis of all earthly beauty' (191). To counteract this world of flux, Margaret makes a point of strolling 'slowly' up the avenue, 'stopping to watch the sky' and fingering 'the little horseshoes on the lower branches' (249) of the chestnut at Howards End. Watching a narrow road in the pleasant English countryside, she finds that 'its little hesitations pleased her. Having no urgent destiny, it strolled downhill or up as it wished, taking no trouble about the gradients, nor about the view, which nevertheless expanded' (249). The car is indicted through the contrast between Margaret's consciousness of individualized phenomena arranged within a determinate space and the sensation of car journeys that disconnect places only to merge them into an indistinguishable sameness. Her reactions foreground the paradoxical logic of efficiency, which simultaneously fragments and totalizes social spaces.

Beyond Margaret's dislike of car journeys, Forster holds the car responsible for a sense of flux and hurry that is already invading urban spaces and threatens eventually to engulf the rural landscape. The city of London is said to suffer the effects of the car in that 'month by month the roads smelt more strongly of petrol, and were more difficult to cross, and human beings heard each other speak with greater difficulty, breathed less of the air, and saw less of the sky' (102). Not unlike Henry Ford, who ended up deploring the modern world his car had created, Margaret feels overwhelmed by urban spaces designed to allow for the efficient flow of rapidly moving vehicles. Destructive of the leisurely pace of her genteel existence, the car is blamed for creating superficial social connections that do not bind people to each other or to the earth. Expressing her sense of alienation, she contends that the city has 'a heart that certainly beats, but with no pulsation of humanity' (102). London is a city bustling with nervous energy devoid of both meaning and purpose; it is 'a tract of quivery grey, intelligent without purpose, and excitable without love' (102). Disturbed by 'the architecture of hurry' and 'the language of hurry' in London, she comments that 'month by month things were stepping livelier' and asks herself rather pertinently, 'But to what goal?' (103). Like Henry, London is increasingly efficient but also spiritually impoverished. The car is indicative of a utilitarian purpose at odds with the aesthetic sensibilities of the Schlegel sisters. In this urban space, human beings exist as disconnected atoms whose movements are random and meaningless. The dark underside of the car's promise of freedom and independence is the anxiety that human consciousness will be irrevocably severed from the traditional social ties that Margaret cherishes.

The Problem: Fragmentation and Reification

Always on the move, Henry Wilcox is a successful businessman who is intent on increasing his wealth through colonial expansion and who accumulates houses, solves problems, manages people, and restlessly seeks opportunities to maximize the gains he hopes to procure. In metaphorical conformity with the internal combustion engine, he pursues efficient outcomes for their own sake. Although the car-crazy Wilcoxes are unmistakably the primary target of the social critique in *Howards End,* Forster is not without sympathy for their aspirations and accomplishments. As even Marx acknowledges, capitalism was initially admirable in its transformative power. In its early dynamic phase, for instance, capitalism was a revolutionary force that emancipated society from feudal oppression. It is the desire of the Wilcoxes to change the world that briefly attracts Helen to Paul and compels Margaret to marry Henry. The Wilcoxes exemplify a devotion to work that appears to the independently wealthy Schlegel sisters in the romantic light of 'modernity as adventure.' Born into a class that had for centuries considered work to be a mark of social inferiority, Margaret convinces herself that work is for the Wilcoxes not an economic necessity but a creative outlet for their enterprising energies. According to her construction, Paul sought employment in Nigeria because he decided to engage in an ennobling human activity. Ignoring the fact that men have laboured and suffered for centuries under various forms of socio-economic oppression, she confidently claims that 'in the last century men have developed the desire for work, and they must not starve it. It's a new desire. It goes with a great deal that's bad, but in itself it's good' (*Howards* 104). Elevated to the position of modern adventurer, Paul is said to be intent on realizing not only his own human potential but also the nation's destiny: 'He doesn't want the money, it is work he wants, though it is beastly work – dull country, dishonest natives, an eternal fidget over fresh water and food. A nation who can produce men of that sort may well be proud. No wonder England has become an Empire' (105). Even if Paul experiences working conditions that are unpleasant and routinized, he is redeemed by the larger purpose of Empire he serves.

In Margaret's opinion, enterprising men like the Wilcoxes engage in work that ensures the efficient management of the world according to British standards of decency. Although she declares herself bored by the notion of 'Empire,' she nevertheless can 'appreciate the heroism that builds it up' (106). If Henry strikes her as a 'real man,' it is because he impresses her not with 'youth's creative power' but with 'its self-confidence and optimism' (162, 152). She appreciates, for instance, that Howards End, which had been mismanaged by the former owners, was saved through Henry's decisive actions and efficient interventions. Knowing full well that 'the days for small farms are over,' Henry nevertheless saved the house, 'without fine feelings or deep insight, but he had saved it, and she loved him for the deed.' Efficiency is here portrayed not just as a necessary evil; the Wilcoxes are 'deliverers' who had 'worked and died for England for thousands of years' so that Margaret and Helen could 'sit here without having our throats cut' (192, 164). The novel offers at least a grudging acknowledgment of the contribution that efficiency makes to the standard of living the cultured Schlegels enjoy.

But this economic contribution brings with it cultural costs that the

Schlegel sisters are unwilling to pay. Through their reactions to the bustling Wilcoxes and through the narrator's comments, Forster leaves little doubt about his dislike of the Wilcoxes' moral callousness, careless destructiveness, self-confident complacency, thoughtless acquisitiveness, and control over others. Henry Wilcox is the ideological carrier of the culturally and socially detrimental effects that Forster associates with the new economic order of global capitalism. The narrator seems to sum up the novel's primary theme when he describes Margaret's eventual rebellion against her husband as a 'protest against the inner darkness in high places that comes with a commercial age' (309). Unlike Henry Ford, he is a non-productive finance capitalist who trades in commodities he may never see or own. Even in his everyday world, he is so divorced from the world of objects and people that a home is for him real estate and his housekeeper and chauffeur are tolerated 'because he could get good value out of them' (248). In his consciousness, he has internalized the total-output model of efficiency; he and his sons divide everything into manageable units that can be calculated, accumulated,[1] and exchanged. Through this process of fragmentation, they assume control over a mathematized environment and transform human beings into docile bodies to be managed. Focused on efficient outcomes, they are not concerned with the meaning of their actions. Typifying the Age of Commerce, Henry Wilcox is singled out for a tendency to objectify others and for an inability to recognize his own reified consciousness.

Henry and his son Charles are repeatedly conflated into the efficient modern businessman whose attitudes and behaviour are marked by a nexus of categories – speed, organization, fragmentation, objectification, reification, control – reminiscent of the logic introduced at the Ford Motor Company. In the opening scene we are introduced to Charles Wilcox, who is described as 'dark, clean-shaven,' and 'accustomed to command' (17). Exuding a self-possession that strikes Mrs Munt as 'extraordinary,' he is shown complaining that the train station is 'abominably organized' and pointing out that his 'time's of value' (17, 18). Scaring Mrs Munt with his fast driving, he is above all associated with speed and hurry. Representing a lack of imagination and individuality, he is, like his father, at his 'best when serving on committees' (93). In the observing eyes of the Schlegels, the Wilcoxes are guilty of commodifying human beings. For the Wilcoxes, a tenant like Mr Bryce is someone 'who had no right to sublet' (184); a man in a horse-drawn buggy, who is injured by Henry's car, is assessed according to insurance terms 'against third-party risks' (83); and a grieving girl, whose cat is killed by a Wilcox car, can be left to be dealt with by the 'insurance company' (198). Like Taylor's efficiency expert, Henry is a specialist[2] who focuses on steps that result in immediate solutions to narrowly defined problems. The Wilcox men generally avoid making 'the mistake of handling human affairs in the bulk, but disposed of them item by item, sharply' (93). For them, life resembles an assembly line; it can be lived 'sharply' as long as everything confronting them can be reduced to manageable fragments that can be counted, accumulated, exchanged, or discarded.

Henry is a danger not only to others but also to himself. Responding to the ideology of efficiency, he is the character whose consciousness is so thoroughly reified that he cannot recognize his own victimization under the socio-economic system that seems to favour him. It is through Margaret's reactions to Henry that the reader is alerted to the self-destructive aspects of her husband's investment in efficiency. Without him 'realizing that she was penetrating to the depths of his soul' (172), she recognizes, for instance, that his narrow commitment to efficiency cuts him off from his 'inner world' and makes him a poor judge of people. At the time of Henry's proposal of marriage, the conflict between his outer and his inner life strikes Margaret most forcefully: 'Outwardly he was cheerful, reliable, and brave: but within, all had reverted to chaos, ruled, as far as it was ruled at all, by an incomplete asceticism' (174). Both the Schlegel sisters and the critics insist that his focus on being an efficient man who gets things done leads him to neglect his emotional life and repress his passions. This failing is emphasized in his marriage proposal to Margaret; she has to delicately avert her eyes so as not to embarrass him 'as he struggled for possessions that money cannot buy' (155). Concentrating on what can be counted, arranged, and exchanged, he ignores personal relations that count so heavily with the Schlegel sisters. 'While his investments went right,' observes Margaret, 'his friends generally went wrong' (194). For most of the novel, Margaret convinces herself that Henry has to be protected from knowledge of his inner emptiness. He is an efficient man of action for whom self-contemplation might be psychologically debilitating and dangerous.

Henry's approach to his daughter's wedding and his collusion in the suppression of Ruth Wilcox's testamentary desire to leave Howards End to Margaret speak most forcefully to Forster's objections to the efficient commercial age. Margaret acknowledges that Henry knows how to organize and manage his daughter's wedding so that 'everything went like one o'clock' (207); however, she could not help but notice that 'Henry treated a marriage like a funeral, item by item, never raising his eyes to the whole' (205). Accused of routinizing life, Henry is shown to be draining it of all vitality. Organized like the perfectly planned assembly line, the wedding meant to be a celebration of life is transformed into a deadly funeral procession. Dividing events into fragments, he is able to streamline and control them but loses sight of their deeper meaning. A similar indictment of efficiency as a spiritual and moral death surfaces in the refusal of the Wilcoxes to honour what Ruth intended to convey in her will. They resist the meaning of the note by inspecting it for its authenticity, its legality, and the soundness of the writer's mind. This division of the note into different areas of specialization allows the Wilcoxes to repress the note's 'personal appeal' and to ignore evidence of Ruth's 'desire for "a more inward light"' (94, 85). As the narrator comments, 'Had they considered the note as whole it would have driven them miserable or mad. Considered item by item, the emotional content was minimized, and all went forward smoothly' (93). Disconnecting parts from the whole, the Wilcoxes dehumanize Ruth Wilcox by distancing themselves from the meaning of the message they do not wish to honour. As efficient committee men, the Wilcoxes exemplify Marx's contention that the commodity form of production converts social relations into relations between objects. Translating the human suffering they inflict into abstract terms, they are only capable of forming mediated rather than direct social interactions. What Forster deplores above all is that his society's new emphasis on efficiency tends to fragment society into areas of specialization that prevent people from seeing it as a harmoniously connected whole. The division of labour that Adam Smith celebrates and Marx holds responsible for human alienation creates increasingly atomized and narrowly defined subjects. In conformity with the disassociation of personality typical of today's CEO, Henry Wilcox rationalizes and compartmentalizes the contradictory demands of his corporate interests and his personal or moral concerns.

It is in his treatment of Leonard Bast that Henry reveals the full extent of his reified consciousness and his ability to rationalize the suffering he causes others as the inescapable cost of an efficiently functioning market economy. A debate on the market ensues when the Schlegel sisters realize that they had contributed to Leonard's unemployment by passing on to him Henry's chance remark that the insurance company the clerk worked for was in financial trouble. It turned out that this information was false, and it was the bank, Leonard's new employer, that went bankrupt. In an attempt to appease her own feelings of guilt, Helen blames Henry for Leonard's dire financial straits. When it finally dawns on Henry that Helen holds him personally responsible for Leonard's fate, he defends what happened as simply 'part of the battle of life' (178); the clerk is the hapless victim of powers beyond anyone's control. Implicitly appealing to Adam Smith's 'invisible hand' argument, Henry contends that 'no one's to blame' (178). In the grand historical scheme, it seems, Leonard unfortunately found himself in the wrong place at the wrong time. 'As civilization moves forward,' Henry lectures Helen, 'the shoe is bound to pinch in places, and it's absurd to pretend that anyone is responsible personally' (179). In other words, the 'free' market is as indifferent to the nature of individual exploiters as it is to the merits of individual victims.

Seeing himself as an agent of history, Henry asserts that civilization has been 'moulded by great impersonal forces' to which, sadly, some individuals have to be sacrificed so that the rest of humanity can progress. Although it is doubtful that Leonard would find this reassuring, Henry complacently affirms that 'the tendency of civilization has on the whole been upward' (179). Henry's self-justification implicitly invokes Hegel's 'ruse of Reason' argument, the contention that apparently 'evil' events, when seen from a retrospective viewpoint, will reveal themselves as a necessary precondition for a higher 'good' to emerge. Margaret falls into the same dialectical reasoning when she reflects that some day 'there may be no need for [Henry's] type,' while 'at present, homage is due to it from those who think themselves superior, and who possibly are' (152). Far from sharing her sister's sanguine resignation to market forces, Helen vows to 'stand injustice no longer' and to 'show up the wretchedness that lies under this luxury, this talk of impersonal forces' (209). She is clearly outraged by Henry's self-serving suggestion that 'in some mystical way the Mr Basts of the future will benefit because the Mr Basts of today are in pain' (180). Responding to her socialist sympathies, Henry smugly contends that 'one sound man of business did more good to the world than a dozen of your social reformers' (24). In his view, the market rather than the state is the most efficient provider of the public good. Poverty is not a social or moral issue, but a measure of the market's efficacy. In his complacent bourgeois-capitalist eyes, social inequality can be elevated to a law of nature: 'The poor are poor, and one's sorry for them, but there it is' (179). Such Social Darwinism is presumably meant to account for Henry's inner emptiness and his inability to enter into meaningful personal relationships. But it also implies a social vision based on the supposedly self-regulating mechanisms of the market; for Henry, the market inexorably seeks a state of equilibrium between input and output predicated on competitive players making best responses to each other. In this indifferent totality, social justice is treated as 'waste' to be dismissed or eliminated.

Interestingly enough, Margaret's reactions to the imperial dimensions of Henry's financial capitalism have their source not so much in Helen's moral outrage as in the threat the global economy poses to unique individual and local traditions. It is only in her dismay at Henry's plan to 'capture' her sister at Howards End that she indicts imperialism on moral grounds. Her own initial complicity with this plan makes her call it 'clever and well-meaning,' but also an action drawing its 'ethics from the wolf-pack' (263). As Henry sprang into action to 'save' Helen from herself, Margaret accuses him of patriarchal paternalism and naked colonial conquest: 'The genial, tentative host disappeared, and they saw instead the man who had carved money out of Greece and Africa, and bought forests from the natives for a few bottles of gin' (264). But Henry's crime is less the exploitation of colonialism's victims than the violent imposition of a standardizing order on the whole world. If imperialism 'always had been one of her difficulties,' it is because Margaret objects to it as a force making everything seem 'just alike in these days' (183). Visiting the 'offices of the Imperial and West African Rubber Company,' she complains that the maps on the walls emphasize the globalized 'Imperial side of the company' over the more local 'West African' one (183). Bringing into focus 'the formlessness and vagueness that one associates with Africa,' these maps speak to the will to power that imperialists like Henry impose on other continents. Henry's imperialist forays are thus violent conquests and homogenizing incursions rather than nurturing connections with different cultures. 'Building up empires,' men like Henry Wilcox are held responsible for 'leveling all the world into what they call common sense' (223). Although Margaret appreciates the enterprising colonial spirit that accounts for Henry's wealth and England's stature as 'an Empire' (105), she fears that the traditional England she cherishes will be imperilled by such homogenizing tendencies.

Margaret's opposition to the standardization of the globe through global capitalism finds an unlikely echo in Leonard Bast's reflections on imperialism as a 'motor passed him' (301) while he made his way towards Howards End for his fateful encounter with Charles Wilcox. On the positive side, the 'Imperial' type strikes Leonard as 'healthy, ever in motion' and poised 'to inherit the earth' (301). Yet this type is not the civilizing force he thinks he is. Although the imperialist 'carries his country's virtue overseas,' he reveals himself to be 'a destroyer' who 'prepares the way for cosmopolitanism, and though his ambitions may be fulfilled, the earth that he inherits will be grey' (301). Linking imperialism and cosmopolitanism, Forster exhibits a deep-seated anxiety about the disappearance of unique cultural and geographical markers under the spread of the globalized efficiency model promoted by the Wilcoxes. From Margaret's perspective, the 'continual flux' of London, which in Henry's world is 'good for trade,' strikes her as an 'eternal formlessness; all qualities, good, bad, and indifferent, streaming away – streaming, streaming for ever' (171). This formlessness does not inspire a creative desire to impose an aesthetically pleasing or socially coherent order. It is, instead, affiliated with a soulless homogeneity that she fears will spread throughout the world. As she puts it, London 'is only part of something else, I'm afraid. Life's going to be melted down, all over the world' (316). She objects to imperialism as an efficient totality that paradoxically privileges both the fragmentation of previously cohesive social communities and their homogenization under the impact of global capitalism.

A Failed Alternative: Connections and Proportion

In contrast to Henry Wilcox's concentration on efficient organization in his business and everyday life, Margaret seeks to preserve what is distinctive, unique, and particular within a harmonious organic whole. The values she embraces are almost term-by-term opposites of the categories that the novel associates with the car and the Wilcoxes. Against compartmentalization and fragmentation, she promotes temporal continuities with a cultural tradition predicated on deep roots and organic wholeness. Against standardization and homogenization, she celebrates the unique individual and the particularized object both in nature and in society. In her vision of an organically coherent world, the inner and the outer life ought to correspond in the neo-Romantic sense of spiritual freedom being guaranteed by a social community reflecting the spontaneous order of nature. The conflict between the inner and the outer life is to be resolved through an appeal to a quasi-mystical notion of reconciliation: 'Our business is not to contrast the two, but to reconcile them' (*Howards* 98). According to this formula, the subject is not encouraged to withdraw from the world into itself or to the pastoral idyll of Howards End; on the contrary, the subject and the object are meant to transform each other so as to form a harmonious unity. This essentially Hegelian notion of a 'unity of opposites' re-emerges in what is undoubtedly Margaret's most famous philosophical statement: 'Only connect! That was the whole of her sermon. Only connect the prose and the passion, and both will be exalted, and human love will be seen at its height. Live in fragments no longer' (174–5). The passion is not meant to obliterate the prose but to find an accommodation with it. To 'connect' is to 'transform' the two opposites so as to sublate them into a higher unity identified here as 'love.' Margaret's sermon implies that the quasi-mystical moment of sublation can be achieved through a quasi-rational program designed to 'connect' the fragments. It seems to be up to the well-intentioned individual to find and execute the healing connections between contradictory forces. When things are 'so contradictory,' she preaches at another point, all we can do is to try and 'live by proportion' (70). Proportion is presumably what emerges when opposites have been reconciled or sublated. But, contrary to the injunction to will 'connections,' Margaret now maintains that 'proportion' cannot be planned: 'Don't *begin* with proportion. Only prigs do that. Let proportion come in as a last resource, when better things have failed' (70). What 'better things' she has in mind are not clear. She may well be thinking of proportion as inferior to the loftier notion of sublation. In any case, once 'better things' have failed, proportion is equivalent to the spontaneous emergence of an equilibrium between competing forces. In her logic, then, Margaret shares with Henry a belief in the mysterious ability of systems – the market, nature, society – to configure themselves into spontaneously balanced totalities.

The philosophical underpinnings of Margaret's supposedly wise sermons on 'connection' and 'proportion' have been traced to a 'Coleridgean/Hegelian' appeal to 'reconciliation, a creative synthesis of antithetical types' as well as to a Shelleyan search for 'a transcendent union of the prosaic and the poetical' (White 53). More accurately, perhaps, she reprises a tension between Kantian rationalism and Romantic expressivism to which Charles Taylor points in his seminal *Hegel* (1975). Seeing Hegel himself as a conflicted figure, Taylor accentuates that the philosopher seeks to reconcile the contradictory aspirations of his age, 'to combine the fullest moral autonomy of the subject with the highest expressive unity within man, between and with nature' (49). His privileging of reason gestures in the direction of Kant, while his call for the subject to actualize itself through union with the quasi-mystical notion of 'cosmic spirit' (*Geist*) acknowledges the Romantics. Margaret's 'inner life' and 'outer life' reprises Hegel's opposition between the subject (the individual) and the object (nature, society). Preoccupied with the notion of personal freedom or autonomy, the Romantics contended that the subject is to define itself from within by creating meaning out of its own essence. For Taylor, the Romantic generation aimed 'to bring man back to unity with nature within and without, while maintaining his highest spiritual achievements, consciousness and moral freedom, intact' (39). It is through harmony with nature that 'man' gains his freedom: 'He must adapt to unconscious, unreflective forces which are beyond the call of reason; and he can only reach harmony with them by listening to what is most unconscious, unreflective in him, the voice of instinct' (39). Margaret's emphasis on the 'inner life' corresponds to the Romantic ideal of self-creation, while her investment in organic social relations references the Romantic celebration of nature.

Yet, unlike the Romantics, Hegel does not encourage a withdrawal from the world into a private, spiritual 'inner life.' For him, the subject's self-actualization could not take place in social isolation as a personal catharsis. The question for him, as for Margaret, is how to connect subject and object without obliterating either of them. In his dialectical logic, the subject needs the object to realize itself as subject just as the object needs the subject to recognize itself as object. This mutually interdependent dynamic retains the distinctiveness of both terms of the opposition. Subject and object are here not collapsed into each other but retain their singularity. In its ideal formulation, Hegel's 'unity of opposites' means that the subject can no more dominate the object than the object can control the activities of the subject. The key to the ultimate reconciliation or 'sublation' of subject and object is reason; it is through reason that the subject recognizes the shortcomings of the present and is able to imagine a better world. Hegel's narrative conceives of increasingly rational subjects intent on reconciling the actual or immediately given reality with the potential or ideal reality he envisages. The philosopher thus calls on the subject to transform the objective world so as to make it conform to the dictates of reason. In this scenario, reason enjoins the subject to subordinate its short-term interests to the long-term interests of the collectivity. However, if the subject's freedom is predicated on its willingness to 'freely' submit to the collectivity, then freedom is difficult to distinguish from coercion. It is this undecidability at the heart of Hegel's conception of the sublation of opposites that plagues Margaret's own attempt to reconcile (Romantic) expressivism and (Kantian) rationalism. If the organic wholeness she endorses can only be achieved through her active intervention, then her attempt to connect 'the prose and the passion' is no longer a spontaneous process but a forced or 'violent' yoking together of contradictory elements. In his critique of Hegel, Theodor W. Adorno famously deconstructs the Hegelian aspiration for a sublation of opposites as an illusion that glosses over the reality of society as an inescapably antagonistic space. Marked by the conflicted material conditions of late capitalism, modern society is no longer open to the reconciliation of opposites on which both Hegel's philosophy and Margaret's 'wisdom' are predicated. The ideal of organic wholeness is for Adorno not only a false or 'untrue' conception of present social reality, but a dangerous illusion poised to perpetuate society's violence by disguising it with the pleasing image of harmony.

The inheritor of Hegelian conundrums, Margaret is shown to invest in a reconciliation of opposites that proves not only unattainable but raises questions about its very desirability. Although there are critics who continue a long interpretative tradition of accepting at face value Margaret's self-identified power as a mediator and synthesizer, recent critical attention has tended to scrutinize her efficacy. The organic ideal she opposes to the efficiency calculus is, in the first place, deeply implicated in the very modernity she seeks to resist. In the second place, the unifiers she embraces consistently fail to connect the antagonistic fragments to be reconciled. And, finally, she and her sister are at times just as guilty of objectifying others as are the Wilcoxes. Neither the Schlegels nor Howards End function as convincing symbols of the unified community with which Forster concludes the novel. Leslie White is undoubtedly right when he contends that 'Forster *does* want connection, but the utopian vision of *Howards End* is destabilized by his ambivalence (as well as his inchoate aesthetic positions) regarding the utility of the exceptional person in a culture threatening or unsympathetic to the dynamical, imaginative life' (White 57). It is as if Forster were so successful in his depiction of the efficient modern age that he forecloses on the possibility of imagining an effective alternative to the 'loss of proportion' he often saw as 'the problem of modernity' (Levenson, 'Liberalism' 305).

Margaret's nostalgic privileging of Howards End as a harmonious whole is deeply implicated in the modern world from which the house is meant to offer a site of resistance or, at the very least, an avenue of escape. Unlike Ruth Wilcox, who is 'naturally' rooted in the countryside, Margaret is an urban figure intent on 'artificially' recreating lost connections to the land. In 'Buying Time: *Howards End* and Commodified Nostalgia,' Elizabeth Outka links the appeal of Howards End as a rural retreat to the 'Arts and Crafts Movement' and to the success of Edwin Lutyens's architectural style, blending 'nostalgia and commerce' (Outka 342). She argues convincingly that Margaret's celebration of the English country house exemplifies the 'commodified nostalgia' (331) representative of moneyed city dwellers in Forster's time. Having grown up in London, Margaret imbues country houses with nostalgic emotions that were typically being 'sold and consumed' (333) as real estate in Edwardian England.[3] It is because Forster cannot rid the house of its commercial taint that he has to shift the nostalgic vision culminating at Howards End into the aesthetic or symbolic register. It is not only the Wilcoxes who 'collect houses as your Victor collects tadpoles' (*Howards* 159); Margaret is in fact much preoccupied with real estate and the question 'Where to live?' (Born 142). Her decision to marry Henry has been seen as evidence of 'economic shrewdness' and appears to be 'fiscally opportunistic' (153). It is certainly the case that Henry's money had earlier 'saved' Howards End at Ruth's request. Moreover, the 'pleasing prospects' Margaret appreciates from the house consist of land belonging to 'the people at the Park' who had 'made their pile over copper' (*Howards* 192). Forster leaves no doubt that the English countryside is financed by commerce just as Margaret depends on Henry's money to keep Howards End economically afloat. Howards End and the English countryside survive only as a sentimental luxury on the part of enterprising businessmen. When Margaret thrills to the beauty not only of Howards End but also of Oniton Grange, she acts as one of those 'urban tourists' who imagine 'living an idealized country life' (Outka 339, 340). In her enthusiasm for Oniton Grange as 'a permanent home' that was, 'like herself,' rather 'imperfect' (*Howards* 242), she assumes that she can 'buy, ready-made, a nostalgic and romantic atmosphere' (Outka 340). Oniton thus 'reflects the ugly side of commodified nostalgia' (340). In the case of Howards End, the ghost of Ruth Wilcox is meant to conceal the logic of commodification. When Henry eventually respects his first wife's wishes and wills Margaret the house, the commercial taint is meant to be attenuated through the imprint of Ruth's spiritual legacy.

Although it will be Margaret's own money that will from then on pay for the expenses at Howards End, even this money is already contaminated by the efficient outer world she seeks to keep out. Her social ideal of a community of individuals who draw on a spiritually rich inner life in their relations with others remains inescapably implicated in the utilitarian social relations that reify the consciousness of the Wilcoxes and, upon closer inspection, of the Schlegels themselves. Forster is said to acknowledge 'the vital role of money earned from commercial ventures,' money that 'supports the cultural life of the Schlegels' and revitalizes Howards End as a 'marker of authenticity' (Outka 340). While the Wilcox money is 'not sufficient to maintain the house's spirit,' it is a 'necessary prerequisite' (341). As Turner nicely puts it, Forster's 'recognition of capital's indispensability and his profound desire to be rid of it' means that he 'seems fascinated by the power of money to change things and people but at the same time distrustful of its superficiality, of the structures that produce it and that are required to manage it, and of the world of impermanence it ushers in' (334). Margaret is prepared to acknowledge that her openness to the 'inner life' and to high culture is predicated on her economically privileged social position. Belonging to the unproductive class of rentiers who owe their leisured existence to the dividends made available by capitalism, she exhibits a politically anemic social self-consciousness that critics often trace to Forster's own unease with his economically privileged background. Speaking of his access to culture through education, he famously mused in 'The Challenge of Our Time': 'But though the education was humane it was imperfect, inasmuch as we none of us realized our economic position. In came the nice fat dividends, up rose the lofty thoughts, and we did not realize that all the time we were exploiting the poor of our own country and the backward races abroad, and getting bigger profits from our investments than we should' (68; in Delany, 'Islands' 285; Turner 328; Pinkerton 236). This sentiment surfaces in the novel when Margaret points out that 'all our thoughts are the thoughts of six-hundred pounders' (59) and compels Helen to 'admit that independent thoughts are in nine cases out of ten the result of independent means' (120). As Delany explains, the problem for Forster was 'how to uphold the civic and cultural virtues intrinsic to the rentier way of life, yet avoid complicity with commerce or technology' (291). Typifying Forster's own liberal guilt, the Schlegel sisters try to assuage their consciences with Margaret proffering sermons on money and Helen wanting to give away a large sum to Leonard.

In contrast to Helen's grandiose gesture, Margaret contents herself with a self-conscious acknowledgment of her privileges. Observing that 'the very soul of the world is economic,' so that the 'lowest abyss is not the absence of love, but the absence of coin' (58), Margaret seems to be echoing Karl Marx's famous contention that 'it is not the consciousness of men that determines their being, but on the contrary, their social being that determines their consciousness' (Marx 4). It is clearly to Margaret's credit that she points out to her aunt, 'You and I and the Wilcoxes stand upon money as upon islands. It is so firm beneath our feet that we forget its very existence' (58). Countering Mrs Munt's caution that taking risks in life is 'most dangerous,' her niece further insists that 'there's never any great risk as long as you have money' (58). Moreover, it is often financial security that accounts for a generous heart. Making arrangements for her marriage to Henry, Margaret encourages him to look after his children, to 'give away all you can, bearing in mind I've a clear six hundred.' Her virtue is not social responsibility but mere gratitude for her privileges: 'What a mercy it is to have all this money about one!' (169). Since a fine mind and a generous disposition are presumably indispensable to the cultivation of the 'inner self,' then the economic stability on which this superiority is predicated depends on the mode of production made possible by the efficient Wilcoxes. The consciousness of the Schlegels can thus not escape the reifying impact that capitalism has on a social class parasitic on those labouring for the preservation of a society's higher cultural values.

In spite of their rhetoric of 'personal relations,' the Schlegel sisters are often shown to be either failing to connect with others or dangerous to those they touch. Where Margaret's promotion of a synthesizing ideal leaves her complicit with the moneyed classes, Helen's socialist sympathies lead her to embrace 'social engineering' as a strategy to contest the free-market position Henry occupies. What prevents the establishment of harmonious connections is the tendency of both sisters to objectify others. The novel's opening scene alerts us to the possibility that Helen succumbs to the objectifying logic of Wilcox efficiency at the very moment when she declares herself unequivocally hostile to everything they stand for. In her own self-understanding, Helen confidently asserts that her temporary fascination with Paul Wilcox ends by confirming the value she has always put on the inner life and on personal relations. Explaining her experience to Margaret, Helen contrasts 'a great outer life that you and I have never touched' with a world of 'personal relations' that the two sisters consider to be 'supreme' (27). At first Helen conceded that this 'outer life, though obviously horrid, often seems the real one – there's grit in it. It does breed character'; in contrast, 'personal relations' may well lead to 'sloppiness' (27). Before the arrival of Paul, Helen had sent two letters to her sister in which she called the Wilcoxes 'the jolliest family you can imagine' (7). Impressed by the 'energy of the Wilcoxes,' she was struck by Henry's ability to impose his opinions on her 'without hurting' (23, 7) her feelings. Against her own genteel cultural upbringing, Helen acknowledges that the modern world represented by the Wilcoxes may in fact be the 'real' world. By the end of the scene, though, Helen dismisses Paul and his family for leading meaningless, empty lives. In the cold light of the morning after her engagement, she knew that 'it was no good' (25). Noticing how frightened Paul was that she might say 'the wrong thing' while he listened to his brother talking 'about Stocks and Shares,' she realizes that Paul 'had nothing to fall back upon' (26, 25, 27). Paul's fear made her feel 'for a moment that the whole Wilcox family was a fraud, just a wall of newspapers and motor-cars and golf-clubs, and that if it fell I should find nothing behind it but panic and emptiness' (26). During breakfast, then, Helen decides that 'personal relations are the real life, for ever and ever' (27). From now on, the Wilcoxes are for Helen symbolic of the despised world of efficient hollow men.

But the 'supreme' value Helen places on meaningful personal relations is undermined by her treatment of Paul as a 'cog' in the family economy. Arriving in the middle of her fascination with the Wilcoxes, Paul kindles in her a passion that is mediated by a social context within which she objectifies Paul. As the narrator editorializes, 'she had fallen in love, not with an individual, but with a family' (23). Playing his part in her romantic construction of the Wilcox family, Paul mediates her momentary desire for participation in the outer world. This most 'intense' moment of Helen's life is revealed as the 'embrace of this boy who played no part in it' (25). Helen's retreat into the inner life and personal relations is from the beginning compromised by her own inability to see Paul as a person in his own right. This same inability will later inform her disastrous attempt to 'save' Leonard. Helen's principled stand against Henry's dismissal of Leonard as a victim of market forces makes her adopt an interventionist socialist agenda that is far more damaging to the clerk than Henry's inability to see him as a human being. Like Paul Wilcox before him, Leonard is for Helen not an individual but a social construct to be either patronized as a cultural inferior or managed as social-engineering project. It is for good reason, then, that the narrator pointedly asks: 'Did Leonard grow out of

Paul?' (292). If anything, her objectification of Leonard is far more blatant. Although he had earlier appeared at Wickham Place to retrieve his umbrella, he impinges on her consciousness only retrospectively as she argues with a group of affluent women at a dinner party about the best way for upper-class ladies to assist the poor. It is this moment of theoretical socialism that initiates her well-intentioned but highly destructive interference in the lives of the Basts. Expressing her own guilt in the clerk's tragedy for which she blames Henry, she at one point acknowledges, 'We upper classes have ruined him' (209). To reinforce this point, Forster makes the situation of the Basts worse when Helen carelessly forgets to settle their hotel bill and accidentally takes their return train tickets with her. Leonard is indeed as objectified by Helen as by Henry and is at greater risk from her personal interference than he is from his impersonal neglect.

That Leonard has no individual existence for the Schlegel sisters is further borne out by the condescending paternalism with which they treat him. They want him to be 'the naive and sweet-tempered boy for whom Nature had intended him' (114) rather than the cultural equal he would like them to recognize in him. Margaret reflects insultingly that while 'culture had worked in her own case,' she 'doubted whether it humanized the majority,' and concludes that 'so many of the good chaps' who are trying to cross the gulf that 'stretches between the natural and the philosophic man' are in fact 'wrecked' in the attempt (109). The logic of Leonard's liminal position between civilized human being and natural man makes him the 'exception' that affirms the 'normal' status of both the Schlegels and the Wilcoxes. It is Henry who sees through the paternalism of the Schlegel sisters, asking Margaret: 'What right have you to conclude [Leonard's life] is an unsuccessful life, or, as you call it, "grey"?' (137). Henry here draws attention to the hypocrisy of liberals in their condescension towards the less fortunate while he also implicitly rationalizes his own part in the exploitation of Leonard by denying his suffering. But it is Margaret's refusal to help Leonard and her advice to forget him that most clearly undermines her sermon on connecting through her objectification of others. As Born concludes, 'For all her talk about connection, Margaret seems rather ill-equipped, and not at all predisposed, to connect with people' (151–2). At the same time, Leonard himself is in turn guilty of objectifying the Schlegels by elevating them to abstract embodiments of the cultural capital he covets: 'Seeing them as art objects and as grail symbols, he cannot see them as human beings, just as they, for entirely different reasons, have difficulty seeing him as a human being' (Hoy 226). As an antidote to the coldly rational efficiency calculus of capitalist modernity, Margaret's celebration of personal relations as a Hegelian reconciliation of social antagonisms fails to transcend the reification of her own consciousness as she succumbs to the temptation of objectifying others.

Margaret's ideal of connecting 'the prose and the passion' fails even more spectacularly in the novel's treatment of class conflicts. Class markers are a crucial aspect of the characters depicted in *Howards End,* often determining their attitudes and behaviour. As J. Hillis Miller points out, 'England was in 1910 still a class society in a way the United States has never been' (471). For a long time, critics followed in Lionel Trilling's footsteps, arguing that 'the novel's characters all belong to the middle class' (Born 146). More recent criticism calls for finer class distinctions, challenging the 'flaccid term, "middle class"' by differentiating between 'the poverty-line Basts, the independently wealthy Schlegels, and the rapidly rising Wilcoxes' (146). In Delany's reading of the novel's class topography, the Schlegels are sandwiched between two other classes: 'Below the Schlegels are the Basts, representing the half-submergd yet aspiring lower middle class; above them are the richer Wilcoxes, go-ahead business people' ('Islands' 286). The depiction of each sister is an invitation to judge the claims of the rentier class in relation to the other two classes: 'The older sister, Margaret, concentrates on trying to understand the class above her; the younger, Helen, on understanding the class below. Each takes her sympathy to the point of sexual connection – Margaret's willed and reasoned, Helen's impulsive' (286). What these topographies overlook is the ghostly presence of the 'Age of Property,' the essentially feudal class represented by Ruth Wilcox, and the 'unthinkable' working class of which the Basts are the tip of the iceberg. The conflicting ideological claims of the Schlegels and the Wilcoxes can be examined in light of their respective responses to the 'unthinkable' material conditions of the emergent lower classes and the 'unseen' spiritual values of the residual propertied class. Although the Basts are not entirely representative of the unthinkable 'People of the Abyss,' just as Ruth Wilcox is not quite emblematic of the powerful aristocracy, they constitute the lower and upper limits of a range of possible moral responses of the bourgeoisie to its 'other' on a class continuum. As we will see, Forster's symbolic reconciliation of the Schlegels and the Wilcoxes at Howards End is destabilized by the commodification of Ruth Wilcox and the suppression of the Basts. At the core of the novel's inability to 'connect' the classes are the dead figures of Ruth Wilcox and Leonard Bast.

It is only fitting that Ruth Wilcox should symbolize the 'unseen' in the novel as she hovers as a ghost over the other characters. The tradition she embodies can no longer survive in the efficient modern world embraced by her husband and children. As the 'proper' owner of Howards End, she becomes, 'at least in Margaret Schlegel's mind, a ubiquitous, all-knowing, all-encompassing spirit' (J.H. Miller 470). The Schlegel sisters and Henry are, in Margaret's view, 'only fragments of that woman's mind. She knows everything. She is everything. She is the house, and the tree that leans over it' (*Howards* 292). Her ghost symbolizes the 'unseen,' a word, Miller reminds us, that appears frequently in positive contexts and is 'especially associated with Margaret Schlegel' (467). What is most noticeable about *Howards End* is that it 'measures the characters by their responses to the unseen' (469); those 'who are sensitive to the unseen are unequivocally superior to those who are not' (478). Margaret's 'optimistic slogan' (480) to 'connect the prose and the passion' (*Howards* 174) is crucially not 'a demand to do this or that, but a demand to be and to feel in a certain way that then leads spontaneously to right action. That the action is right, however, cannot be verified by any preexisting code of ethics or moral behavior' (J.H. Miller 480). Whatever forges the 'connections' necessary to a social community that honours 'proportion' is an 'expressive' force beyond rational conceptualization. If the 'unseen is unknowable, not amenable to reason or understanding' (481), then the injunction to connect is less an articulation of liberal humanist ideals than of a nostalgic yearning for traditional values rooted in premodern times.

It has been well documented that Ruth Wilcox functions as the remnant of a residual ideology that the sociohistorical moment of modernity no longer makes unproblematically available to Margaret. Having been born at Howards End, Ruth simply *is* what Margaret desires to *become* through urban tourism. By stressing that Howards End had been Ruth's 'own property' (69), the narrator situates it outside the commercial orbit of the Wilcoxes. Firmly rooted in both the land and older social traditions, Ruth is dismayed to hear that Margaret will be evicted from Wickham Place to make room for a modern apartment building. To be 'parted from your house' is for her 'worse than dying' (79); affirming permanence as her supreme value, she asks: 'Can what they call civilization be right, if people mayn't die in the room where they were born?' (79). What is most significant about Ruth is that 'she worshipped the past, and that the instinctive wisdom the past can alone bestow had descended upon her – that wisdom to which we give the clumsy name of aristocracy' (22).[4] She is a figure not of enlightened humanism but of natural aristocratic blood and breeding. Like Edward Ashburnham in *The Good Soldier,* she retains a 'nobility' and 'spiritual existence' that can be neither lost nor appropriated; her claim to an aristocratic system of values circulates outside the economy of capitalist production and consumption. However, while her refusal to participate in an economy of exchange leaves her spiritual claims inviolate, it makes her vulnerable to material dispossession. In the 'battles' her son 'had fought against her gentle conservatism,' Charles proved victorious, replacing the pony paddock that 'she loved more dearly than the garden itself' (88) with a garage for the new motorcars. Ruth Wilcox and Howards End clearly represent for Margaret the imperilled traditional world resistant to Henry's affirmation of impermanence and disconnection from the past; it is against her sensitivity to the unseen that the Wilcox men are judged and found wanting. But Margaret also senses that resistance to speed and flux as well as to the accumulation and efficient management of possessions as a meaningless end in themselves rests on the weak shoulders of a dead woman's ghost, her own often compromised voice, and a small farm threatened by the spread of urban sprawl. With 'everyone moving,' Margaret wonders if it is, in fact, 'worth while attempting the past when there is this continual flux even in the hearts of men' (129). In spite of the novel's insistence on the spiritual bond between Ruth and Margaret, their actual interactions tend to be strained, indicating that the two women are ultimately not able to 'connect.' Ruth strikes Margaret as 'inclined to hysteria,' and there was a 'growing discomfort' between them, with Margaret fearing at one point that she 'had been snubbed' (79, 80). Although Margaret is singled out for her sensitivity to the 'unseen' suggested by this 'shadowy woman' of 'undefinable rarity' (81), her own relationship to the past is so marked by commodified nostalgia that she is never entirely at ease in the older woman's company.

At the other end of the class spectrum, Leonard Bast and his wife Jacky test Margaret's ideal of 'connection' and 'proportion' beyond its breaking point. Although Leonard is not the oppressed cog of the assembly-line worker at Ford Motor Company, he is representative of the white-collar clerk subjected to Taylor's scientific management practices. Forster clarifies that the proletariat is beyond his powers of conceptualization: 'We are not concerned with the very poor. They are unthinkable, and only to be approached by the statistician or the poet. The story deals with gentlefolk, or with those who are obliged to pretend they are gentlefolk' (44).[5] Whether we condemn this statement as Forster's 'shocking and surprising acceptance of class-based social exclusion that contradicts the ethos of connection that the novel would seem to endorse' (Armstrong 318) or see it as 'an honest admission that he knows nothing at all about [the Basts]' (Widdowson 90) and their class, we are at the very least invited to judge the Schlegels and the Wilcoxes according to their responses to the 'unthinkable' typified by Leonard Bast, just as we were asked to assess them in their responses to the 'unseen' represented by Ruth Wilcox. It is men such as Leonard who silently labour to create the material conditions on which the opposition between the economically parasitic Schlegels and the hardworking Wilcoxes operates. In contrast to their assessment of the entrepreneurial spirit exhibited by the work performed by the Wilcoxes as an 'adventure,' the Schlegels portray Leonard's job as the demeaning drudgery exemplifying modernity as 'routine' from which they intend to save him.

The critical chorus castigating Forster for his inability to make Leonard a convincing character[6] suggests that the class markers clinging to him were in fact barely 'thinkable' for Forster. Hovering just above the poverty level, he points towards the 'unthinkable' in the narrative economy. At the bottom of the novel's social stratification, Leonard himself comes to understand his own place in the insurance company as a 'cog' in a machine beyond his powers of conceptualization and control. Clarifying that he can only grasp 'his own corner of the machine, but nothing beyond it' (131), he draws attention to the increasingly specialized nature of routinized work: 'I could do one particular branch of insurance in one particular office well enough to command a salary, but that's all' (212). Not unlike the assembly-line laborer, this white-collar employee is assigned a specialized task serving a larger corporate purpose he cannot see. Leonard is encouraged neither to think for himself nor to act on his own initiative. In the first place, then, he seems to be victimized by the efficiency calculus that reifies his consciousness by reducing him to a commodity to be used, controlled, exchanged, and finally dispensed with. In Turner's terms, Forster invents 'certain characters who stand as figures for surplus and the human cost of capitalism – the Basts – but who are rendered as mere symbols, "flat" characters on which to hang, in passing, a certain amount of remorse over the system and its expense' (339). The victimization of the Basts in

Howards End is not about their plight but about the destructive power of the Wilcoxes and the self-serving hypocrisy of the Schlegels.

Although Forster criticizes the class system that bars Leonard from realizing his sociocultural ambitions, he offers him no escape from the fate of being a cog in the economic machine. In an effort to escape from his narrowly constrained world and to counteract his sense of disconnection from the whole social context, Leonard attends concerts and reads books in order to improve himself and his socio-economic prospects, demonstrating a naive belief in the liberal-humanist 'ideals of Culture and Equality' that Forster 'associates with both Ruskin and Arnold' (Hoy 222). Forster makes him typical of the increasingly impoverished lower middle classes who continued to identify culturally with the social aspirations of the bourgeoisie at the same time as their actual buying power often sank below that of the working classes. Invested in the bourgeois-capitalist ideology holding out the promise of social mobility to the enterprising individual, Leonard assumes that he can improve his socio-economic position by acquiring the 'cultural capital'[7] he sees in the possession of economically privileged people like the Schlegels: 'I care a good deal about improving myself by means of Literature and Art, and so getting a wider outlook' (52). Identifying with a social class that excludes him, Leonard feels 'superior' (49) to those people who blame their lack of success on bad luck; unlike such hapless victims of circumstance, 'he did believe in effort and in a steady preparation for the change that he desired' (49). Believing himself to be a self-originating or 'free' agent, he proudly contends that he does not 'take heed of what anyone says' (52). Refusing to resign himself to the status of a cog in a machine, he tries to regain his human dignity through cultural accomplishments that he pursues not for their own sake but for the socio-economic advantages they offer as cultural capital.

It gradually dawns on Leonard that his efforts at self-improvement will not enhance his socio-economic situation. A visit to Wickham Place provides him with the unwelcome insight that no amount of self-improvement would open the door to 'some ample room' into which the Schlegels disappeared, a place 'whither he could never follow them, not if he read for ten hours a day' (53).[8] No matter how enterprising and efficient in his acquisition of cultural capital he may prove to be, he can never be more than a temporary distraction and exotic curiosity for the Schlegels. For them, he can only become 'thinkable' in terms that exclude his actual 'unthinkable' conditions of existence. While the Schlegels and the Wilcoxes are socially mobile, Leonard Bast is, like

Ruth Wilcox, 'naturally' situated in the class into which he was born. Although heavily inflected by Forster's editorial irony, the narrator's comments on Leonard's 'continual aspiration' show him to be condemned to inescapable frustration because 'some are born cultured; the rest had better go in for whatever comes easy. To see life steadily and to see it whole was not for the likes of him' (53). Whatever 'comes easy' turns out to be work that 'stops one thinking,' as it does for the assembly line 'cog.' The books Helen reclaims for him from the bailiff cannot compensate for 'something regular to do' that would allow him to 'settle down again.' Forster signals his character's resignation to his lot when he has him acknowledge that 'the real thing's money, and all the rest is a dream' (222). By the end of the novel, before he meets his death buried under books, Leonard had joined the ranks of the unemployable or 'unthinkable,' presumably a fitting punishment for his audacity to question his socially predetermined destiny. If only he could find routinized work again, he could 'settle down' to live out his life in the state of happy stupor reserved for docile bodies who enjoy the benefits of Frederick Winslow Taylor's attempt to create a perfect fit between worker and task.

It seems, then, that Leonard's fate was sealed by the indifference of the capitalist system to its exploited cogs and by the failure of the Schlegel sisters to bridge the class divide in spite of their rhetoric of 'connection' and 'proportion.' However, the final blow to his chances of survival may well have been delivered by the pre-modern tradition embodied by Ruth Wilcox. For Judith Weissman, it is the capitalist system that is clearly at fault for Leonard's inability to find employment after he had lost his job through the interference of Henry Wilcox and the Schlegel sisters: 'Henry might have taken him in, but he will not, for the ethics of capitalism do not allow for charity' (437). But the 'ethics of capitalism' are so indifferent to individual cogs that Henry was in fact inclined to employ the clerk. As Margaret reminds Helen, the unemployed Leonard 'wants work, and that we can't give him, but possibly Henry can' (213). After Henry had agreed to interview Leonard, she exults: 'Henry would save the Basts as he had saved Howards End, while Helen and her friends were discussing the ethics of salvation' (215). It appears, then, that Henry was initially prepared to 'save' the man who poses a moral dilemma not for his own capitalist assumptions but for the liberal self-understanding of the Schlegel sisters. It is possible that Leonard would regain employment through the same combination of market forces and personal connections that had cost him his job in the first place. But the Schlegel sisters are made to carry their share of guilt for Leonard's death. Helen's dogmatic socialism proves more damaging to Leonard than any market forces. Turning him into the recipient of her social-engineering ambitions, she is the catalyst for the events that eventually lead to his poverty and death. Her brief sexual connection to assuage her liberal guilt proves far more detrimental to Leonard than Henry's refusal to connect with him. Finally, for Pat C. Hoy, it is Margaret herself who is responsible for 'failing to get work for [Leonard]' (234). Distancing herself from Leonard, she certainly falls short of her ideal to connect the 'prose and the passion.' In the end, though, Henry rescinds his offer of an interview because he discovers that Leonard's wife had at one time been his own mistress. It is thus not hostility to Helen's socialism or a change in market forces that condemns Leonard. The conflict between Henry's contention that the market is an objective equalizer and Helen's insistence on social reform is ultimately resolved not in the socio-economic sphere but in the novel's sexual economy. Where the market is indifferent to moral questions, the social world of personal relations remains dominated by the patriarchal values of a residual ideology treating women as the property of their husbands. It is this ideology that compels Henry to condemn Helen for a sexual indiscretion with Leonard Bast of which he himself had earlier been guilty with Jackie Bast. Although Margaret is prepared to forgive her future husband his past sin, she is infuriated by the realization that Henry has internalized a patriarchal value system that sanctions a sexual double standard. In sexual relations, then, Henry acts as a 'protectionist' in contrast to his endorsement of the 'free market' in his business dealings. Since it was his wife's sexual 'connection' with Henry that blocks Leonard's access to financial survival, it looks suspiciously as if he is ultimately doomed neither by market forces nor by socialism, but by the residual aristocratic paternalism of the Age of Property of which Ruth Wilcox and Margaret Schlegel are the ideological carriers.

If the novel's guiding wisdom is Margaret's emphasis on connections and proportion, then the spontaneous cooperation of decent individuals based on the organic order of nature assumes a social equilibrium that proves susceptible to collectively self-defeating scenarios typified by the Prisoner's Dilemma. Although the Schlegel sisters make well-intentioned and apparently appropriate decisions in given circumstances, these contribute to unintended consequences detrimental to the social community in question. Helen's rescue mission not only exacerbates the material conditions of the Basts, but also steers Leonard into the path of Charles Wilcox. Margaret's genuine desire to connect the Schlegels and the Wilcoxes culminates in the compromise of her own principles and the destruction of Henry and his son. Recent critical attention has focused precisely on the failure of the 'ethos of connection,' emphasizing that it is being called 'into question' as it is being 'evoked' (Armstrong 317). Margaret is accused of having 'betrayed her values' (White 56) and of eventually having abandoned 'the attempt to articulate a unifying vision for her private and public discourse' (Born 144). In the novel as a whole, the 'prose and the passion remain in conflict' and 'attempts to connect them are rarely successful' (J.H. Miller 480).[9] It is generally accepted that Forster resorts to the 'personal' and the 'unseen' to compensate for his inability to deal with the problems of modernity in economic or political terms. For Forster, then, 'the effort to "link the individual to the whole" is a matter not of social organization but of imaginary figuration' (Levenson 1985, 308). The novel's 'lingering attachment to a political ideal' is accompanied by 'its refusal of a political program' (Levenson, 'Liberalism' 308). In spite of her rhetoric to the contrary, Margaret cannot step outside of her class self-interest in order to connect with the unfathomable Basts; she is not prepared to reconfigure the equilibrium between benefits and costs in the socio-economic arena. Her sermons on connections and proportion reveal themselves as self-serving illusions designed to conceal the social antagonisms that are at least foregrounded in Henry's market logic.

A Dubious Ideal

Aside from Margaret's inability to put her ideal into practice, the ideal itself may be ideologically suspect. Her retreat into illusion saves her from having to acknowledge her own recourse to coercion to ensure the pastoral idyll on which *Howards End* concludes. It is only by breaking Henry and forgetting Leonard that Margaret was able to achieve the reconciliation that Helen acknowledges: 'But you picked up the pieces, and made us a home' (315). However, the community over which Margaret presides falls short of a Hegelian sublation of equally strong contradictory forces. Both Helen and Henry are diminished characters. In contrast to the earlier zeal Helen exhibited for social reform, she now seems contented with the domesticated position of expectant mother taking refuge from the world on the farm. More disturbingly, perhaps, Margaret assumes control over a Henry who has been physically and mentally broken. The ideal of spontaneous synthesis conceals the violence that went into the reversal of the master position in the relationship between Margaret and Henry. By the end of the novel, Henry's 'possession of male mastery,' the 'virility' that had been 'part of his attraction for Margaret,' has been supplanted by Margaret's 'triumphant matriarchal possession of Howards End, firmly in control of her broken and subdued husband' (J.H. Miller 477). She herself expresses astonishment at this unintended reversal of power positions: 'She, who had never expected to conquer anyone, had charged straight through these Wilcoxes and broken up their lives' (*Howards* 318). The ideal of a harmonious whole consisting of cooperating social agents has become perverted into an enforced totality. Her triumph over Henry is encapsulated in the scene when she upbraids him about his sexual double standard: 'You shall see the connexion if it kills you, Henry!' (287). By violently imposing on him a philosophy of connection that can logically operate only if entered into freely, she confirms Frederick Winslow Taylor's insight that cooperation always has to be enforced. If there is 'something uncanny in her triumph' (318), it is that the unintended consequences of her noble intentions hint at the violent or dark underside of the ideal of synthesis.[10] This underside is reinforced in the exclusion of those who cannot be assimilated into the utopian vision of organic wholeness. As the closest representative of the 'unthinkable' proletariat the narrator rules out from the start, Leonard Bast is by the end of the novel identified as 'the personal' that must be 'forgotten' if it proves inconvenient. Addressing Helen, who carries Leonard's child, Margaret proclaims: 'Then I can't have you worrying about Leonard. Don't drag in the personal when it will not come. Forget him' (314). In order to set up Howards End as a purified place, both Henry's efficient commercial investments and Leonard's economic plight 'have been ruthlessly and effectively purged' (Outka 346). If Leonard is sacrificed to Henry's dehumanizing indifference in the focus on efficient outcomes, he is also kept out of Margaret's comfortable bourgeois retreat and symbolically repressed so as not to disrupt her self-deceptive conception of herself as a synthesizing unifier. The organic alternative that Margaret espouses to counteract the mechanistic efficiency calculus thus 'exposes the will-to-power in efforts to merge perspectives and forge unity out of differences' (Armstrong 317).

Armstrong's suggestion that Margaret's 'will to harmonize and unify is not only unrealistic but also potentially dangerous' (321) is further borne out when we consider the ideological implications of the specifically English landscape as a source of her nostalgic yearning for a synthesis. Howards End may well be implicated in the kind of ideological rhetoric that could, retroactively, be considered to have been particularly conducive to the emergence of the fascist regime of terror in Germany. The celebration of the 'unseen' in the novel is incontrovertibly linked to what is deemed 'natural' and 'spontaneous.' As the legacy the first Mrs Wilcox bequeaths to the second, Howards End is the repository of neo-Romantic yearnings that are most potently symbolized by the special spiritual powers with which Margaret imbues the wych-elm and chestnut tree. Fingering the 'little horseshoes' (*Howards* 249) of the chestnut with loving absorption, she appreciates each part of the tree for its self-sufficiency as well as for its contribution to a pleasing overall effect. Identified as a specifically 'English tree,' the wych-elm at one point overwhelms her with its 'peculiar glory,' a glory she traces to the 'strength and adventure in its roots' and the 'tenderness' in its 'utmost fingers,' concluding that the 'girth, that a dozen men could not have spanned, became in the end evanescent, till pale bud clusters seemed to float in the air' (192). The paradoxical combinations of strength and tenderness, adventure and rootedness, and permanence and evanescence create an emotional effect that is clearly beyond rational analysis. The organic metaphor of the tree is symptomatic of Margaret's privileging of authentic social bonds mysteriously rooted in English cultural traditions and English soil. For her, the 'free' social community ought to express a natural state of equilibrium not unlike the efficient distribution of wealth by means of an 'invisible hand' in the 'free' market. Social relations are for her not to be regulated through calculation; they are best left to the 'rule-of-thumb' good sense of individuals who combine the paradoxical values she associates with the wych-elm. What forges 'connections' is thus largely 'unseen' and can only be felt. Margaret's sermons on connections and proportion are thus less an articulation less of liberal humanist ideals than of a nostalgic yearning for traditional values rooted in pre-modern times.

The privileging of Howards End over London has been well documented. The deeply felt values of the English countryside are held up against the cosmopolitan superficiality of London. Elevating Howards End as a refuge from London, Margaret nostalgically privileges the age of 'feudal ownership of the land.' Imbuing the Age of Property with 'dignity,' she compares it favourably to the 'modern ownership of movables' that is 'reducing us again to a nomadic horde. We are reverting to the civilization of luggage, and historians of the future will note how the middle classes accreted possessions without taking root in the earth, and may find in this the secret of their imaginative poverty' (141). London was for her 'but a foretaste of this nomadic civilization which is altering human nature so profoundly' (243), evidence of an irreversible process to which her only response can be withdrawal to Howards End. The house itself is for her symbolic of the English landscape of small farms that Henry considers to be economically no longer viable. As a refuge from the age of industrial efficiency, the countryside is sentimentally associated with what is best in England. In the country, she reflects, 'men had been up since dawn' (301), men who are 'England's hope. Clumsily they carry forward the torch of the sun, until such time as the nation sees fit to take it up. Half clodhopper, half board-school prig, they can still throw back to a nobler stock, and breed yeomen' (301). Even Leonard Bast is imbued with 'hints of robustness that survived in him' and 'more than a hint of primitive good looks' (109). Projecting onto him a neo-primitivism they consider to be a site of resistance to the emergent triumph of the Wilcoxes, the Schlegel sisters align themselves with the authentic blood of the yeoman against the Wilcoxes' culture of luggage. Margaret operates under the nostalgic assumption that feudal ownership of land allowed human beings to live harmoniously within nature. Ignoring the suffering of the landless masses throughout history, she associates land ownership with the cultivation of the imagination. In contrast with this firm connection with the land, the bourgeois-capitalists' accumulation of mutable possessions is devoid of emotional bonds.

In addition to such class-based elitism, Margaret's commitment to England as an organic whole leads her to indulge in a perhaps rather facile nationalism that has her prefer local 'Stilton' to foreign 'Gruyère' (*Howards* 145). Gibson argues that *Howards End* contains a thorough 'critique of cosmopolitanism' that is marked by Forster's approval of the 'earlier meaning' of the cosmopolitan as a 'citizen of the world' and his dislike of a later meaning arising with the 'advent of nationalism' that involves a 'contrast between patriotism or "old home feeling" and international, often financial interests' (106, 107). Where Tibby Schlegel is a 'self-described cosmopolitan' who 'resembles the "citizen of the world"' in limited ways, Henry Wilcox falls somewhere between an 'international financier and the typical British merchant or manufacturer' (108). However, it is the father of the Schlegel siblings, the naturalized Englishman who had been born in Germany, who is representative of the universal or cosmopolitan currency Forster attaches to cultural ideals. Forster goes to some length to stress that the father was neither 'the aggressive German' nor 'the domestic German,' but 'the countryman of Hegel and Kant,' that is, he was an 'idealist, inclined to be dreamy, whose Imperialism was the Imperialism of the air' (28). Rejecting the 'clouds of materialism obscuring the Fatherland,' the father hopes that eventually 'the mild intellectual light' of his own disposition would 're-emerge' (29) in the land of his birth. Indeed, what the Schlegel children inherit from the German side of their lineage is their cultural sophistication and idealistic intellectualism. In contrast, Henry Wilcox is a 'citizen of the world' not in his openness to ideas from abroad but in his financial ambitions to dominate the economic structures of British imperialism. When Margaret shares with the Wilcoxes a dislike of cosmopolitanism, she seeks to protect the English cultural values that they consider to be obstacles to their imperial economic expansionism. What emerges from the contrast between cosmopolitanism as the circulation of ideas forming our cultural identity and cosmopolitanism as the circulation of capital in the arena of economic activity is Margaret's deep investment in specifically *English* cultural values. In response to Helen, who threatens to 'throw the treacle' at her if she starts to argue in her 'honest-English vein,' Margaret counters that it is 'a better vein than the cosmopolitan' (149). Her retreat to Howards End is also her affirmation of a nationalism squarely located in the English country house, the English tree, and the English landscape. As a culmination of this ideology, the estranged Schlegel sisters are reunited at Howards End, this place with 'wonderful powers' conducive to them regaining 'the knowledge that they never could be parted because their love was rooted in common things' (279, 278). That these 'common things' are embedded in the English tradition is finally affirmed by the narrator's comment that where 'explanations and appeals had failed,' the two sisters were saved by 'the past sanctifying the present' (278). Resistance to the modern efficient world of the Wilcoxes is here located in a pre-modern past that is both unattainable and open to appropriation by dangerous political forces.

In addition to being unattainable, Margaret's synthesizing ideal has her wade into the politically dangerous territory of an irrational neo-Romanticism that had its parallel in the cultural attitudes of Wilhelmine and Weimar Germany, attitudes that can be seen as a fruitful seedbed for the emergence of the Nazis. It is only in retrospect that the seemingly innocent pleasures Margaret takes in the contemplation of English trees, the English landscape, and the English 'yeoman' reveal dubious associations with a neo-Romantic rhetoric later surfacing in

Hitler's speeches. In his acclaimed analysis of the cultural roots of fascism, George Mosse singles out the 'linden tree' as a typical icon of 'the peasant strength of the Volk, with roots anchored in the past while the crown aspired toward the cosmos and its spirit' (26). Proto-fascist intellectuals like Paul de Lagarde, Ludwig Klages, and Oswald Spengler regularly objected to the 'parasitical city dweller, traditionless, utterly matter-of-fact, religionless, clever, unfruitful, deeply contemptuous of the countryman' (Spengler 32), encouraging young people to embrace 'the small towns, the villages, the peasant' as symbols of the authentic 'connection between the history of the Volk and its fusion with the landscape' (Mosse 16). During the Wilhelmine and Weimar years, German students (*Wandervögel*) in particular advocated a revitalization of sterile modern existence through a return to more organic or primitive ways of life. In the modern metropolis, young people saw their 'spiritual roots dislodged through industrialization and the atomization of modern man' (Mosse 8–9). Not unlike the Schlegel sisters, young Germans treated their feelings of alienation as a metaphysical or spiritual crisis rather than as a socio-economic problem. Exploiting the German students' exaltation of the 'organic nature of the Volk' (Mosse 102) and the sacredness of the German soil, the Nazi movement was later to transpose 'Volk' into race; indeed, Hitler specifically locates 'the foundation of the new renascence' in the 'maintenance of the German peasant' (Hitler, in Baynes, 250). Widdowson is undoubtedly right when he complaints that Forster's desire to retain 'the victorious "vision"' (87) of an idealized England in the novel's final scenes is responsible not only for a forced plot, but also for the endorsement of (perhaps suspect) nostalgic longings.

A Seemingly Inescapable Dilemma

A reading that attributes to *Howards End* a perhaps slight and certainly unwitting complicity with fascist totalitarianism may appear to foreclose all lines of escape from the pervasive hold of efficiency on modernity. On the one hand, this reading indicts the efficient Wilcoxes for participating in the dehumanizing and reifying tendencies of modernity that Bauman considers to be a necessary, if not sufficient, condition for the emergence of Nazism. On the other hand, Margaret's neo-Romantic counter-discourse appears as the potentially dangerous seedbed for an ideological rhetoric that enabled Hitler to promote a racist agenda. Neither an investment in efficiency nor resistance to its increasing pervasiveness leads inexorably to Auschwitz; however, an unquestioning investment in the efficiency calculus authorizes Henry's dehumanization of Leonard, just as an unexamined endorsement of irrational longings for authentic community encourages the clerk's exclusion by the Schlegels. Both tendencies are in their *logic,* rather than in their actual manifestations in the novel, allied with the possibility of opening the door to unintended consequences. Henry's unregulated market is as susceptible to collateral damage as is Margaret's equally unregulated organic community. What Forster's novel forecloses is a specifically *political* analysis of the evils of modernity he decries and then resolves in the personal register of individual choice and family relations as well as in the aesthetic register of symbolic reconciliation. What is conspicuously absent in the novel is a political engagement with liberalism. Although Forster is consistently identified as a 'liberal,' the ideal he promotes in *Howards End* is not so much liberalism as neo-Romanticism. The social connections Margaret privileges are fantasized as mysteriously arranging themselves into an organic whole rather than being recognized by her as the negotiated alliances on which liberal consensus depends.

Levenson is thus quite right to stress that Forster 'frequently remarked upon the obsolescence of the liberal ideal,' but always did so 'from the standpoint of an obsolescent liberal' ('Liberalism' 303). Drawing on L.T. Hobhouse's *Liberalism* (1911), Levenson argues that Forster must thus be associated with an 'older liberalism' devoted to 'the removal of constraints' in the belief that 'once individuals were allowed to develop freely, an "ethical harmony" would ensue' (303). This 'negative' sense of freedom was followed by the 'positive aspect of the liberal movement' that focused on 'the regulation of behavior, the intervention in markets, the exercise of legal restraints and "social control,"' which struck Hobhouse as 'the complete subordination of individual to social claims' (in Levenson, 304). The earlier 'harmonic conception' assumed a belief 'in steady progress, a slow course of mutual adjustment in which the self and the state would move gradually toward equilibrium' (304), much as the best responses of players in the free market were thought to result in a state of equilibrium. But this older liberalism is in fact closer to the spiritual neo-Romanticism embraced by the Weimar students than to any political program, except perhaps John Stuart Mill's defence of property rights.

The rival equilibrium model to this older liberalism was Bentham's 'greatest happiness principle,' which 'looked to the state to harmonize competing interests' (Levenson, 'Liberalism' 303). The highest value is now no longer attached to 'the individual but to the community and its collective will' (303). In his steadfast defence of the individual, Forster decried social regulation as a violation of freedom. It is, after all, Helen who is shown to be the most misguided character in the novel; her socialism is proof not only of her impetuosity but also of the threat she poses to others. Instead of relying on Margaret's neo-Romantic idealism to counter the Social Darwinism that Henry parades as the social concomitant of free-market economics, Forster might have considered the possibility of regulating the market through political and legal constraints. But the economic and the political are in Forster's novel incommensurate, so that the reifying impact of efficiency on the individual consciousness and its homogenizing social tendencies could only be resisted in the aesthetic register of the imagination.

10 Efficiency and the Perfect Society: Aldous Huxley's *Brave New World*

In the narratives examined up to this point, efficiency as an emergent ideology tended to be dramatized in its impact on the consciousness of individuals whose personal and cultural self-understanding had been thrown into crisis. In contrast to both direct and indirect indictments of the specifically social and personal costs of an investment in efficiency, Aldous Huxley's *Brave New World* (1932) draws out the broader social, cultural, and philosophical implications of the industrial revolution that the novel metaphorically attributes to Henry Ford. Displacing Christ as the central cultural figure, 'Our Ford' is revered as the founder of a 'civilization' projected into the future, a society that erects statues to him, circulates his autobiography 'with large golden T's' (218) stamped on it as the new Bible, pays tribute to him with 'Ford's Day celebrations' (52), remembers him in mock church services, and thinks of itself as taking place in 'this year of stability, A.F. 632' (4). Not surprisingly, critics have linked the standardization and rationalization of all aspects of social life in *Brave New World* to Ford's obsession with the efficient production and mass consumption of the car as a technology dedicated to speed and motion. Carrying the mantle of the 'founding father' is the current World Controller, Mustapha Mond, who is modelled on both the Turkish political reformer Mustapha Ataturk and the British industrialist Sir Alfred Mond.

These economic and political associations are framed by philosophical debates between Mond and three dissidents with the symbolic names of Bernard Marx, Helmholtz Watson, and John Savage. Born on a 'primitive' Indian reservation, Savage is quite obviously representative of Jean-Jacques Rousseau's celebration of the natural man not yet corrupted by civilization. This figure has generated critical debates focused primarily on Huxley's rejection of the nostalgia with which D.H. Lawrence imbued the Pueblo peoples of the American southwest. Although *Brave New World* targets efficiency in the name of Ford, Huxley's indictment of the assembly line is in many ways a more straightforward, and hence more superficial, confrontation with the efficiency calculus than his extrapolation from the factory to the 'perfect' society over which Mond exercises total political control. The novel sets in motion two competing explanations for the homogenized social order, which is for Huxley destructive of the unique or 'free' individual posited by the liberal tradition. On the most obvious level, economic determinism reduces human beings to cogs in the Fordian machine. More provocatively, Huxley has Mond justify his political position through a denigration of Jean-Jacques Rousseau's social-contract theory in *The Social Contract* (1762) at the expense of his espousal of Thomas Hobbes's alternative philosophical proposal in his *Leviathan* (1651). Leaping from Ford's factory floor to Mond's 'perfect' society, Huxley depicts a socially engineered society that is as indebted to Frederick Winslow Taylor as it is to Henry Ford.[1] More importantly, Mond in effect functions as an 'efficiency expert' who plans and organizes an objective system from a supposedly impartial external position. In political terms, this external position identifies him as the embodiment of the sovereignty principle embraced by Thomas Hobbes that Rousseau proceeded to revise and oppose. Mond's philosophical debates with the novel's dissidents are thus informed by competing social-contract theories at issue in utopian dreams of the 'perfect' social order. Under the gaze of Henry Ford, *Brave New World* explores the logic of efficiency no longer as a diffuse concern, but as a concentrated confrontation with its most radical economic, political, and philosophical ramifications.

'Our Ford': The Efficient Social Machine

Brave New World began as Huxley's anti-utopian satire of H.G. Wells's *Men Like Gods*, his fantasy of a society owing its 'perfection' to technology, rationalization, and bureaucratization. However, as he revised the manuscript, the 'future according to H.G. Wells grew increasingly anti-Fordian' (Meckier 427). Reconstructing the genesis of *Brave New World*, Jerome Meckier documents Huxley's Americanization of London as the centre of civilization by charting the 'additional insults to Henry Ford' (427) the author inked in as he amended earlier manuscript versions. Although the metaphorical centrality of Ford has never been in doubt,

Meckier confirms it by discovering 'at least 110 references to Ford and things Fordian' (432). In addition to direct invocations of Ford's name, Huxley depicts a society that worships technical efficiency in all areas of life. Taking obvious delight in the mock-precise technical and pseudoscientific terms he invents to suggest the triumph of rational efficiency, he has the brave new worlders speak expertly of 'Bokanovsky's Process' and 'Podsnap's Technique' (*Brave* 6, 8), procedures that are being 'stimulated,' 'supplied,' and done 'automatically' (12). Using the language of rationalization, they speak of 'calculations,' cite 'a few figures,' conduct a 'test,' consider the 'margin of safety,' and employ a 'system of labelling' to keep track of things (10–13). The 'hum and rattle of machinery' inside the Central London Hatchery and Conditioning Centre is typical of the society as a whole. Inhabitants are moved along in escalators, 'light monorail trains,' helicopters, and the 'Blue Pacific Rocket' that takes off from 'Charing-T Tower' (11, 72, 58). This 'hive of industry' in which 'every one was busy, everything in ordered motion' (146), is dominated by a sense of time that is both precise and constantly accelerating. Typifying the tyranny of time in the Ford system, Huxley specifies that the 'hands of all the four thousand electric clocks in all the Bloomsbury Centre's four thousand rooms marked twenty-seven minutes past two' (146). Worshipping Ford as the founder of their civilization, the brave new worlders exhibit an obsession with technical efficiency that makes them accept their subordination to the machine without question or protest.

The brave new worlders are not only chained to machines but are literally produced by them and for them. In an absurd extension of Ford's desire to integrate workers into the factory system as a closed totality, Huxley opens *Brave New World* with the famous introduction to the Central London Hatchery and Conditioning Centre in which children are no longer born but are 'decanted' from bottles. Biological reproduction has been superseded by eugenic engineering. The embryo is placed in a bottle and is then 'assembled' at a series of workstations. Passing 'Metre 320 on Rack 11,' the unnamed narrator visiting the Hatchery encounters a young Beta-Minus mechanic who 'was busy with screwdriver and spanner on the blood-surrogate pump of a passing bottle. The hum of the electric motor deepened by fractions of a tone as he turned the nuts. Down, down ... A final twist, a glance at the revolution counter, and he was done. He moved two paces down the line and began the same process on the next pump' (14). The eugenics laboratories solve the problem of unpredictable workers inhibiting the efficient production process at the Ford Motor Company. Instead of bribing or threatening workers, the brave new world adjusts the workers to machines by mechanically engineering them for the functions they are destined to perform. In Huxley's assembly-line laboratories, the 'principle of mass production' is applied to the creation of 'standard men and women' in 'uniform batches,' so that Fordian London is peopled by 'millions of identical twins' (7). In a nightmare extension of Ford's assembly-line logic, Huxley paints a picture of a totally homogenized world in which 'crowds of lower-caste workers' are 'queued up in front of the monorail station – seven or eight hundred Gamma, Delta and Epsilon men and women, with not more than a dozen faces and statures between them' (164). The diversity of individuals is considered an inefficient remnant of a time before rationalization was embraced as the precondition of prosperity. The appeal of standardization introduced with the assembly line is for Huxley an affront to the creative human spirit and a denial of individual and political freedom. It could be said that *Brave New World* draws out the totalitarian dimensions of an efficient system from which all wasteful residues have been eliminated.

Social control over docile bodies extends beyond eugenics to include ongoing and highly sophisticated psychological conditioning techniques. In the nursery, infants undergo various forms of 'neo-Pavlovian conditioning' (*Brave* 19). They are exposed to deafening noises in order to develop an aversion to such non-utilitarian pursuits as the appreciation of books or flowers. If Pavlovian conditioning through repetition proved insufficient, it was periodically followed up with electric shock therapy. Such early conditioning was above all intended to eliminate all extraneous human emotions or wasteful individual aspirations. Everything that Huxley considers to be humanly meaningful is here treated as waste material destined for elimination. The 'love of nature,' for instance, is a 'gratuitous' indulgence (23). Similarly, 'lower-caste people' are taught to hate books because there is no point in them 'wasting the Community's time' (22) on frivolous reading. Pavlovian conditioning is most specifically aimed at preventing individuals from thinking for themselves. Huxley demonstrates exhaustively how methods like 'sleep-teaching' (25) and the mindless repetition of slogans produce automatic and compliant reactions. In many ways, the 'god' of the new world state could have been Taylor rather than Ford: Taylor's principles of scientific management infuse the description of the eugenics factory so that the 'Deltas and Epsilons are the equivalents of Taylor's gorillas and human oxen' (Firchow, *End* 108). Predestined to execute assigned tasks, these 'gorillas' are kept docile with rewards (*soma* drugs) for the completion of work, as Taylor recommended. Unlike the managers at Ford Motor, the Human Element Manager of Huxley's brave new world can justifiably boast that 'we hardly have any trouble with our workers' (160). Where the sociological department at Ford Motor struggled to make workers internalize docility, the managers in Huxley's novel are presented with already tamed workers who had compliance injected into their embryonic blood and later reinforced through social-engineering practices. Mass-produced according to standard categories, the brave new worlders are infinitely replaceable and interchangeable. Most disturbingly, they are no longer traumatized by their dehumanization, as was the case for workers at Ford Motor; radically extending Babbitt's social compliance, they have so internalized the requirements of the system that they are 'happy' performing their assigned tasks. Eugenics and social engineering here combine to create a homeostatic state that recursively reproduces itself. Projected six hundred years into the future, the creative spirit animating the Crystal Palace has inexorably deteriorated into Dostoevsky's frozen tomb, and 'modernization as *adventure*' has transformed itself with a vengeance into 'modernization as *routine*' or Weber's 'iron cage.'

The Efficient Mustapha Mond: Fordism, Capitalism, Totalitarianism

The shocking picture of the rigorously rationalized life in the future that emerges with the visit to the Hatchery and the lecture of the Director of Hatcheries and Conditioning (D.H.C.) on history remains to this day the most memorable aspect of *Brave New World*. Once Huxley has detailed the dehumanizing conditions he traces to Ford's assembly line, he turns his attention to Mustapha Mond, one of the seven World Controllers in charge of the highly Americanized London. It seems appropriate that the homogenized Fordian civilization should be governed by a totalitarian ruler. It is through this figure that Huxley examines the economic, political, and philosophical sources of the dehumanized future he imagines in such graphic terms. Addressed as 'his fordship, Mustapha Mond' (33) when he first appears during the D.H.C.'s lecture, Mond is unambiguously the representative of the industrial revolution Ford introduced with the assembly line and the Model T. Where his title ('his fordship') links him to Ford, his name conjures up economic and political associations intended to consolidate the standardizing impact of Fordism (as well as Taylorism). On the one hand, his first name alludes to 'a notorious seventeenth-century Turkish sultan' and, more recently, to 'Mustafa Kemel (Kemel Ataturk), president of Turkey since 1923' (430). For Meckier, this perhaps unexpected intrusion of the Orient into this Euro-centric novel can be explained as follows: 'Huxley feared a worldwide dystopia governed by sophisticated Western dictators wielding greater power than an oriental despot' (Meckier 430). Mond's first name thus draws attention to the political despotism that Huxley fears will arise with economic rationalization. His last name is French for 'world,' conjuring up the World Controller's political aspiration for global domination. But the name also alludes to the British industrialist Sir Alfred Mond, symbolizing the economic usurpation of politics the novel demonstrates. According to Meckier's research into the novel's genesis, Huxley had initially intended to caricature 'H.G. Wells and Sir Alfred Mond in the composite figure of Mustapha Mond because he considered both men proponents of antihumanistic rationalization – the reorganization of society on an allegedly more scientific, more efficient, more technological basis' (432–3). While Huxley did not visit the United States before the completion of *Brave New World,* he had 'toured Mond's Billingham plant for producing sodium and synthetic ammonia' (433). Although Ford as 'America's archetypal technocrat' (427) grew in metaphorical importance, Sir Alfred Mond remained a palpable target for Huxley's hostility to industrial capitalism.

The political impetus in the brave new world is to ensure the economic control on which political power rests. As the D.H.C. stresses, the social engineering project is designed to adapt 'future demand to future industrial supply' (48). Compulsory production and consumption are carefully calculated so as to maintain an efficient equilibrium between input and output. Comparing the fictional Mond's attitudes with those expressed by Sir Alfred Mond, James Sexton indicates how much Huxley was indebted to the British industrialist who defined rationalization as follows: 'The application of scientific organization to industry, by the unification of the processes of production and distribution with the object of approximating supply to demand' (in Sexton, 93). What Huxley adopts from Mond's position is a 'fascination with rationalization and mechanization' that can be traced to the 'materialist legacy left by F.W. Taylor and Ford' (89). But in Sir Alfred Mond, Huxley discovered a figure committed to rationalization not only as an industrial but also as a sociopolitical principle. For the historical as for the fictional Mond, rationalization presented itself as the solution to the threat of chaos. Faced with social turmoil and financial crises during the interwar years in Britain, Sir Alfred Mond responded by proposing an efficiently organized economic order: 'To Mond the solution to the failing market system was a growing rationalization of economic units – he hoped to see a kind of economic League of Nations, where industrialists could, effectively and co-operatively, shape the economic destiny of most of the world' (94). The emphasis on cooperation echoes Taylor's assumption that rational social agents, who are prepared to rationally pursue a rational purpose, would logically choose cooperation over competition. However, as Taylorism proved, this kind of cooperation easily deteriorates from embracing voluntary to taking on coercive forms. In 'Machinery, Psychology and Politics,' Huxley draws out the coercive dimensions of a society modelled on the logic of the factory. In a reference to Alphonse Séché, whose *La morale de la machine* (1929) he admired for its brutal frankness, he comments on the extrapolation from the factory to society at large in terms that apply directly to the conflation of economic and political power in the figure of the fictional Mond: the 'modern industrial State ... should be organized like a very efficient factory, or group of factories, with hierarchically graded experts in charge of every department and a single Henry Ford at the head to co-ordinate their activities and dictate the policy of the whole concern' ('Machinery' 749; in Sexton, 92). For Huxley, then, the economic and the political spheres are not incommensurate but continuous.

Brave New World dramatizes overlapping but not necessarily converging understandings of the social ramifications of Henry Ford's obsession with efficiency. As the first part of the novel indicates, Huxley associated efficiency with the technical efficiency of the machine. In this sense, efficiency is blamed not only for its reifying impact on individuals and its homogenizing social tendencies, but also for its elevation of efficiency as a value or an end in itself. The brave new world is inhuman in that it eliminates meaningful human activity or purpose as 'waste' detrimental to society's stability. Responsible for protecting the brave new world's 'faith in happiness as the Sovereign Good,' Mond outlaws 'explanations in terms of purpose' (*Brave* 177) as a threat to the instrumental rationality of the perfect society. No longer allowed to serve a meaningful purpose beyond its own static reproduction, the efficient organization of the Fordian social 'machine' becomes an end in itself: 'Just as in the Ford factory, each step in production is carefully timed and each task is assigned to just the right kind of worker. The whole operation is supervised by men like Henry Foster and the DHC, people dedicated to the cult of statistics and efficiency. Their minds – like Ford's and Tayor's – are completely taken up (and in) by the task, with no room left over to inquire what the point of the task might be' (Firchow, *End* 108). Fearing that 'efficiency has become absolutely necessary' (Sexton 96), Huxley agreed with Séché who observed: 'Once started the machine demands (under threat of economic ruin) that it shall never be unnecessarily stopped, never thrown out of its stride. Production and yet more production – that is the fundamental law of the machine's being. The necessary corollary to this law is consumption and yet more consumption' (in Sexton, 91). Credited with a revolutionary production process and the concept of mass consumption, Ford was both an innovative industrialist and an enabler of the economic power base of corporate capitalism. One of the first automakers to embrace global expansion, he merges in Huxley's figurative territory with Sir Alfred Mond's ambition to concentrate corporate power. In 1926 Mond 'personally pushed through the amalgamation of the British chemical industries into one massive, multi-national corporation: Imperial Chemical Industries Limited IC)' (Sexton 94). The historical Mond's desire for the consolidation of economic power in the hands of a few CEOs finds its radical political extension in the fictional Mond's membership in a seven-man conglomerate of World Controllers.

Although one might ask why the chemically preconditioned brave new worlders are in need of Pavlovian conditioning, let alone ideological manipulation, it is clear that Huxley is not particularly concerned with narrative or logical consistency. As is the case with utopian projections, *Brave New World* is really about the present social conditions rather than the future. And the present was for him threatened by the rationalizing tendencies of economic and political forces that conjured up the spectre of totalitarianism. In his new 'Foreword' to the 1946 edition of *Brave New World,* he envisaged a totalitarian future consisting of a group of despots who control a mass of hapless consumers: 'A really efficient totalitarian state would be one in which the all-powerful executive of political bosses and their army of managers control a population of slaves who do not have to be coerced, because they love their servitude' (xiv). Anticipating Althusser's analysis of the 'Ideological State Apparatuses,' he points to the role of 'ministries of propaganda, newspaper editors and schoolteachers' (xiv) in the reproduction of the totalitarian ideology. In *Brave New World,* Mond's totalitarian control operates not by means of violent repression but by means of ideological seduction. Huxley shares with Sinclair Lewis an insistence on the church and the school as the primary sites of Althusser's ISAs. If eugenics and Pavlovian conditioning can be considered the absurd equivalents of natural aptitude reinforced by training or education, then mass consumerism and pseudo-religion masquerade as society's ideological cement. Aside from keeping the new brave worlders contented through drugs (*soma*) and the immediate satisfaction of every desire through consumption, sex, technology, and entertainment (the 'feelies'), they are safe and secure in the knowledge that they are never alone. The brave new world is the radical extension of the group tyranny that Whyte outlines in *The Organization Man* and Lewis dramatizes in *Babbitt.* The 'Ford's Day celebrations, and Community Sings, and Solidarity Services' (*Brave* 52) that Mond touts as proof of social harmony is for the dissidents evidence of group tyranny. Through the absurd exaggeration of the 'happy group' Whyte associates with bureaucratization and suburbanization, Huxley expresses his outrage against the subordination of the individual to the collectivity.

The Philosophical Roots of the Efficient Society

In *Brave New World,* Mustapha Mond is figuratively linked to industrialization (Ford) and to corporate capitalism (Sir Alfred Mond); however, in the novel he functions primarily as a political figure with the power to decide the fate of dissidents or 'enemies' of the state. Often considered the novel's most interesting character, the World Controller stands out as a unique, perhaps even 'charismatic,' figure whose intellectual curiosity makes him engage in philosophical debates with the dissidents Bernard Marx, Helmholtz Watson, and John Savage. Although commentators have discussed John Savage as a representative of Rousseau's legacy, the philosophical roots of Mond's totalitarian control have not been analysed from the perspective of the 'perfect' society as an extrapolation from Taylor's efficiently managed social space. The critical literature tends to dwell on the total-outcome model of efficiency associated with the assembly-line production of human beings and the many references to Ford as the source of the standardized society. But efficiency is also implicated in Mond's open hostility to Rousseau's *Social Contract* and unacknowledged approval of Thomas Hobbes's *Leviathan. Brave New World* re-enacts an old conflict between competing social-contract theories: on the one hand, Rousseau's reliance on ideology to cement group solidarity and, on the other, Hobbes's invocation of the Sovereign as the despotic arbiter of disputes that Taylor reanimates in his privileging of the 'efficiency expert.'

After Huxley's depiction of the chemically predetermined and thoroughly managed bodies populating the brave new world, the reader might well ask what need there is for a repressive World Controller. Since Huxley is no radical historical materialist, he also, and perhaps contradictorily, traces the material 'iron cage' he describes to the *ideas* of philosophers and social theorists. The symbolic significance of names – such as Bernard Marx, Lenina Crowne, Polly Trotsky, Sarojini Engels, Herbert Bakunin, Morgana Rothchild – has given rise to the justified complaint that the 'apparently odd assortment of names' tends to 'have very specific and often contradictory overtones' (Firchow, *End* 83). The same could be said for political philosophies referenced not only in the allusion to Rousseau in the figure of John Savage but also in recognizable maxims like Jeremy Bentham's 'the greatest happiness of the greatest number' and Karl Marx's 'from everybody according to his ability' and 'to everybody according to his need.' For Firchow, Huxley's point is to show that these apparently 'divergent' forces all 'share the claims of "totality"' and are 'fundamentally materialist, in that they envision man's salvation here on earth.' Moreover, they 'all glorify machines and modern technology' and tend to 'subordinate the individual to the claims of a collective whole' (83). But beyond such confusing political symbolism, Mond can be seen to occupy a philosophically more coherent political position. It is as if Huxley wanted to compensate for the novel's nightmare vision of material and social determinism by providing Mond with political power as a gesture denoting the continued viability of political agency. Mond's most important role is to affirm the power of *politics* as such at the very historical moment when Huxley sees it being usurped by *economic* interests. By blaming philosophical influences for Mond's totalitarian position, Huxley suggests that the brave new world was not the *inevitable* outcome of technical efficiency; it was also the sedimented accumulation of political decisions reached by social agents. If Mond is contradictorily the symbolic representative of both material determinism and political decisionism, it is because Huxley could no more abandon his liberal investment in the power of ideas and individual action than he could refrain from dramatizing the material conditions of the Fordian society as an inescapable 'iron cage.'

The solidarity of the happy group in the brave new world is captured by the political slogan 'Community, Identity, Stability' displayed on a shield as the 'World State's motto' (*Brave* 3). In this smoothly functioning system, cooperation is prized above all for facilitating social and economic integration. Having at his disposal standardized men and women whose talents correspond exactly to the tasks society assigns them, it is Mond's responsibility to make certain that his society continues to achieve the 'optimum population' (223) necessary to maintain a perfect equilibrium between economic production and consumption. This Malthusian preoccupation echoes the eugenics debates in which Wells played such a prominent role.[2] The desired equilibrium is paradoxically reached by means of a hierarchical power structure that is simultaneously radically egalitarian. As Mond explains, the undifferentiated masses of workers are dominated by a small elite of Alpha beings. 'Modelled on the iceberg,' he elaborates, the hierarchical distribution amounts to 'eight-ninths below the water line, one-ninth above' (223). But this power asymmetry does not create the kind of suffering and resentment expressed by the workers at Ford Motor. Maintaining the exact balance between high-caste Alphas and lower-caste Betas, Deltas, Epsilons, and Gammas, Mond inserts each member of each caste into a predetermined slot in the totality. Declaring complete satisfaction with their place in the system, the workers are not tempted into competitive behaviour that might disrupt the process of production. In a world in which 'men are physico-chemically equal' (74), they neither envy each other's social role nor covet each other's possessions. In this efficiently arranged society, it would make no sense for predetermined Delta-workers to aspire to the ranks of the equally predetermined 'Alpha Plus Intellectuals' (17). Since an 'Alpha-decanted, Alpha-conditioned man would go mad if he had to perform Epsilon Semi-Moron work,' it is clearly preferable to ask an Epsilon to 'make Epsilon sacrifices' (222). In this homeostatic environment, all the ranks happily embrace who they are and what they do. As one of the inferior castes rejoices, 'I'm *so* glad I'm a Beta' (27).

On the surface, it appears as if Mond simply affirmed a repressive caste system: 'You cannot pour upper-caste champagne-surrogate into lower-caste bottles.' To counteract this impression, he relates 'an experiment in rebottling' that took place in A.F. 473 on 'the island of Cyprus' (223). This Cyprus experiment is meant to confirm that 'a factory staffed by Alphas – that is to say by separate and unrelated individuals of good heredity and conditioned so as to be capable (within limits) of making a free choice and assuming responsibility' is an "absurdity"' (222). It is an 'absurdity' because, as the Prisoner's Dilemma proves, free individuals find it impossible to cooperate spontaneously. Mond stresses that the experiment took place under optimal conditions. Cyprus had been 're-colonized with a specially prepared batch of twenty-two thousand Alphas. All agricultural and industrial equipment was handed over to them and they were left to manage their own affairs.' Under these best possible material conditions, one would expect that intelligent beings would spontaneously organize themselves into a Pareto-efficient social order. However, as Mond cynically exults, the 'result exactly fulfilled all the theoretical predictions. The land wasn't properly worked; there were strikes in all the factories; the laws were set at naught, orders disobeyed; all the people detailed for a spell of low-grade work were perpetually intriguing for high-grade jobs, and all the people with high-grade jobs were counter-intriguing at all costs to stay where they were.' Instead of cooperating to optimize productivity, the Alpha society degenerated into a 'first-class civil war.' In the terms of the Prisoner's Dilemma, each Alpha chose the 'I-win/you lose' option, thereby producing an inevitable race to the bottom. The Cyprus experiment concluded with the Alphas begging to have the totalitarian regime reinstituted: 'When nineteen out of the twenty-two thousand had been killed, the survivors unanimously petitioned the World Controllers to resumed the government of the island' (223). This story is meant to convince the dissidents that Mond's paternalistic social engineering is justified, no matter how totalitarian its disciplinary control mechanisms have become. The Alphas are willing to sacrifice individual freedom in exchange for the collective security provided by a figure empowered to arbitrate disputes from a position external to the space occupied by the disputants.

The Cyprus experiment tests Rousseau's ideal of a harmonious social community predicated on Pareto-efficient principles. Each social agent would voluntarily sacrifice some short-term self-interest to further the common good, which, in turn, would result in long-term benefits to each individual. Endowed with reason, citizens would obey the laws of society because they understood these to be based on the rational principles outlined in Rousseau's *Social Contract.* However, as Mond observes, left to their own devices, the Alphas soon regressed to the state of war described in Hobbes's *Leviathan.* Unlike Rousseau, Hobbes assumed that human beings were not capable of acting rationally to achieve the common good. Fearing for their survival, they were willing to enter into a social contract with a sovereign committed to the smooth functioning of the social group. Catering to the desire for security expressed by the Alphas seeking readmission to the Fordian society, Mond constructs himself as a Hobbesian sovereign who knows best what his subjects need. He is also in the position of the efficiency expert in Taylor's system who assesses the means for achieving an optimally efficient outcome from an external and hence supposedly disinterested perspective. Where Rousseau relied on voluntary cooperation, Hobbes and Taylor sought to enforce it in order to prevent free-rider problems from arising.

Mustapha Mond and Hobbes's Principle of Sovereignty

The only relatively 'free' or 'individualistic' character of the novel, Mond controls the Fordian society from a position of power located outside the system.[3] His legitimacy depends on his ability to organize the society without at the same time being implicated in the system he administers. 'But as I make the laws here,' he declares, 'I can also break them' (219). On closer inspection, though, his power to act as he pleases is premised on, and hence curtailed by, the very political position that empowers him. In Weber's terms, his despotism is bureaucratic rather than charismatic because his leadership is not based on the loyalty he inspires in subjects responding to his *personal* appeal. For Weber, charismatic authority is not hampered by either tradition or bureaucratization. Addressed as 'his fordship,' Mond upholds the legacy of Henry Ford as the founding moment of the brave new world. This feudal form of power seems to function primarily on symbolic levels. The actual source of his power lies in the processes of Fordian rationalization and Taylorized normalization criticized by Weber and Huxley. As long as the laws Mond 'makes' reinforce the 'security, identity, and community' of the brave new world, he is presumably free to indulge his whims. The totality is indifferent to the means employed to achieve its efficient outcomes. When he boasts that he can 'break' the laws he makes, he is referring specifically to his secret reading of the prohibited plays of Shakespeare. Mond can indulge in such hidden transgressions because they are no threat to the hegemony. But the need for secrecy indicates that it is only by subordinating his *personal* preferences to the demands of the *impersonal* system that Mond can maintain his grip on power. 'Making' and 'breaking' the laws is thus a *political* boast without any real political consequences. His main function in the brave new world is to 'manage' rather than 'lead' the efficiently rationalized social order. He is in the position of the pragmatic, utilitarian efficiency expert whose task is to ensure the efficient functioning of the socially engineered totality. Although Huxley never explicitly references Thomas Hobbes, Mond's self-understanding has its philosophical roots in *Leviathan*.

Mond's eccentric position in relation to the subjects he absolutely controls enacts Hobbes's contention that the security of men and women requires a sovereign principle that is autonomous, indivisible, and transcendental. In *Leviathan* Hobbes conceptualizes a social-contract theory intended to resolve the kind of antagonistic social relations prevalent among the Alphas on Cyprus. In his opinion, men naturally live in a state of war; in their pursuit of brute self-interest, they will use violence against each other for gain (competition), for safety (self-defence), and for reputation (glory). 'Hereby is manifest,' he contends, 'that during the time men live without a common Power to keep them all in awe, they are in that condition which is called Warre; and such a warre, as is of every man against every man' (Hobbes 88). Human subjects are for him irrational creatures whose violent and capricious instincts have to be curbed for their own good. Hobbes understood that as long as human beings are left to exist in a 'state of nature,' they are inclined to choose the lose-lose scenarios outlined in the Prisoner's Dilemma. In game-theoretical terms, Hobbes noticed that 'our attempts to secure our own self-interest are collectively self-defeating' (Heath 49). If left to their own devices, fears Hobbes, human beings will try to freeload by stealing food from their neighbours rather than growing it themselves. If everybody acted strictly according to self-interest, then the society would be faced not only with lawlessness but also with starvation.

In an effort to curb this free-rider syndrome, Hobbes introduced a system of rules and laws the enforcement of which he entrusted to an autonomous and indivisible locus of sovereign power. According to Hobbes's myth of society's origins, 'free' individuals initially entered into a social contract with a sovereign; they relinquished their individual freedom in exchange for protection. In theory, at least, the Sovereign's power would be constrained by the moral responsibility he accepted for the collectivity's overall well-being. Naturally antagonistic human beings would be forced to cooperate because, apparently working against their immediate self-interest, an efficient rational program of cooperation would in the final analysis ensure security and prosperity for all. The idea of a social contract presupposes, of course, that there are terms to be negotiated between parties. But, after having exchanged freedom for security, individuals had forever forfeited the possibility of renegotiating the initial terms; they have once and for all entrusted their well-being to a supposedly benign sovereign who knows what is best for the commonwealth. In the same way, neither workers nor bosses could argue with Taylor's efficiency expert once he had been put in place. Hobbes elevates the right to the securing of life *as such* above all other considerations. His sovereign is a social engineer entrusted with the legal enforcement of the most efficient organization of social agents, an organization designed to minimize 'wasteful' antagonism and maximize 'productive' cooperation. Like Taylor after him, Hobbes sought to regulate the relations between 'naturally' irrational and antagonistic individuals so as to compel them to act collectively in the common interest of the right to life or security. While exerting top-down force on their subjects, Hobbes and Taylor were interested in regulating *bodies* without attempting to change human nature.

By populating the Fordian society with characters who have been reduced to mere instrumental bodies, *Brave New World* extrapolates the absurd dimensions of the pragmatic solutions of Hobbes and Taylor to social conflict. The standardized Alphas, Betas, Deltas, Epsilons, and Gammas reflect the depersonalized and depoliticized bodies of the rationalized society dedicated to the material and social security Lewis satirizes in the pathetic figure of Babbitt. In this thoroughly regulated society, even the World Controller is in fact controlled by the system over which he nominally presides. Far from being able to act as a wilful individualist capable of arbitrarily pursuing his self-interest or whims, Mond subordinates his own desires to the needs of the collectivity. By his own account, he acts not out of megalomaniac ambition but out of a Hobbesian sense of responsibility for the common good. As Huxley repeatedly insists, the World Controller's decrees often go against his own inclinations. In his conversations with the dissidents, for instance, he reveals a nostalgic longing for the pre-modern world he so rigorously seeks to repress. He regrets that he had to 'let the science go' (227), and he secretly enjoys reading the books he prevents others from accessing. It even seems to pain him that he will have to exile the dissidents whose intellectual conversations he will miss. If he draws the line between the good citizen and the dissident, it is not because he prefers one to the other, but because he considers himself bound to ensure the Fordian system's security. In fact, the rosy picture he paints of the community of exiled dissidents conveys a sense of longing for what he feels compelled to prohibit. Instead of dreading exile, Bernard Marx ought to appreciate that he is 'being sent to a place where he'll meet the most interesting set of men and women to be found anywhere in the world. All the people who, for one reason or another, have got too self-consciously individual to fit into community-life. All the people who aren't satisfied with orthodoxy, who've got independent ideas of their own. Every one, in a word, who's any one' (227). When asked why he refrained from joining the exiled community he so admires, Mond claims to have 'preferred' the responsibility of serving the community to the self-indulgent pursuit of scientific curiosity, intellectual satisfaction, or aesthetic contemplation. According to him, his control over others should be understood as the most genuine self-sacrifice. Arguing that he paid his dues by 'choosing to serve happiness,' he clarifies that he is speaking of 'other people's – not mine' (229). In a self-denying gesture, he forbids what nurtures his own interest in 'truth' and 'science' because he has concluded that 'truth's a menace, science is a public danger' (227). According to this logic, his sovereign power is the well-intended outcome of an admirable moral choice. Not unlike Hobbes's benign sovereign or Taylor's objective efficiency expert, Mond paternalistically oppresses others for their own good. Carried out from a superior position of knowledge, the constant surveillance of his subjects is meant to protect them from straying from unquestioningly performing the tasks that make them happy. Restricting their freedom is really an act of love.

Game theory allows us to understand why it is logically advantageous for the brave new worlders to allow Mond to constrain their freedom. In a game-theoretical summary of Hobbes, it has been pointed out that 'each individual in the state of nature can behave peacefully or in a war-like fashion' (Hargreaves and Varoufakis 174). Peace is, of course, the better choice since it allows people to go about their daily lives. Paradoxically, though, 'bellicosity is the best response to both those who are peaceful (because you can extract wealth and privileges by bullying those who choose peace) and those who are bellicose (because might can only be stopped by might). In short, "war" is the strictly dominant strategy and the population is caught in a *Prisoner's Dilemma* where war prevails and life is "nasty, brutish and short"' (174). According to Hobbes, self-interested individuals will see that it makes sense to endow the sovereign or state with the authority to enforce peace. In exchange for peace, the individual agrees to hand over some of his or her freedom to the state endowed with the legitimate power to use force against lawbreakers. We need to stress that Hobbes 'thought the scope of [the state's] interventions in this regard would be quite minimal' (175). However, game theorists have little difficulty in uncovering 'interactions that resemble *Prisoner's Dilemma*' (175) in broad areas of today's social world. Liberal political theory thus suggests 'that the State (or some similar collective agency) will be called upon to police a considerable number of social interactions in order to avoid the suboptimal outcomes associated with this type of interaction' (175). *Brave New World* conveys Huxley's justified fear that disciplinary regimes would increasingly present themselves as the optimal outcome anticipated in Hobbes's *Leviathan.*

Taylor's Hobbesian attempt to convert a race to the bottom into a race to the top finds in *Brave New World* its most logical and most terrifying realization. Mond resolves the Prisoner's Dilemma afflicting all social-contract theories by constructing not only a stable but also an egalitarian world in which all the people are *equally* happy. Happiness may take different forms for Alphas and Epsilons, but the happiness of the former does not diminish the happiness of the latter. It appears that the brave new world is not only rationally efficient but also socially just. Yet this apparently just or egalitarian society satisfies preferences that have been emptied of all human meaningfulness. The insistence on the right to life and happiness is in the brave new world translated into the imposition of the *same* life and the *same* happiness on everybody. The ideal of a Pareto-efficient balancing of preference-satisfactions has been betrayed by the predetermination of desires calculated to correspond to society's capacity to mass-produce consumer goods and entertainment. Mond's sin against humanity is not the exercise of absolute power but the homogenization of the efficiently planned totality that leaves no room for *individual* preferences or degrees of satisfaction.

As an efficient manager, Mond dictates how the brave new world functions. But as a political figure, he can at best engage in sham displays of power. His only specifically political power is to make the decision as to who is a compliant citizen and who is a dissident. Concluding that the material conditions of his day were against the possibility of 'politics as a "vocation"' (Weber, 'Politics' 114), Weber singles out 'physical *force*' as the proper domain of politics: 'Of course, force is certainly not the normal or the only means of the state – nobody says that – but force is a means specific to the state' (78). Along the same lines, Carl Schmitt later argues in *The Concept of the Political* (1932) that the 'specific political distinction to which political actions and motives can be reduced is that between friend and enemy' (26). Political power thus depends on the state's legitimate right to declare war. It is not necessary that war be waged; however, if this possibility were to be removed, the concept of the political would dissolve along with it. Unlike other political thinkers, Schmitt analyses the fundamental logic of the political independent of ideological investments:

> The political enemy need not be morally evil or aesthetically ugly; he need not appear as an economic competitor, and it may even be advantageous to engage with him in business transactions. But he is, nevertheless, the other, the stranger; and it is sufficient for his nature that he is, in a specially intense way, existentially something different and alien, so that in the extreme case conflicts with him are possible. (Schmitt 27)

Embodying the principle of sovereignty, the nation state requires for its legitimacy an enemy; if no actual enemy exists to threaten the community, the state has to create an alien other to justify its very existence. Huxley's *Brave New World* dramatizes not only a thoroughly depersonalized but also a highly depoliticized social order. Mond's power to decide the fate of the dissidents is thus his only specifically political function. The dissidents are the 'waste' necessary for the efficient totality to affirm its political legitimacy. Mond's authority to exile the dissidents constitutes a political act that is simultaneously being invalidated by his scientific management of the consumer society. Dissidents are subject to the political logic of 'us versus them'; they are judged to be either friend or enemy of a *collectivity*. In contrast, consumers participate in a market economy that recognizes competitors but not enemies;[4] acting *as atomistic individuals* pursuing their own self-interest, they do not easily coalesce into antagonistic collectivities. Huxley's scientifically managed brave new world is so rigorously efficient that political dissent can at best be a mere diversion for the World Controller. Moreover, in the brave new world, there are, of course, no more real competitors than there are real enemies. For competition to arise, there would have to be *individuals* motivated by conflicting desires. Coerced cooperation eliminates sites of resistance to the totalized system. As their fate indicates, the dissidents are no threat to the efficiently functioning society; their resistance simply strengthens Mond's display of sham political power. The political despotism with which Mond is tarnished reveals itself to be illusory and superfluous. His threat to civilization lies in his scientific management of the bodies required for the partnership of production and consumption that Huxley represents in the symbolic association of Mond with both Henry Ford and Sir Alfred Mond.

Rousseau's Radical Egalitarianism

If Mond is the paternalistic Hobbesian figure enforcing security and happiness, John Savage is his Rousseauistic counterpart agitating for freedom and self-development. In the history of philosophy, Hobbes is seen as the one who lacks faith in human nature and Rousseau as the one who assumes that human beings are naturally 'good' and hence capable of governing themselves. Where Hobbes curbs human instincts without making any attempt to change human nature, Rousseau seeks to improve human nature through education so as to make social subjects fit to govern themselves. In Mond's Hobbesian view, Rousseau's notion of liberty is simply the right 'to be inefficient and miserable' (*Brave* 46). Through John Savage, the 'free' man whom Mond calls 'a round peg in a square hole' (46), Huxley examines and condemns Rousseau's social-contract theory as yet another social-engineering project designed to homogenize society and standardize individuals. In *Brave New World,* Hobbes's benign paternalism is indicted for condoning the suppression of the free individual by the collectivity. But, as Huxley's treatment of John Savage indicates, Rousseau's revision of Hobbes's contract theory is equally condemned for offering no more than an alternative mechanism for facilitating the enforcement of agreements, a mechanism that underpins the primacy accorded to collective action. In the novel's emotional economy, Huxley seems, if anything, even more hostile to Rousseau than to Hobbes. First of all, Hobbes openly acknowledges the violent enforcement of security he advocates, whereas Rousseau hypocritically conceals the ideological violence lurking in the democratic process he outlines. In the second place, where Hobbes advocates external control over bodies, Rousseau sets his sights on the individual's mind and soul.

As critics keep pointing out, Savage is an ambiguous figure in the novel. Far from being Rousseau's 'natural man,' he was the son of 'civilized' parents. His mother Linda was accidentally left behind by the D.H.C. on a visit to the Reservation. Although it is not clear by what logic an embryo starting out in a bottle finds its way into a woman's womb, John Savage is said to have been born 'naturally' to his mother. This hybrid state explains why he feels no more at home in the Pueblo community than in the Fordian society to which he is later returned by Bernard Marx. No matter how unconvincing John may be as a man of nature, Huxley needs him to be in a 'savage' state for at least two crucial reasons. On the one hand, Huxley heightens the absurd dimensions of the brave new world by having John comment on what he observes from an 'innocent' outsider's perspective. On the other hand, the Rousseauistic 'savage' acts as an obvious philosophical counterpoint to the 'civilized' Hobbesian Mond. Moreover, the 'savage' state in which Bernard discovers him on the Reservation suggests that eugenic predetermination is not sufficient to ensure the docility on which the brave new world depends. Eugenic predisposition has to be reinforced through processes of psychological manipulation and social indoctrination, just as Rousseau's 'natural' man has to be moulded and improved through education. Interestingly enough, Huxley does not want to pursue this parallel, most likely because he needs to disavow the ideological implications of high culture to which Althusser has alerted us. In contrast with the 'feelies' and other hypnotic mass cultural phenomena, the acquisition of cultural capital is championed in both John Savage's reading of Shakespeare and in Mond's enjoyment of forbidden books. This investment in high culture as an attribute of the unique individual sidesteps the crucial function of cultural education in the reproduction of the dominant ideology. The concept of the unique individual is itself an ideological move privileging a certain social theory or even social-engineering project. Although his cultural conservatism and elitism undermine his critique of Rousseau, Huxley proved an astute critic of Rousseau's social-contract theory and the popular tendency to celebrate it as the source of liberalism and democracy.

In his revision of Hobbes's myth of social origins, Rousseau contends that Hobbes arrived at his pessimistic conclusions about human nature because he failed to trace human history to its earliest beginnings. 'Above all,' Rousseau contends in *Discourse on Inequality* (1754), 'let us not conclude with Hobbes that man is naturally wicked because he has no idea of goodness or vice-ridden because he has no knowledge of virtue' (44). In his initial solitary state, Rousseau's hypothetical natural man was strong, self-reliant, peaceful, and free. In a state of nature, far from being at war with others, 'man's concern for his survival least encroaches on that of others' (44). Even after an increase in their number had driven men out of their solitary state, their social relations were not marked by the aggressive selfishness that Hobbes attributed to them. Men are saved from being the 'monsters' depicted in *Leviathan* because 'nature' has given them 'pity to bolster their reason' (46). Hobbes was wrong to assume that men act primarily out of self-interest; for Rousseau, 'pity is a natural sentiment' that moderates 'the action of self-love' (47). In other words, Hobbes failed to see that 'it is pity that in the state of nature takes the place of laws, moral habits, and virtues' (47). If men can be said to be both rational and naturally virtuous, then their social relations could be regulated according to a quasi-Pareto-efficient maxim: 'Do what is good to yourself with as little possible harm to others'

(47). This initial win-win scenario of natural men living harmoniously in small social groups was in the course of history transformed into the lose-lose situation of civilized men competing for property: 'The true founder of civil society was the first man who, having enclosed a piece of land, thought of saying, "This is mine," and came across people simple enough to believe him' (55). It is this power-asymmetry that Rousseau holds primarily responsible for 'improving human reason while worsening the human species,' making 'man wicked while making him sociable' (53). It was when the 'human race became more sociable' (60) that competition for material possessions and social esteem made men jealous and vicious. The result is 'competition and rivalry' that drive men to 'gain some advantage at other people's expense' (66). Rousseau's social-contract theory is thus his Pareto-efficient attempt to solve the free-rider problem he so acutely identifies.

The state of war that is for Hobbes man's natural inclination is for Rousseau the consequence of the disorder sown by the 'right of property' (*Discourse* 67) that both the 'mightiest' and the 'poorest' claimed for themselves. Faced with the spectacle of the race to the bottom that ensued with 'the encroachments of the rich' and 'the thievery of the poor' (67), men decided to form associations to protect themselves from common enemies. But the political institutions they created were open to abuses, putting 'new shackles on the weak' and giving 'new powers to the rich' (69). The leaders that the people have instituted 'to defend their freedom' abuse their power to 'enslave' (72) them. Civilized man has exchanged the freedom that the 'savage' enjoyed for the security he gains through subservience. Rousseau considers this trade-off an unacceptable bargain with the devil: 'I know that enslaved peoples do nothing but boast of the peace and repose they enjoy in their chains' (73). It is to free man from these chains that Rousseau devises a social contract that targets inequality as the source of all human misery. His contract is meant to correct the false trail on which men embarked in their rush to secure their liberty through political association. At the 'origin of society,' there were some ignored 'wise ones' who 'saw that men must resolve to sacrifice one part of their freedom in order to preserve the other' (69). The social contract is an attempt to achieve this Pareto-efficient win-win scenario. For Hobbes, men were either absolutely free and living in chaos or absolutely ruled and enjoying security. In contrast, realizing that the absolute freedom of the 'savage' is no longer available to the 'civilized' man, Rousseau aims for a Pareto-efficient solution that seeks to maximize every man's freedom without thereby decreasing the freedom of every other man.

The Social Contract (1762) begins with the famous statement that 'man was born free, and he is everywhere in chains' (49). Rousseau argues in the *Discourse on Inequality* that the right of 'life and freedom' (75) is inalienable; it would be 'an offense against both nature and reason' (75) to renounce life and freedom. It would consequently be illegitimate for subjects to give away their freedom to a sovereign, no matter how benign, as Hobbes suggests. Rousseau thus proposes the 'establishment of the body politic as a true contract between a people and the leaders they choose' (*Discourse* 76). The sovereign authority lies with the people, who have, 'with respect to their social relations, combined all their will into a single one' (76). The people will create articles of law that 'obligate every member of the state without exception' (76) and bind the powers of the magistrates who will be chosen by the people. It is this contractual reciprocal arrangement that established Rousseau as the 'father of democracy.'

In conformity with the Tit-for-Tat strategy of game theory, Rousseau assumes that citizens begin to trust each other as they learn from previous experience. As game theorists stress, the introduction of repetition or time into the analysis of free-rider problems weakens the need for a Hobbesian sovereign (or the state) to arbitrate disputes. It is at least *possible* for cooperation to arise spontaneously: 'The moment time comes into the picture (and the interaction ceases to be static), repetition allows the players, in effect, to enforce an agreement *themselves.* Players are able to do this by being able to threaten to punish their opponents in future plays of the game if they transgress now' (Hargreaves and Varoufakis 206). Players are encouraged to cooperate in the first place and, if they defect, they are punished and hence compelled to cooperate again in the future. Revising Hobbes's social-contract theory, Rousseau assumes that social agents will rationally embrace the benefits of Tit-for-Tat strategies. The democratic process simply codifies how social agents might behave spontaneously in order to enjoy the benefits of a reciprocal cooperative strategy. When I vote on a public benefit, I do so by calculating not my immediate self-interest but the benefits that might accrue to me from collective action. As long as I can trust my neighbour to honour my right to a piece of property, I am willing to reciprocate by allowing him to claim another piece of property. It is in our mutual interest to enter into an agreement that establishes and protects property or other rights.

However, Tit-for-Tat strategies foreground the paradox that spontaneous cooperation depends on the enforcement of agreements. The moment Rousseau introduces legal obligations that apply to everyone 'without exception,' he shifts the weight of the argument from spontaneous cooperation to social control. In *The Social Contract*, Rousseau clarifies that the people cannot govern themselves without a contract that empowers the lawgiver and the government to make decisions that are in the interests of the commonwealth. He insists that the sovereign people 'ought to be the author' of the laws that govern them: 'The right of laying down the rules of society belongs only to those who form the society' (*Social* 83). But how, Rousseau asks himself, will the people exercise this legislative power? 'How can a blind multitude, which often does not know what it wants, because it seldom knows what is good for it, undertake by itself an enterprise as vast and difficult as a system of legislation? By themselves the people always will what is good, but by themselves they do not always discern it. The general will is always rightful, but the judgment which guides it is not always enlightened' (83). Like Hobbes before him, then, Rousseau needs to introduce a 'lawgiver' who 'is, in every respect, an extraordinary man in the state' (83, 85). But unlike Hobbes's sovereign, Rousseau's lawgiver is to have 'command over laws' but not 'command over men' (85). The lawgiver establishes rules that are binding on the executive power. Rousseau's social contract operates on a balance of power that theoretically prevents abuses of power.

A problem arises when Rousseau cannot entirely trust the people to sacrifice their immediate self-interest to the interests of the commonwealth. It is therefore imperative that the people 'subordinate their will to their reason' so that the community can be taught to 'recognize what it desires' (83). It turns out that the naturally 'free man' has to be manipulated into voluntarily accepting limits on his individual freedom. Rousseau's social contract reveals itself as a social-engineering project: 'Whoever ventures on the enterprise of setting up a people must be ready, shall we say, to change human nature, to transform each individual, who by himself is entirely complete and solitary, into a part of a much greater whole, from which that same individual will then receive, in a sense, his life and his being' (84). Rousseau does not advocate a return to a state of nature; he proposes to improve on nature by designing 'perfect' social institutions based on moral precepts: 'The founder of nations must weaken the structure of man in order to fortify it, to replace the physical and independent existence we have received from nature with a moral and communal existence' (84–5). In short, human subjects must be coerced to be moral and hence free. Rousseau's 'perfect' society can no more tolerate dissent than Hobbes's openly despotic alternative. Although I may voluntarily consent to limits on my freedom, I may, nevertheless, feel oppressed by the 'General Will' of the majority. Rousseau tried to get around this problem by arguing that nonconformist impulses can be adjusted so as to reintegrate dissenters into the community. Where Hobbes was not in the least interested in changing or improving human beings, Rousseau sought to shape or manipulate them through education. Through education, individuals would learn not only how to become more rational but also how to act collectively through cooperation. Where Hobbes provides a framework within which individuals are subjected to external constraints, Rousseau instals disciplinary mechanisms designed to help subjects internalize their submission to the system. *Brave New World* draws out the coercive implications of Rousseau's reliance on education when Huxley satirizes the elaborate manipulative apparatus of Pavlovian conditioning, propaganda, and mass hypnosis.

In the perfect society of A.F. 632, Rousseau's ideal of social community becomes the nightmare of inescapable conformity. For Huxley, Rousseau's social contract represents a cure for the Prisoner's Dilemma that seems more poisonous than the sickness it is meant to heal. In the first instance, Huxley seems irked by Rousseau's hypocritical stance as a defender of individual freedom when his social contract actually empowers the collectivity. Most pointedly, Rousseau promotes a radical egalitarianism that strikes Huxley as the antithesis of freedom. Although both the French Revolution and the American Declaration of Independence appeal to Rousseau's democratic motto of *liberté, égalité, fraternité,* they have to sacrifice freedom if they are to satisfy the demands of equality and fraternity. Huxley thus accuses Rousseau of sacrificing *liberté* on the altar of *égalité* and *fraternité.* Satirizing Rousseau's motto, Huxley indicates the defeat of freedom by placing the brave new world under the new 'planetary motto' of 'Community, Identity, Stability.' Carrying the imprint of Hobbes, these categories foreground the repressive potential in Rousseau's appeal to what resurfaces in Hegel as an 'identity thesis' predicated on the sublation of subject and object. Where Rousseau considers the group to be the individual's 'friend,' Huxley impugns it as a 'tyrant.' In the Fordian society, 'community' has become claustrophobically all-inclusive in that 'every one belongs to every one else' (*Brave* 40). Since identification with the group is here paradoxically both total and egalitarian, Huxley ridicules Rousseau's claim that his social contract will remove the individual's chains. The emphasis on 'Identity' similarly deconstructs Rousseau's insistence that the individual's sense of self is secured but not seriously constrained by the group. In the brave new world, the identity of individuals is predetermined and their desires conditioned to coincide with the needs of the collectivity. In a society dedicated to 'stability' as the 'primal and the ultimate need' (43), each individual is endowed with a rock-solid but hardly 'free' sense of self. Huxley makes it clear that Mond exploits ideas like community and identity as useful fictions to be exploited for the reproduction of the material conditions of the efficient social order. The planetary motto 'Community, Identity, Stability' reconverts Rousseau's emphasis on liberty, equality, and fraternity into the Hobbesian register of securing the right to life at all cost. Tracing the dehumanized brave new world to Hobbes, Huxley draws out the totalitarian dimensions of a pragmatic social contract privileging security rather than individuality. But it is in Mond's favour that he, like Hobbes, does not promise freedom.

By ostentatiously dropping 'liberty' from his planetary motto, Mond restores Rousseau's revision of Hobbes's social contract to its original preoccupation with social harmony through enforced cooperation. However, he replaces Hobbes's reliance on openly repressive state control with ideological strategies that Rousseau advocates to induce individual subjects to voluntarily embrace group solidarity. As Mond points out, violence is a less effective instrument for the achievement of social control than the tactics of social engineering introduced by Rousseau and perpetuated by Ford and Taylor. He recalls only two historical instances when violence was used to put down resistance. At one point, 'eight hundred Simple Lifers were mowed down by machine guns at Golders Green,' and later there was the 'famous British Museum Massacre' in which 'two thousand culture fans [were] gassed with dichlorethyl sulphide.' But the World Controllers soon adopted the 'slower but infinitely surer methods of ectogenesis, neo-Pavlovian conditioning and hypnopaedia' (50). These methods convey the coercive element in Rousseau's insistence that the social contract promises the 'citizen' liberty at a higher 'rational' level than is available to the 'natural man.' Huxley seems particularly irritated by Rousseau's appeal to what constitutes a 'rational action model,' the belief that rational social agents will act rationally if they are presented with a rational plan.

Huxley satirizes this 'rational action model' on many levels. By replacing moral education with Pavlovian conditioning, for instance, Huxley implies that education, far from sharpening critical reason, functions as an ideology intent on adjusting the 'free' individual to the repressive collectivity. *Brave New World* demonstrates exhaustively how methods like 'sleep-teaching' and the mindless repetition of slogans produce automatic and compliant reactions designed to prevent individuals from thinking for themselves. With satiric irony, Huxley refers to such indoctrination as an example of Kant's 'categorical imperative,' a moral precept indebted to Rousseau's philosophy of enlightenment. Mockingly inverting Kant's privileging of reason, Huxley has the D.H.C. intone that moral education 'ought never, in any circumstances,' be 'rational' (Huxley 1998, 26). 'Hypnopaedia,' the use of 'words without reason,' is praised as the 'greatest moralizing and socializing force of all time' (28). Undermining both Kant and Rousseau, Huxley contends that moral behaviour is no longer a human attainment but the predetermined reflex resulting from repeated Tit-for-Tat strategies. In Huxley's Fordian world, the 'citizen' who replaces 'natural man' in Rousseau's *Social Contract* is further reduced to the artificially conditioned automaton. In Huxley's novel, the 'perfection' of the human being through eugenics and Pavlovian conditioning leads Firchow to conclude that 'the dream of the *philosophes* that man can be made perfect has become horribly true, and it has become true precisely because the new world is superlatively gifted at genetic engineering' (*End* 92). Chemical predetermination and Pavlovian conditioning thus satirize Rousseau's conviction that the spontaneity of the 'savage' can be reclaimed and enhanced by recourse to our rational capacities and the benefits of education. In the brave new world, the voluntary internalization of society's rules in Rousseau's social contract has itself become subject to external coercion. Huxley thereby deconstructs the distinction between Hobbes's 'violent' or external imposition of social order and Rousseau's 'non-violent' or voluntary internalization of society's rules; Rousseau's will of the people has become indistinguishable from Hobbes's will of the sovereign.

Huxley's fear that Rousseau's democratic alternative to Hobbes's despotism is no more protective of the free or unique individual finds its most hysterical expression in the conflict between Mustapha Mond and John Savage. The logic of the harmonious organic community in Rousseau's *Social Contract* can no more tolerate the 'exceptional' human subject or dissenting voice than any of the openly totalitarian systems Huxley saw encroaching on his world in the 1930s. The timid forays of Bernard Marx and Helmholtz Watson into eccentric behaviour find their full-blown exemplification in John Savage's refusal to conform. While his breaches of the code governing the brave new world were at first unintentional, he eventually engages in acts of open rebellion. In an early scene he unwittingly offends by withdrawing to his room to read Shakespeare instead of giving a scheduled lecture. By the end of the novel, he has retreated from 'civilization' into nature and finally disrupts the efficient social order by committing suicide as a spectacular public display. His open rebellion is triggered by his mother's death. Violating the disavowal of mortality in the brave new world, John Savage flaunts emotions that 'wholesome-death conditioning' was supposed to have eradicated. His 'scandalous exhibition' (*Brave* 206) of grief draws the unwanted attention of 'an interminable stream of identical eight-year-old male twins' (201). Feeling so overwhelmed by the standardized homogeneity confronting him, he is invaded by 'a sinking sense of horror and disgust.' Haunted by 'the recurrent delirium of his days and nights, the nightmare of swarming indistinguishable sameness,' he affirms his individuality, appointing himself, in a charismatic gesture, the prophet of freedom: 'I come to bring you freedom' (209, 211). However, in this efficiently rationalized society, he is met by the 'blank expression of dull and sullen resentment' (212) in his audience. In the end, he paradoxically resorts to the very coercive behaviour that he claims to be opposing: 'I'll *make* you free whether you want to or not' (213). The dissenter thereby enacts the 'famous/infamous Rousseauean doctrine that there can be no slaves, not even willing ones, for such perverse creatures would have to be "forced to be free"' (Firchow, *End* 90). In *Brave New World,* Rousseau is thus indicted for philosophically reinforcing the standardization of society through his promotion of egalitarianism, for his intolerance of the intolerant, and for the violence concealed in his rhetoric of freedom.

Sites of Resistance or Removal of Waste

In *Brave New World* Huxley examines the detrimental impact of society's commitment to efficiency in several overlapping and often conflicting registers. At the same time, he refers to sites of resistance or alternatives to the efficient society as social practices that are both superseded and no longer tolerated. These alternatives echo nostalgic yearnings for lines of flight from the 'iron cage' evident, as we have seen, in the novels of Conrad, Lawrence, Ford, and Forster. Unlike other modernists, Huxley regrets the passing of liberal-humanist values, but refuses to indulge in fantasies of recuperating these as viable oppositional strategies. While Mond prides himself on presiding over an efficiently organized society that is stable, predictable, and comfortable, he also knows that the 'happiness' he secures for his subjects comes at a price. What he is willing to sacrifice are all the moral and human principles that Huxley implicitly privileges. The satiric edge of *Brave New World* owes its poignancy to the reversal of values that forces readers to re-examine the threat posed to the humanist notion of the individual by the ideology of efficiency. Treated to history lessons, the 'civilized' can only shake their heads in merriment as they listen to stories of the strange beliefs and behaviours of their pre-Fordian ancestors. Huxley mocks the socially engineered society of A.F. 632 by subjecting it to a pseudo-favourable comparison with the 'past' that Mond openly denigrates and secretly prefers. At the same time, *Brave New World* holds no brief for the possibility that the past can be resurrected.

Throughout *Brave New World*, Huxley holds up high art as a mark of elitist distinction threatened by the vulgar masses. Although Mond himself prefers *Othello* to the 'feelies,' he sacrifices high art as 'the price we have to pay for stability' (220) and happiness. Beyond aesthetic satisfaction, the price to be paid embraces a broad spectrum of cultural, social, and political values. Mond speaks derisively of the historical backwardness of values embedded in such institutions as 'Christianity,' 'liberalism,' 'Parliament,' and 'democracy' (46–7). In his effort to preserve social stability, he mounts propaganda campaigns 'against the Past' in general and, more specifically, against 'viviparous reproduction,' the 'pyramids,' 'Shakespeare,' 'scientific education,' 'God,' the 'soul,' 'immortality,' 'old age,' the 'ethics and philosophy of under-consumption' and 'under-production,' and against such activities as 'reading' and 'thinking' (51–5). It is not just critical thought but also deeply felt emotion that is a threat to social stability. When Bernard would like 'to know what passion is,' Lenina reminds him of the slogan 'When the individual feels, the community reels' (94). Obeying the social injunction to be sexually 'a little more promiscuous,' Lenina fails to connect with Bernard and offends John Savage who rejects her as a 'whore' (43, 194), a term that conveys nothing to her. Although the novel privileges the aesthetic and intellectual satisfactions symbolized in the works of Shakespeare, it is perhaps most successful in its dramatization of the difference between love and empty sexual coupling that speaks to Huxley's investment in emotional depth and meaningful personal relationships. Alternative values to the cohesion of the efficiently organized social order are being erased, redeployed, or appropriated by the totalized system.

In the brave new world, the laudable intention of making society an inclusive sphere is presented as an oppressive totalitarian order. The three dissenters deviate from the norm for different reasons. Bernard is chemically 'defective' in that alcohol was accidentally mixed into his blood-surrogate; he is consequently lacking an essential ingredient in the process of predetermination. Suffering from 'the consciousness of being separate,' he is unable to join the mass hysteria of the Solidarity Service, thus feeling 'utterly miserable' for remaining 'separate and unatoned' (67, 86). In contrast, Helmholtz is 'all alone' on account of a 'mental excess' endowing him with 'too much ability' (67) to think for himself. Brought up on the Reservation, John Savage has been deprived of the 'benefits' of Pavlovian conditioning. Where Bernard and Helmholtz share the speculative 'knowledge that they were individuals' (67), John unselfconsciously or 'naturally' defies cultural norms. He is the embodiment of the longings for an individual personality that haunts the other two. Bernard's unhappiness goes deeper than feeling excluded from exhibitions of mass hysteria; he suffers acutely the weight that standardization imposes on him. He would like to be 'more on my own, not so completely a part of something else. Not just a cell in the social body' (90). Contradicting Lenina's claim that 'everybody is happy nowadays,' Bernard would prefer to be 'free – not enslaved by my conditioning,' thereby siding with John, who would 'rather be unhappy than have the sort of false, lying happiness you were having here' (91, 179). The dissenters attract the attention of Mond's disciplinary gaze because, in the Fordian society, 'no offence is so heinous as unorthodoxy of behavior' for it 'threatens more than the life of a mere individual; it strikes at Society itself' (148). In a society that privileges inclusiveness as an absolute social good, individuals have to be so thoroughly integrated into the totality that solitude, inwardness, and meaningful relationships with others have to be eliminated. Not surprisingly, then, Mond mounts an assault on 'family, monogamy, romance' as evidence of an unhealthy tendency towards 'exclusiveness' (40). In a typical reversal of values, the 'civilized' are taught to shudder in horror at the spectacle of their ancestors engaging in 'suffocating intimacies' and 'dangerous, insane, obscene relationships between the members of a family group' presided over by 'viviparous mother' (36). The World Controller's self-sacrificing concern for the subjects under his paternalistic jurisdiction cannot tolerate individual eccentricities and social allegiances that offend against the normalizing disciplines of the Fordian society.

In Huxley's novel, then, the various manifestations of the concept of efficiency – in its overlapping technical, social, cultural, and private dimensions – are brought together to capture, though in perhaps a somewhat incoherent fashion, its complex and conflicted ideological ramifications as it radiates from the assembly line to infiltrate the consciousness of individuals in their social and private activities. The social homogenization that Huxley traces to the standardization of the production process at the Ford Motor Company is for him a scourge not only for the exploitation of workers it makes possible but for its detrimental effects on a meaningful human life. The 'totalitarianism' of *Brave New World* arises alongside the disciplinary mechanisms that Taylor's 'scientifically managed' bodies happily internalize; the 'happy' citizens of the Fordian future are no more than an exaggerated version of the 'happy' inhabitants of the American suburb satirized in Lewis's *Babbitt*.

It is not so much the yoke imposed on the worker chained to machines in the production process but the voluntary submission to the injunction to 'consume' evident in late capitalism that Huxley holds responsible for our dehumanization. We sacrifice what makes life meaningful for the sake of economic security; the 'totalized' or 'totalitarian' society emerges as the comfort of Weber's 'iron cage' rather than the threat of a world modelled on Hitler's concentration camps. The brave new world is thus a depoliticized space given over to the requirements of economic imperatives. Mond's display of power is not unlike the murderous efficiency of Auschwitz; in both instances, a political spectacle arises to disavow the victory of economic expediency over the principles of political agency.

Taking its cues from its initial technical sources in the engineering of machines, efficiency retains its utilitarian focus even in its social dimensions. The 'perfect society' predicated on the smoothly functioning engine is invariably envisioned either as a total-output model dedicated to the eradication of waste or a rational-choice model intent on creating an optimal distribution of costs and rewards. Both models incite social engineering projects that foreground a highly problematical complicity of a desire for efficient outcomes with more or less unpalatable social-control mechanisms.

Not surprisingly, the literary representations of the impact of the efficiency calculus on society manifest anxieties about the disciplinary aspects implicit in the pursuit of efficiency from publicly debated eugenic solutions to the most personal intersubjective dynamics. Characters obsessed with efficiency for its own sake tend to be vilified for exhibiting everything from psychopathological traits to a pathetically reified consciousness. Whether such characters are seemingly destroyed (Kurtz, Gerald Crich, Henry Wilcox) or seemingly successful (Leonora Ashburnham, Babbitt), the efficiency calculus they embrace invariably triumphs over all alternatives.

Investment in the culture of efficiency has become so complete that it is no longer visible; it has become an ideology that works by itself. In spite of its undeniably desirable dimensions, the pursuit of efficiency is deeply implicated in disciplinary mechanisms and social control. It is this aspect of efficiency that needs to be kept in mind as we are tempted to endorse efficiency as a universal panacea.

Notes

Introduction

1 With the exception of Charles Darwin, Alexander takes her historical evidence from relatively unfamiliar sources. In her first chapter, she compares tests conducted on waterwheels by the engineer John Smeaton in the 1750s and the Franklin Institute in 1830. In other chapters, she analyses the industrial philosophy of Gérard-Joseph Christian, director of the 'Conservatoire des Arts et Métiers,' to discuss the relationship between machines and human labour in early-nineteenth-century France. She also focuses on technical as well as personal efficiency as forms of management by analysing a technical journal, *Engineering Magazine,* and a weekly for general readers, *The Independent,* both published during the progressive era of the United States. Disputes roused by Robert William Fogel's *Time on the Cross* (1974) and *Without Consent or Contract* (1989) allow her to consider the morally troubling issue of efficiency and exploitation so pointedly at stake in the case of slave labour in the United States.

2 In *The Gay Science* (1887), Friedrich Nietzsche already anticipates the premium that Ford and Taylor will be putting on thinking 'with a watch in one's hand': 'Virtue has come to consist of doing something in less time than someone else' (Nietzsche 259).

3 For a discussion of the relationship between form and ideology, see Evelyn Cobley, *Temptations of Faust: The Logic of Fascism and Postmodern Archaeologies of Modernity,* in which she analyses the ideological implications of experimental aesthetic form in the context of German National Socialism. Where the earlier study approached categories such as fragmentation, totality, rigorous integration, and decentred system from a formal perspective, the current study departs from this preoccupation with representation to subject a social 'theme' to theoretical analysis.

4 Although my discussion focuses on the debate between Stephen Heath and Janice Gross Stein, other scholars deal in some ways with the moral problem of equity in relation to efficiency. See Walter J. Schultz, *Moral Conditions of Economic Efficiency* (Cambridge: Cambridge UP, 2001), Kelvin Lancaster, *Variety, Equity, and Efficiency* (New York: Columbia UP, 1979), Colette Bowe, ed., *Industrial Efficiency and the Role of Government* (London: H.M. Stationery Office, 1977), Edward E. Zajac, *Fairness or Efficiency* (Cambridge, MA: Ballinger, 1978), Tommaso Padoa-Schioppa, *Efficiency, Stability, and Equity* (Oxford: Oxford UP, 1987), Allen E. Buchanan, *Ethics, Efficiency, and the Market* (Totowa, NJ: Rowman & Allanheld, 1985), Akhtar A. Awan, *Equality, Efficiency, and Property Ownership in the Islamic Economic System* (Lanham, MD: University Press of America, 1983), Yeh-Fang Sun, *Social Needs versus Economic Efficiency in China* (Armonk, NY: M.E. Sharpe, 1982).

5 See, for instance, Andrew H. Miller's *Novels behind Glass: Commodity Culture and Victorian Narrative* (1995) and Thomas Richards's *The Commodity Culture of Victorian England: Advertising Spectacle, 1851–1914* (1990).

1. Efficiency and the Great Exhibition of 1851

1 The importance of the Exhibition of 1851 is attested by the number of books that it immediately generated: Henry Babbage's *The Exposition of 1851* (1851), John Gilbert's *The Crystal Palace that Fox Built* (1851), John Tallis's three-volume *History and Description of the Crystal Palace, and the Exhibition of the world's Industry in 1851* (1851), and George Clayton's *Sermons on the Great Exhibition* (1851). This first display of human ingenuity and industrial might continues to attract scholarly interest; in the 1950s, there seems to have been a revival of interest with C.H. Gibbs-Smith's *The Great Exhibition of 1851* (London: H.M.S.O., 1981; 1950) and Yvonne Ffrench's *The Great Exhbition: 1851* (1950); more recent studies include Jeffrey Auerbach's *The Great Exhibition of 1851: A Nation on Display* (New Haven: Yale UP, 1999), Hermione Hobhouse's *The Crystal Palace and the Great Exhibition: Art, Science, and Productive Industry* (2002), and a brief analysis in Walter L. Arnstein's *Britain Yesterday and Today – 1830 to Present* (1996). More recently, Victorian scholars have analysed the Great Exhibition as ushering in the consumer society (notably Richards, *The Commodity Culture of Victorian England* [1990], Miller, *Novels behind Class* [1995], and Isabel Armstrong, *Victorian Glassworlds: Glass Culture and the Imagination, 1830–1880* [Oxford: Oxford UP, 2008].) This list does not include separate discussions of the Crystal Palace building. In contrast, the Paris 1900 Exposition, for instance,

seems to be discussed in just one book-length study: Richard D. Mandell, *Paris 1900: The Great World's Fair* (1967).

2 In his study *Britain Yesterday and Today: 1830 to the Present,* Walter L. Arnstein discusses the Great Exhibition of 1851 under the chapter heading 'Prosperity, Propriety, and Progress.'

3 In an ideologically interesting move, she attributes the poor aesthetic showing to the emergence of the bourgeoisie: 'It is merely depressing to see the result of middle-class taste in the mass let loose in such quantities and all at once. It proves the importance of the aristocratic element as the ultimate arbiter of taste. 1851 is the first collective opportunity for assessing the damage done to the arts in one generation alone of philistine freebooting; it marks the high-water of hideosity' (Ffrench 231).

4 See his on-line article 'Internationalism and the Search for National Identity: Britain and the Great Exhibition of 1851,' http://www.stanford.edu/group/ww1/spring2000/exhibition/paper.htm.

5 On a smaller scale, it is instructive to look at the list of agricultural implements exhibited at the Pan-American Exposition of 1901 in Chicago. Improvements in agricultural implements through mechanization seem to apply to all areas farm life; the Chicago exhibits testify to the pervasive spirit of invention and ingenuity in the United States. The list of exhibits encompasses ploughs, corn planters, harrows, cultivators, hay presses, other haying tools, diggers, subsoilers, pumps, grapple forks, bailing presses, fertilizer and manure spreaders, traction engines, threshers, wagons, carts, buggies, grinding mills, gasoline engines, cream separators, continuous pasteurizers, incubators, and a variety of hand tools. Inventions large and small competed in the efficiency stakes by advertising themselves as being bigger, better, faster, more durable, and more convenient than devices produced by others.

6 In *Novels behind Glass,* Andrew H. Miller quotes from some nineteenth-century novelists and poets, especially from Thackeray's *Vanity Fair.* However, most of his documentation derives from popular sources, especially from *Punch.*

7 Chernyshevsky visited in 1859 and Dostoevsky in 1862.

8 The Great Exhibition underscores this shift from creative production to passive consumption in that it 'represented the material world as an unchanging configuration of consumable objects – as a kind of still life' (Richards 46).

9 There were, for instance, 'large numbers of bicycles and automobiles ' and the first demonstrations of a 'practical steam turbine,' of 'X-rays, of wireless telegraphy, and of sound synchronized with movies' (Mandell 67).

10 With the Gallery of Machines being dominated by the giant wheels of a 40-foot-high dynamo generating electricity for the fair, exhibits included an 'industrial crane that lifted twenty-five tons,' a 'quietly hissing dynamo that generated 5,000 horsepower and took just two men to operate it' (Mandell 82). It seems only fitting that the Exposition's Grand Prix was awarded to Rudolph Christian Carl Diesel's engine, this adaptation of the personal automobile for the rugged requirements of trucks and other machinery used in industry and agriculture.

11 In a repetition of 1851, the architectural concept for the building destined to house the 1900 Exposition was again an issue of great debate. And, once again, art was to be showcased along with scientific advances. The plan called for two 'new palaces of fine arts' (Mandell 45) and the arts were generally well represented. The new Petit Palais, for instance, 'contained a retrospective display of older French art,' while the larger Grand Palais contained 'the best modern painting and sculpture of all nations' (Mandell 63, 64), including examples of the new movement of Art Nouveau.

2. Efficient Machines and Docile Bodies

1 A library keynote search for both Henry Ford and the Ford Motor Company turned up over 10,000 entries for each.

2 According to Kanigel, the 'canonical scholarly distinction between Taylor and Ford' (496) oversimplifies the oppositon between theory and practice. Fifteen years before Ford, Taylor had developed a tool grinder at Midvale that 'was expressly to make an unskilled job out of a skilled one' (496). By the same token, Ford used elements of time study early in his career and, by 1913, he had already instituted a 'time-study department of about eighty men' who 'canvassed the factories, recording the number of minutes required to complete each process' (Brinkley 282).

3 The efficiency expert was gently satirized in *Cheaper by the Dozen* (1948), a book written by Frank Gilbreth, Jr, and Ernestine Gilbreth Carey, the children of time-and-motion-study proponent Frank Gilbreth, a man who had collaborated with F.W. Taylor and proved to be an enthusiastic promoter of efficiency in all spheres of public and private life. Frank and Ernestine were efficiency experts and the parents of twelve children; two of the children recall with amusement how the household was organized according to principles of efficiency. In 1950 the book was turned into a movie of the same name, starring Clifton Webb and Myrna Loy as the parents. This first film remained more or less faithful to the book. But in 2003, 20th Century Fox brought out a comedy of the same name, starring Steve Martin; how-

ever, aside from featuring a family with twelve children, this remake bears little resemblance to the original movie or to the book.

4 We know that he attended the centennial exhibition of 1876 in Philadelphia, admiring tributes to machine technology that included the giant Corliss steam engine. By 1900, at the Paris Exposition, Taylor was in charge of Bethlehem Steel's high-speed steel lathe, which was capable of cutting metal at speeds never seen before. Taylor's exhibit drew the awe of the crowd and was received with admiration by competitors. It is certainly fitting that Taylor came to everyone's attention for having increased the speed of a steel-cutting tool.

5 Management was now constrained by the following equilibrium-efficient principles:

First. They [management] develop a science for each element of a man's work, which replaces the old rule-of-thumb method.

Second. They scientifically select and then train, teach, and develop the workman, whereas in the past he chose his own work and trained himself as best he could.

Third. They heartily cooperate with the men so as to insure all of the work being done in accordance with the principles of the science which has been developed.

Fourth. There is an almost equal division of the work and the responsibility between the management and the workmen. The management take over all work for which they are better fitted than the workmen, while in the past almost all of the work and the greater part of the responsibility were thrown upon the men. (Taylor 36–7)

6 The dehumanizing conditions in factories such as the Ford Motor Company have been famously satirized in Charlie Chaplin's film comedy *Modern Times* (1936). Efficiency is most memorably targeted when the Chaplin character desperately seeks to keep up with an accelerating assembly line that requires him to screw nuts into fast-moving machine parts with which he cannot keep pace.

7 In his seminal essay 'Ideology and Ideological State Apparatuses (Notes towards an Investigation) (*January–April 1969*)' Louis Althusser famously differentiates between two manifestations of power in modern societies: the Repressive State Apparatus (RSA) and the Ideological State Apparatus (ISA). The RSA refers to such institutions as the police or the military and works primarily through the threat of open violence. But Althusser is more interested in the analysis of such ISAs as the school or the church; such social institutions are for him sites of power working almost invisibly to dominate individuals through ideology. The educational apparatus,

for instance, inculcates in students behaviour that serves to reproduce the dominant socio-economic structures.

8 In *Choice, Rationality, and Social Theory,* Barry Hindess stresses and criticizes 'three fundamental assumptions' he traces to an essay by M. Hollis ('Rational Man and Social Science,' *Rational Action,* ed. R. Harrison [Cambridge: Cambridge UP, 1979] 1–16): 'First, actors are rational and their rationality is understood in strictly utilitarian terms. Actors have a given set of ends, they choose between them in a consistent fashion, and they select from the available means of action those most appropriate to the realization of their chosen ends. In this sense of rationality, the ends themselves are neither rational nor irrational, they are simply there. Secondly, actors are assumed to be narrowly self-interested. Thirdly, they are social atoms: "they could be picked at random from their groups, because it made no difference *who* they were"' (Hollis 6). They are human individuals, but they are not regarded as essentially located within a social structure of positions and roles' (Hindess 29).

9 Hargreaves and Varoufakis illustrate this kind of disjunction between an egalitarian and a just choice in their analysis of a thought experiment suggested by the philosopher John Rawls. Not surprisingly, perhaps, they conclude their analysis by suggesting that 'some inequality in society makes the society as a whole more productive' (167).

3. An Experiment in (In)Efficient Organization and Social Engineeering

1 According to Kazimierz Smolen, the English title of Höss's memoir is *My Soul, Evolution, Life and Ordeals;* however, he provides no bibliographical reference and I have not been able to locate this version. Smolen informs us that 'the original manuscript of these memoirs can be found in the archives of the State Museum at Oswiecim (PMO) in the file Reminiscences (Höss) 96, nr. Inw. 49757. They were published as a book in Poland and several other countries, including the Federal Republic of Germany.' In this study I am using the excerpts relating to Auschwitz published as the 'Autobiography of Rudolf Höss' in *KL Auschwitz: Seen by the SS (Rudolf Höss, Pery Broad, Johann Paul Kremer)* (Oswiecim: Auschwitz-Birkenau State Museum, 2007), 28–101. Other English translations are Rudolf Hoess, *Commandant of Auschwitz,* intro. Lord Russell of Liverpool, trans. Constantine FitzGibbon (Cleveland: World Publishing, 1959); Rudolf Höss, *Death Dealer: The Memoirs of the SS Commandant at Auschwitz,* ed. Steven Paskuly, trans. Andrew Pollinger (New York: Da Capo Press, 1996).

2 Originally entitled *If This Is a Man.*
3 This early 'Polish period' was followed by a far more intense 'Jewish period,' from mid-1942 to 1945.
4 Kremer's diary entries were not written retrospectively but as events unfolded.
5 Höss was 'conscious of the fact that the death sentence was the only possibility in his case' (Rawicz 14). If he portrayed himself in a more favourable light than might be justified, it was not because he sought to curry favor with the tribunal that would soon be sentencing him.
6 See *Globe and Mail,* 7 January 2008, A12.
7 See comments by Hans Friedrich (1st SS Infantry Brigade), Michal Kabac (Slovak Hlinka Guard), and Oskar Göning (SS Garrison, Auschwitz) in the television documentary *Auschwitz: Inside the Nazi State* (PBS, 2005).
8 The camps employed up to 47.9% of *Volksdeutsche,* that is, of people of German background from Poland, Romania, Hungary, and Croatia.
9 I borrow this term from a program on the Holocaust, 'Auschwitz: Factories of Death,' aired on the History Channel in January 2006.

4. Efficiency and Disciplinary Power

1 In *Natasha's Dance: A Cultural History of Russia* (2002), Orlando Figes also draws attention to the fact that 'Lenin was a huge fan of Taylorism.' He considered the American's 'scientific' methods as a means of discipline that could remould the worker and society along more controllable and regularized lines' (463).
2 It is probably not incidental that Dos Passos stresses the Puritan roots of Taylor, who 'was born in Germantown, Pennsylvania' (Dos Passos 17) and Ford, whose mother was the 'daughter of a prosperous Pennsylvania Dutch farmer' (47). Both Taylor and Ford lived according to Puritan rules of conduct. Ford's mother, for instance, had 'told him not to drink, smoke, gamble or go into debt, and he never did' (47). The same could have been said about Taylor, a man whose sober and regular life is for Dos Passos symbolized by his dying 'with his watch in his hand' (23). It is this Weberian connection between Protestant rules of conduct and capitalist expectations of workers that is indirectly being targeted in Dos Passos's depiction of the social engineering project at the Ford Motor Company. In a satirical debunking of Taylor's American Plan, Dos Passos refers to Ford's offer of five dollars a day by stressing that he preferred 'cleancut properlymarried American workers' (50) who could be relied upon to be dedicated and ef-

ficient workers. Instead of seeing high wages as an act of benevolence, Dos Passos accuses Ford of exploiting America's spiritual heritage to serve his capitalist obsession with productivity.

3 Although Lewis's animated descriptions of gadgets suggest that he may see the 'romantic possibilities of an industrial-technological world' (Geismar 90), he is unambiguous in his condemnation of the standardization of society to which Babbitt so enthusiastically succumbs.

4 During his period of rebellion, Babbitt initially refused to join the Good Citizens' League, but he eventually reconsidered his options to become one of its most enthusiastic supporters.

5 As Lewis points out, 'for weeks together Babbitt was no more conscious of his children than of the buttons on his coat-sleeves' (214). He was thus 'an average father' who was 'affectionate, bullying, opinionated, ignorant, and rather wistful' (216).

6 See, for instance, Daniel R. Brown's comment: 'Because [Lewis] provides no adequate alternatives to their lives, the characters scurry around pursued by Lewis's scourge, but presented with no escape' (54).

5. Efficiency and Population Control

1 See Peter Firchow, *Aldous Huxley: Satirist and Novelist*, 120n.

2 Published in response to the totalitarian nightmares of Stalinist Russia and Nazi Germany, Orwell's *Nineteen Eighty-Four* was published too late (1949) to fit into my historical timeframe, and its anti-utopian target was the use of political terror rather than an ideological commitment to efficiency.

3 Although chronologically belonging to the earlier generation, Joseph Conrad (1857–1924) spent the first half of his life as a seaman, so that his writing career coincided with that of the later group of modernists.

4 For a history of the Fabians, see Norman and Jeanne MacKenzie.

5 Frank Podmore and Edward Pease, for instance, first met at a spiritualist séance.

6 Stone adds that 'there never were SS stud farms, as we have been encouraged to believe by popular literature' (129).

7 In 1934 Shaw wrote an appeal to chemists 'to discover a humane gas that will kill instantly and painlessly: in short a gentlemanly gas – deadly by all means, but humane, not cruel,' a gas for which 'we shall find a use ... at home' (*Man* 112; in Fritz, 165).

8 Firchow mentions that these anticipations have been catalogued by Mark Hillegas.

9 Wells adds: 'Years ago I slew that sham in "Mankind in the Making"' (*Cor-*

respondence vol. 2, 437). In a footnote referring to Wells's refutations of the claims of eugenics, the editor of the *Correspondence of H.G. Wells*, David C. Smith, indicates that there are 'twenty-six articles on these subjects by H.G. Wells published from 1904 to 1906' (8).

10 *When the Sleeper Wakes* refers to the original version of 1899; Wells published a revised version in 1910 entitled *The Sleeper Awakes*.

11 In his introduction to the 'Oxford World Classics' edition of *Mrs Dalloway* (2000) David Bradshaw also draws attention to the role played by 'eugenicist ideology' (xxv).

12 *The Road to Wigan Pier* was commissioned by the Left Book Club in 1936.

13 Hillegas claims that 'Orwell must have read "The Machine Stops," and there is a possibility that Zamyatin did also' (83).

14 These aspects of the story and the effects of computerization on our lives today have been discussed by Marcia Bundy Seabury.

15 In *A Preface to Forster* (New York: Longman, 1983), Christopher Gillie notes that the immediate stimulus for 'The Machine Stops' was not only 'a reaction to one of the earlier heavens of H.G. Wells,' but a technological achievement. 'On January 13, 1908,' explains Gillie, 'Henri Farman succeeded in flying an aircraft over a circuit of one kilometer in one and a half minutes. The event depressed Forster deeply: "It is coming quickly, and if I live to be old I shall see the sky as pestilential as the roads"' (Gillie 46).

16 In Huxley's essay 'Notes on Liberty and the Boundaries of the Promised Land' (1955) there is a similar passage: 'The more traveling there is, the more will culture and way of life tend everywhere to be standardized and therefore less educative will travel become. There is still some point in going from Burslem to Udaipur. But when all the inhabitants of Udaipur have been sufficiently often to Burslem, there will be no point whatever in making the journey. Leaving out of account a few trifling geological and climatic conditions, the two towns will have become essentially indistinguishable' (86).

17 In 'What I Believe' (1939), Forster makes a similar point: 'What is good in people – and consequently in the world – is their insistence on creation, their belief in friendship and loyalty for their own sakes' (80).

6. 'Criminal' Efficiency

1 See Padmini Mongia, '"Ghosts of the Gothic": Spectral Women and Colonized Spaces in *Lord Jim*,' *Conradiana* 17:2 (1993), 1–16); Peter Edgerly Firchow, *Envisioning Africa: Racism and Imperialism in Conrad's* Heart of Darkness (Lexington: Kentucky UP, 2000); Patrick Brantlinger, 'Victori-

ans and Africans: The Genealogy of Myth of the Dark Continent,' *Critical Inquiry* 12 (Autumn 1985), 166–203); Edward W. Said, *Beginnings: Intention and Method* (New York: Basic Books, 1975); and Benita Parry, *Conrad and Imperialism: Ideological Boundaries and Visionary Frontiers* (London: Macmillan, 1983).

2 The widespread conception that Conrad's denigration of King Leopold's inefficient and immoral administration of the Congo did not mean that Conrad objected to colonialism when 'some real work is done' in British territories falls far short of capturing this underlying capitalist logic.

3 Frances B. Singh, 'The Colonialist Bias of *Heart of Darkness*,' in *Joseph Conrad,* Heart of Darkness: *A Norton Edition*, ed. Robert Kimbrough (New York: W.W. Norton, 1988), 268–80; Edward Said, *Culture and Imperialism*, (New York: Knopf, 1993); Andrea White, *Joseph Conrad and the Adventure Tradition: Constructing and Deconstructing the Imperial Subject* (New York: Cambridge UP, 1993); Linda Dryden, *Joseph Conrad and the Imperial Romance* (New York: St Martin's Press, 2000).

4 Jennifer Karns Alexander uses this quotation as an epigraph opening *The Mantra of Efficiency*, thereby suggesting that Conrad unambiguously endorsed efficiency as a saving grace. For her, Conrad's Marlow 'paid homage to the foundations of the industrial culture he represented in the novella' by dramatizing the disastrous consequences that ensued when Kurtz deviated from the devotion to efficiency she unproblematically attributes to Marlow. Kurtz's tragedy is that he had 'surrendered to savagery and wilderness in horrible pursuit of the "forgotten and brutal instincts" that Marlow believed efficiency had tamed' (1), thereby confirming that Conrad believes that 'efficiency separated savage from civilized.'

5 As has often been pointed out, Conrad is primarily concerned with men. His depiction of women is marked by suspect patriarchal and, in the case of *Heart of Darkness*, racial presuppositions. Andrew Michael Roberts's *Conrad and Masculinity* (2000) disrupts the notion of Conrad's standard models of masculinity, and Susan Jones's *Conrad and Women* (1999) has convincingly complicated the reductive assumption that Conrad neither understood women nor paid much attention to them.

8. Efficiency and Perverse Outcomes

1 It could be argued that *Parade's End* might have been a more logical novel than *The Good Soldier* for inclusion in a study of efficiency. However, as I have indicated in the introduction, *The Good Soldier* allows for a more challenging and hence more interesting examination of the ways efficiency has

infiltrated the social fabric and threatens to reify the consciousness of individuals than would be possible in the reading of a novel such as *Parade's End* that is more explicitly but also more narrowly concerned with the efficiency calculus.

9. Efficiency and Its Alternatives

1 Turner usefully discusses the 'difficulty of representing capital accumulation' by focusing on 'the narrative's persistent attention to the physical objects of everyday life. As concrete objects cluster around the novel's characters to form the fabric of their lives and environments, their accumulation becomes both the narrative's preeminent thematic concern and its primary structuring principle' (329–30).

2 Henry has been linked to Virginia Woolf's negative depiction of medical experts: 'Henry and the specialist he hires exemplify the cosmopolitan perspective in very much the same way Sir William Bradshaw does in Woolf's *Mrs Dalloway*' (Gibson 117).

3 As Outka acknowledges, 'Forster's desire to purify Howards End of commercial taint, to literally take if off the market, is well documented' (333).

4 Christiana Crich in Lawrence's *Women in Love* is similarly associated with aristocratic values.

5 Forster's depiction of Leonard's cultural aspirations illustrates a situation that was even more pronounced in Germany, where industrialization was felt most keenly by the increasingly impoverished lower middle classes, who continued to identify culturally with the bourgeoisie while their buying power was often reduced to a level below that of the working classes.

6 See Born 148 for a discussion of this critical tendency.

7 'Cultural capital' is a term introduced and used by Pierre Bourdieu.

8 While 'piping melodiously of Effort and Self-Sacrifice,' the much-admired Ruskin himself, he realizes eventually, speaks to him with 'the voice of one who had never been dirty or hungry, and had not guessed successfully what dirt and hunger are' (*Howards* 48).

9 Gibson summarizes the position of those who 'find *Howards End* more strained than rewarding' by drawing attention to Forster's inability to offer an 'alternative to liberal values' in terms that stress the failure of Margaret's ideal: 'While the novel's theme is connection, its implications are collapse and recoil from the "human scene"' (106).

10 From a different perspective, Leslie White similarly contends that 'Margaret is the curious seeker ... bent on engaging the world's variety while

working to heal its divisions. But her well-intentioned quest for connection is perverse, and ultimately injurious' (56).

10. Efficiency and the Perfect Society

1 As we will see, the critical consensus has it that Huxley's dystopian civilization carries the imprint of Taylor as well as of Ford.
2 In *Brave New World Revisited* Huxley continued to be preoccupied with the detrimental effects on society of the threat of overpopulation.
3 Mond is caught up in what Derrida describes as that 'contradictorily coherent' situation of the totality that has its centre elsewhere.
4 As Schmitt puts it, 'In the domain of economics there are no enemies, only competitors' (Schmitt 28).

Works Cited

Achebe, Chinua. 'An Image of Africa: Racism in Conrad's *Heart of Darkness.*' *Joseph Conrad, Heart of Darkness: A Norton Critical Edition.* Ed. Robert Kimbrough. New York: W.W. Norton, 1988. 251–62.

Alexander, Jennifer Karns. *The Mantra of Efficiency: From Waterwheel to Social Control.* Baltimore: Johns Hopkins UP, 2008.

Althusser, Louis. 'Ideology and Ideological State Apparatuses (Notes towards an Investigation) (*January–April 1969*).' *Lenin and Philosophy and Other Essays.* New York: Monthly Review Press, 1971. 123–73.

Arendt, Hannah. *Eichmann in Jerusalem: A Report on the Banality of Evil.* New York: Penguin, 1994.

Armstrong, Paul B. 'The Narrator in the Closet: The Ambiguous Narrative Voice in *Howards End.*' *Modern Fiction Studies* 47 (2) (2001), 306–28.

Arnstein, Walter L. *Britain Yesterday and Today: 1830 to the Present.* Lexington, MA: D.C. Heath, 1971.

Bakan, Joel. *The Corporation: The Pathological Pursuit of Profit and Power.* Toronto: Penguin, 2004.

Baker, Stuart E. *Bernard Shaw's Remarkable Religion: A Faith That Fits the Facts.* Gainesville: UP of Florida, 2002.

Bauman, Zygmunt. *Modernity and the Holocaust.* Ithaca, NY: Cornell UP, 1991.

Baynes, Norman H. *The Speeches of Adolf Hitler April 1922–August 1939.* New York: Howard Fetig, 1969.

Berman, Marshall. *All That Is Solid Melts into Air: The Experience of Modernity.* New York: Penguin, 1988.

Born, Daniel. 'Private Gardens, Public Swamps: *Howards End* and the Revaluation of Liberal Guilt.' *Novel* 25 (2) (1992), 141–59.

Brinkley, Douglas. *Wheels for the World: Henry Ford, His Company, and a Century of Progress, 1903–2003.* London: Penguin, 2004.

Britzolakis, Christina. 'Pathologies of the Imperial Metropolis: Impressionism as Traumatic Afterimage in Conrad and Ford.' *Journal of Modern Literature* 29 (1) (2005), 1–20.

Brown, Daniel R. 'Lewis's Satire – A Negative Emphasis.' *The Merrill Studies in Babbitt*. Ed. Martin Light. Columbus, OH: Charles E. Merrill, 1971. 51–63.

Carey, John. *The Intellectuals and the Masses: Pride and Prejudice among the Literary Intelligentsia, 1880–1939*. London: Faber & Faber, 1992.

Chappelow, Allan. *Shaw the Villager and Human Being*. London: Charles Skilton, 1961.

Childs, Donald J. *Modernism and Eugenics: Woolf, Eliot, Yeats, and the Culture of Degeneration*. Cambridge: Cambridge UP, 2001.

Cobley, Evelyn. *Temptations of Faust: The Logic of Fascism and Postmodern Archaeologies of Modernity*. Toronto: Toronto UP, 2002.

Conrad, Joseph. *Chance: A Tale in Two Parts*. New York: W.W. Norton, 1968.

– *The Secret Agent: A Simple Tale*. Harmondsworth: Penguin, 1969.

– *The Collected Letters of Joseph Conrad, 1898–1902*. Ed. Frederick R. Karl and Laurence Davies. Cambridge: Cambridge UP, 1986.

– *Heart of Darkness*. London: Penguin, 1995.

Corwin, Sharon. 'Picturing Efficiency: Precisionism, Scientific Management, and the Effacement of Labor.' *Representations* 84 (2003), 139–65.

Couchman, Gordon W. 'Bernard Shaw and the Gospel of Efficiency.' *Shaw Review* 16 (1973), 11–20.

Czech, Danuta. 'Origins of the Camp, Its Construction and Expansion.' *Auschwitz: Nazi Death Camp*. Ed. Teresa Swiebocka and Franciszek Piper. Oswiecim: Auschwitz-Birkenau State Museum, 2005. 21–39.

Czech, Danuta, and Jadwiga Bezwinska, eds. *KL Auschwitz: Seen by the SS*. Oswiecim: Auschwitz-Birkenau State Museum, 2007.

Daly, Macdonald. 'D.H. Lawrence and Labour in the Great War.' *Modern Language Review* 89 (1) (1994), 19–38.

Delany, Paul. *D.H. Lawrence's Nightmare: The Writer and His Circle in the Years of the Great War*. New York: Basic Books, 1978.

– '"Islands of Money": Rentier Culture in E. M. Forster's *Howards End*.' *English Literature in Transition* 31 (1988), 285–96.

Dos Passos, John. *The Big Money*. London: Constable, 1936.

During, Simon. 'Postmodernism or Post-colonialism Today.' *Postmodernism: A Reader*. Ed. Thomas Docherty. New York: Columbia UP, 1993.

Ebbatson, Roger. *The Evolutionary Self: Hardy, Forster, Lawrence*. Brighton: Harvester Press, 1982.

Fernald, Anne E. '"Out of It": Alienation and Coercion in D.H. Lawrence.' *Modern Fiction Studies* 49 (2) (2003), 183–203.

Ffrench, Yvonne. *The Great Exhibition: 1851*. London: Harvill Press, 1950.

Figes, Orlando. *Natasha's Dance: A Cultural History of Russia*. New York: Picador, 2002.

Firchow, Peter. *Aldous Huxley: Satirist and Novelist*. Minneapolis: U of Minnesota P, 1972.

– *The End of Utopia: A Study of Aldous Huxley's Brave New World*. Lewisburg: Bucknell UP, 1984.

Fleishman, Avrom. *Conrad's Politics: Community and Anarchy in the Fiction of Joseph Conrad*. Baltimore: Johns Hopkins UP, 1967.

Ford, Madox Ford. *The Good Soldier: A Tale of Passion*. New York: Random House, 1955.

– *Parade's End*. 2 vols. New York: Knopf, 1964.

Forster, E.M. *Howards End*. Harmondsworth: Penguin, 1973.

– 'The Machine Stops.' *The Collected Short Stories of E.M. Forster*. London: Sidgwick and Jackson, 1948. 115–58.

– 'What I Believe.' *Two Cheers for Democracy*, 75–82.

Foucault, Michel. *Discipline and Punish: The Birth of the Prison*. New York: Random House, 1979.

– *'Society Must Be Defended': Lectures at the Collège de France, 1975–1976*. New York: Picador, 2003.

Friedman, Philip Allan. '*Babbitt*: Satiric Realism in Form and Content.' *The Merrill Studies in Babbitt*. Ed. Martin Light. Columbus, OH: Charles E. Merrill, 1971. 64–75.

Fritz, Stephen G. 'Reflections on Antecedents of the Holocaust.' *The History Teacher* 23 (2) (1990), 161–79.

Geismar, Maxwell. 'On *Babbitt*.' *The Merrill Studies in Babbitt*. Ed. Martin Light. Columbus, OH: Charles E. Merrill, 1971. 91–7.

Gerth, Hans, and C. Wright Mills. 'Introduction: The Man and His Work.' *From Max Weber*. New York: Oxford UP, 1946. 3–74.

Gibbs, A.M. *Bernard Shaw: A Life*. Gainesville: UP of Florida, 2005.

Gibson, Mary Ellis. 'Illegitimate Order: Cosmopolitanism and Liberalism in Forster's *Howards End*.' *English Literature in Transition* 28 (1985) (2), 106–23.

Gilbreth, Frank B., and Ernestine Gilbreth Carey. *Cheaper by the Dozen*. New York: HarperCollins, 2002.

Gillie, Christopher. *A Preface to Forster*. New York: Longman, 1983.

Goldman, Harvey. *Max Weber and Thomas Mann: Calling and the Shaping of the Self*. Berkeley and Los Angeles: U of California P, 1988.

Gordon, Mary. '"Things That Can't Be Phrased": Forster and *Howards End*.' *Salmagundi* 143 (2004), 89–103.

Grebstein, Sheldon Norman. '*Babbitt*: Synonym for a State of Mind.' *The Merrill Studies in Babbitt*. Ed. Martin Light. Columbus, OH: Charles E. Merrill, 1971. 32–44.

Harford, Tim. *The Undercover Economist*. Toronto: Random House, 2005.

Hargreaves Heap, Shaun P., and Yanis Varoufakis. *Game Theory: A Critical Text*. London: Routledge, 2004.

Harvey, David. *The Condition of Postmodernity*. Oxford: Basil Blackwell, 1990.

Haynes, Roslynn D. *H.G. Wells: Discoverer of the Future*. London: Macmillan, 1980.

Heath, Joseph. *The Efficient Society: Why Canada Is as Close to Utopia as It Gets*. Toronto and London: Penguin, 2001.

Henstra, Sarah. 'Ford and the Costs of Englishness: "Good Soldiering" as Performative Practice.' *Studies in the Novel* 39 (2007) (2), 177–95.

Hilfer, Anthony Channell. 'Lost in a World of Machines.' *The Merrill Studies in Babbitt*. Ed. Martin Light. Columbus, OH: Charles E. Merrill, 1971. 83–91.

Hillegas, Mark. *The Future as Nightmare: H.G. Wells and the Anti-Utopians*. New York: Oxford UP, 1967.

Hindess, Barry. *Choice, Rationality, and Social Theory*. London: Unwin Hyman, 1988.

Hobbes, Thomas. *Leviathan*. Ed. Richard Tuck. Cambridge: Cambridge UP, 1997.

Hobhouse, Hermione. *The Crystal Palace and the Great Exhibition: Art, Science and Productive Industry (A History of the Royal Commission for the Exhibition of 1851)*. London: Athlone Press, 2002.

Hoffman, Karen A. '"Am I no better than a eunuch?": Narrating Masculinity and Empire in Ford Madox Ford's *The Good Soldier*.' *Journal of Modern Literature* 27 (3) (2004), 30–46.

Holderness, Graham. *D.H. Lawrence: History, Ideology and Fiction*. Dublin: Gill and Macmillan Humanities Press, 1982.

Holroyd, Michael. *Bernard Shaw: Volume II, 1898–1918 – The Pursuit of Power*. New York: Random House, 1989.

Höss, Rudolf. 'Autobiography.' *KL Auschwitz: Seen by the SS*. Ed. Danuta Czech and Jadwiga Bezwinska. Oswiecim: Auschwitz-Birkenau State Museum, 2007. 27–101.

Hoy, Pat C. 'The Narrow, Rich Staircase in Forster's *Howards End*.' *Twentieth Century Literature* 31 (2/3) (1985), 221–35.

Hutchisson, James M. *The Rise of Sinclair Lewis, 1920–1930*. University Park: Pennsylvania State UP, 1996.

Huxley, Aldous. *Brave New World*. New York: HarperCollins, 1998.

– *Letters of Aldous Huxley*. Ed. Grover Smith. New York: Harper & Row, 1970.

– 'Machinery, Psychology and Politics.' *The Spectator*, 23 November 1929, 749–51.
– 'Notes on Liberty and the Boundaries of the Promised Land.' *Music at Night and Other Essays*. Harmondsworth: Penguin, 1955. 81–8.
Hynes, Samuel. 'The Epistemology of *The Good Soldier*.' *Sewanee Review* 69 (1961), 226–35.
Iwaszko, Tadeusz. 'Deportation to the Camp and Registration of Prisoners.' *Auschwitz: Nazi Death Camp*. Ed. Teresa Swiebocka and Franciszek Piper. Oswiecim: Auschwitz-Birkenau State Museum, 2005. 54–69.
Jameson, Fredric. *Postmodernism, or, the Cultural Logic of Late Capitalism*. Durham: Duke UP, 1991.
Jarosz, Barbara. 'Organizations of the Camp Resistance Movement and Their Activities.' *Auschwitz: Nazi Death Camp*. Ed. Teresa Swiebocka and Franciszek Piper. Oswiecim: Auschwitz-Birkenau State Museum, 2005. 215–34.
Kanigel, Robert. *The One Best Way: Frederick Winslow Taylor and the Enigma of Efficiency*. Harmondsworth: Penguin, 1997.
Kazin, Alfred. 'The New Realism: Sinclair Lewis.' *The Merrill Studies in Babbitt*. Ed. Martin Light. Columbus, OH: Charles E. Merrill, 1971. 97–105.
Kemper, John. 2000. 'Internationalism and the Search for a National Identity: Britain and the Great Exhibition of 1851.' 2000. At http://www.stanford.edu/group/ww1/spring2000/exhibition/paper.htm.
Kertész, Imre. *Fateless*. Evanston, IL: Northwestern UP, 1992.
Knapp, James F. 'The Discourse of Knowledge: Historical Change in *Women in Love*.' *The Rainbow and Women in Love*. Ed. Gary Day and Libby Di Niro. New York: Palgrave Macmillan, 2004. 153–71.
Kremer, Johann Paul. 'Diary.' *KL Auschwitz: Seen by the SS*. Ed. Danuta Czech and Jadwiga Bezwinska. Oswiecim: Auschwitz-Birkenau State Museum, 2007. 150–215.
Lacan, Jacques. 'The Function and Field of Speech and Language in Psychoanalysis.' *Écrits: A Selection*. Ed. Bruce Fink. New York: W.W. Norton, 2004. 31–106.
– 'The Instance of the Letter in the Unconscious or Reason since Freud.' *Écrits*, 138–68.
Lasik, Aleksander. 'Structure and Character of the Camp SS Administration.' *Auschwitz: Nazi Death Camp*. Ed. Teresa Swiebocka and Franciszek Piper. Oswiecim: Auschwitz-Birkenau State Museum, 2004. 43–53.
Lawrence, D.H. *The Letters of D.H. Lawrence*. Vol. 2: *June 1913–October 1916*. Ed. George J. Zytaruk and James T. Boulton. Cambridge: Cambridge UP, 1981.
– *Women in Love*. Harmondsworth: Penguin, 1983.

Levenson, Michael. 'Liberalism and Symbolism in *Howards End.' Papers on Language & Literature* 21 (3) (1985), 295–316.

– 'The Value of Facts in the *Heart of Darkness.' Heart of Darkness*. Ed. Robert Kimbrough. New York: W.W. Norton, 1988. 391–405.

Levi, Primo. *Survival in Auschwitz*. New York: Touchstone, 1996.

– 'Foreword.' *Death Dealer: The Memoirs of the SS Kommandant at Auschwitz by Rudolph Höss*. Ed. Steven Paskuly. New York: Da Capo Press, 1996. 3–9.

Lewis, Sinclair. *Babbitt*. New York: Penguin, 1998.

Lewisohn, Ludwig. 'Babbitt.' *The Merrill Studies in Babbitt*. Ed. Martin Light. Columbus, OH: Charles E. Merrill,1971. 19–21.

MacKenzie, Norman, and Jeanne MacKenzie. *The First Fabians*. London: Weidenfeld and Nicolson, 1977.

Mandell, Richard D. *Paris 1900: The Great World's Fair*. Toronto: U of Toronto P, 1967.

March-Russell, Paul. '"Imagine, If You Can": Time and the Impossibility of Utopia in E.M. Forster's "The Machine Stops."' *Critical Survey* 17 (2005), 56–71.

Marx, John. *The Modernist Novel and the Decline of Empire*. Cambridge: Cambridge UP, 2005.

Marx, Karl, and Friedrich Engels. *The Marx-Engels Reader*. Ed. Robert C. Tucker. New York: W.W. Norton, 1978.

Meckier, Jerome. 'Aldous Huxley's Americanization of the *Brave New World* Typescript.' *Twentieth-Century Literature* 48 (4) (2002), 427–60.

Merkle, Judith A. 1980. *Management and Ideology: The Legacy of the International Scientific Management Movement*. Berkeley: U of California P.

Michaels, Walter Benn. *The Gold Standard and the Logic of Naturalism: American Literature at the Turn of the Century*. Berkeley: U of California P, 1987.

Mickalites, Carey J. '*The Good Soldier* and Capital's Interiority Complex.' *Studies in the Novel* 38 (3) (2006), 288–303.

Micklethwait, John, and Adrian Wooldridge. *The Company: A Short History of a Revolutionary Idea*. New York: Modern Library, 2005.

Milgram, Stanley. *Obedience*. Ed. Christopher C. Johnson. New Haven: Yale UP, 1965.

Miller, Andrew H. *Novels behind Glass: Commodity Culture and Victorian Narrative*. Cambridge: Cambridge UP, 1995.

Miller, J. Hillis. 'Just Reading *Howards End.' E.M. Forster: Howards End*. Ed. Alistair M. Duckworth. Boston: Bedford Books, 1997. 467–82.

Monta, Anthony P. 'Parade's End in the Context of National Efficiency.' *History and Representation in Ford Madox Ford's Writings*. Ed. Joseph Wiesenfarth. Amsterdam: Rodopi, 2004. 41–51.

Mosse, George L. *The Crisis of German Ideology: Intellectual Origins of the Third Reich*. London: Weidenfeld and Nicolson, 1964.

Nathan, David. 'Failure of an Elderly Gentleman: Shaw and the Jews.' *Shaw: The Annual of Bernard Shaw Studies* 11 (1991), 219–38.

Nietzsche, Friedrich. *The Gay Science*. New York: Vintage, 1974.

Orwell, George. *Down and Out in Paris and London*. London: Secker and Warburg, 1997.

– *The Road to Wigan Pier*. Harmondsworth: Penguin, 1972.

– 'Wells, Hitler and the World State.' *George Orwell: Collected Essays*. London: Secker & Warburg, 1975.

Outka, Elizabeth. 'Buying Time: *Howards End* and Commodified Nostalgia.' *Novel* 36 (3) (2003), 330–50.

Pinkerton, Mary. 1985. 'Ambiguous Connections: Leonard Bast's Role in *Howards End*.' *Twentieth Century Literature* 31 (2/3), 236–46.

Piper, Franciszek. 'Direct Methods for Killing Prisoners.' *Auschwitz: Nazi Death Camp*. Ed. Teresa Swiebocka and Franciszek Piper. Oswiecim: Auschwitz-Birkenau State Museum, 2005. 152–64.

– 'Living Conditions and Work as Methods of Exterminating Prisoners.' *Auschwitz*, 143–51.

– 'The Mass Extermination of Jews in the Gas Chambers.' *Auschwitz*, 165–73.

– 'The Political and Racist Principles of the Nazi Policy of Extermination and Their Realization at KL Auschwitz.' *Auschwitz*, 11–20.

– 'Prisoner Labor.' *Auschwitz*, 103–22.

Pippin, Robert B. *Modernism as a Philosophical Problem*. Oxford: Basil Blackwell, 1991.

Postman, Neil. *Technopoly: The Surrender of Culture to Technology*. New York: Random House, 1993.

Raitt, Suzanne. 'The Rhetoric of Efficiency in Early Modernism.' *Modernism/Modernity* 13 (1) (2006), 89–105.

Rawicz, Jerzy. 'Foreword.' *KL Auschwitz: Seen by the SS*. Ed. Danuta Czech and Jadwiga Bezwinska. Oswiecim: Auschwitz-Birkenau State Museum, 2007. 7–25.

Richards, Thomas. *The Commodity Culture of Victorian England: Advertising and Spectacle, 1851–1914*. Stanford: Stanford UP, 1990.

Ross, Stephen. *Conrad and Empire*. Columbia: U of Missouri P, 2004.

Rousseau, Jean-Jacques. *Discourse on Inequality*. Oxford: Oxford UP, 1999.

– *The Social Contract*. Harmondsworth: Penguin, 1979.

Scaff, Lawrence A. *Fleeing the Iron Cage: Culture, Politics, and Modernity in the Thought of Max Weber*. Berkeley: U of California P, 1989.

Schmitt, Carl. *The Concept of the Political*. Chicago: U of Chicago P, 1996.

Schorer, Mark. 'Sinclair Lewis: *Babbitt*.' *The Merrill Studies in Babbitt*. Ed. Martin Light. Columbus, OH: Charles E. Merrill, 1971. 105–16.

Schriber, Mary Sue. 'You've Come a Long Way, Babbitt! From Zenith to Ilium.' *Twentieth Century Literature* 17 (2) (1971), 101–6.

Sexton, James. '*Brave New World* and the Rationalization of Industry.' *Critical Essays on Aldous Huxley*. Ed. Jerome Meckier. New York: G.K. Hall, 1996. 88–102.

Shaw, Bernard. 'John Bull's Other Island.' *John Bull's Other Island with How He Lied to Her Husband and Major Barbara*. London: Constable, 1931.

– *Man and Superman: A Comedy and a Philosophy*. Baltimore: Penguin, 1964.

– 'Preface to *Geneva*.' *Geneva, Cymbeline Refinished, & Good King Charles*. London: Constable, 1946. 3–27.

– 'Preface to *Major Barbara*.' *John Bull's Other Island, How He Lied to Her Husband, Major Barbara*. New York: Wm. H. Wise, 1930. 207–47.

– 'Preface to *The Millionairess*.' *The Simpleton, the Six, and The Millionairess. Being Three More Plays by Bernard Shaw*. London: Constable, 1936. 105–30.

– 'Preface to *On the Rocks*.' *Too True to be Good, Village Wooing & On the Rocks. Three Plays by Bernard Shaw*. London: Constable, 1934. 143–86.

– 'Preface to *Too True to be Good*.' *Too True to be Good*, 3–25.

– *Shaw on Language*. Ed. Abraham Tauber. London: Peter Owen, 1965.

– 'The Simple Truth about Socialism.' *The Road to Equality: Ten Unpublished Lectures and Essays, 1884–1918*. Ed. Louis Crompton. Boston: Beacon Press, 1971. 155–94.

Smith, Adam. *An Inquiry into the Nature and Causes of the Wealth of Nations*. Vol. 1. Ed. R.H. Campbell and A.S. Skinner. Oxford: Clarendon Press, 1976.

Smolen, Kazimierz. 'The Punishment of War Criminals.' *Auschwitz: Nazi Death Camp*. Ed. Teresa Swiebocka and Franciszek Piper. Oswiecim: Auschwitz-Birkenau State Museum, 2005. 293–304.

Spengler, Oswald. *The Decline of the West: Form and Actuality*. Vol. 1. New York: Alfred A. Knopf, 1926.

Stein, Janice Gross. *The Culture of Efficiency*. Toronto: Anansi, 2001.

Steinbacher, Sybille. *Auschwitz: A History*. London: Penguin, 2005.

Stone, Dan. *Breeding Superman: Nietzsche, Race and Eugenics in Edwardian and Interwar Britain*. Liverpool: Liverpool UP, 2002.

Strzelecka, Irena. 'Experiments.' *Auschwitz: Nazi Death Camp*. Ed. Teresa Swiebocka and Franciszek Piper. Oswiecim: Auschwitz-Birkenau State Museum, 2005. 88–102.

Taylor, Charles. *Hegel*. Cambridge: Cambridge UP, 1975.

Taylor, Frederick Winslow. *The Principles of Scientific Management. Scientific Management: Comprising Shop Management, The Principles of Scientific Management, Testimony before the Special House Committee*. New York: Harper, 1947.

Thacker, Andrew. 'E.M. Forster and the Motor Car.' *Literature and History* 9 (2) (2000), 37–52.

Turner, Henry S. 'Empires of Objects: Accumulation and Entropy in E.M. Forster's *Howards End*.' *Twentieth Century Literature* 46 (3) (2000), 328–45.

Watts, Steven. *The People's Tycoon: Henry Ford and the American Century*. New York: Alfred A. Knopf, 2005.

Weber, Max. 'Bureaucracy.' *From Max Weber*. Ed. Hans Gerth and C. Wright Mills. New York: Oxford UP, 1946. 196–244.

– 'The Meaning of Discipline.' *From Max Weber*, 253–64.

– 'Politics as a Vocation.' *From Max Weber*, 77–128.

– *The Protestant Ethic and the Spirit of Capitalism*. London and New York: Routledge, 1992.

– 'Religious Rejections of the World and Their Directions.' *From Max Weber*, 323–59.

– 'The Sociology of Charismatic Authority.' *From Max Weber*, 245–52.

Weissman, Judith. '*Howards End*: Gasoline and Goddesses.' *E.M. Forster: Howards End*. Ed. Alistair M. Duckworth. Boston: Bedford Books, 1997. 432–67.

Wells, H.G. *Anticipations*. New York: Charles Scribner's Sons, 1924.

– *The Correspondence of H.G. Wells, 1904–1918*. Vol. 2. Ed. David C. Smith. London: Pickering & Chatto, 1998.

– *Mankind in the Making*. London: Chapman & Hall, 1903.

– *The Time Machine*. New York: Random House, 1931.

– *The Shape of Things to Come*. Toronto: Macmillan, 1933.

– *When the Sleeper Wakes*. London: Harper, 1899.

– *The Work, Wealth and Happiness of Mankind*. Vol. 2. New York: Doubleday, Doran & Company, 1931.

White, Leslie. 'Vital Disconnection in *Howards End*.' *Twentieth Century Literature* 51 (1), (2005) 43–62.

Whyte, William H. *The Organization Man*. Garden City, NY: Doubleday, 1957.

Widdowson, Peter. *E.M. Forster's Howards End: Fiction as History*. London: Sussex UP, 1977.

Wisenthal, J.L. *The Marriage of Contraries: Bernard Shaw's Middle Plays*. Cambridge, MA: Harvard UP, 1974.

Woolf, Virginia. *Mrs Dalloway*. Oxford: Oxford UP, 2000.

Worthen, John. *D.H. Lawrence and the Idea of the Novel*. London: Macmillan, 1979.

Zapf, Hubert. 'Taylorism in D.H. Lawrence's *Women in Love*.' *The D.H. Lawrence Review* 15 (1982), 129–39.

Zizek, Slavoj. *For They Know Not What They Do: Enjoyment as a Political Factor*. London: Verso, 1991.

Index